London Mathematical Society L

J. F. ADAMS

Lowndean Professor of Astronomy and Geometry
University of Cambridge

Algebraic Topology – A Student's Guide

CAMBRIDGE AT THE UNIVERSITY PRESS 1972

Published by The Syndics of the Cambridge University Press
Bentley House, 200 Euston Road, London N. W. 1
American Branch: 32 East 57th Street, New York, N. Y. 10022

Library of Congress Catalogue Card No.: 75-163178

ISBN: 0 521 08076 2

First published 1972
Reprinted 1973

Printed offset in Great Britain by
Alden and Mowbray Ltd at the Alden Press, Oxford

Contents

(3) S. Eilenberg, 'La suite spectrale. 1: Construction générale', Séminaire H. Cartan 3 (1950/51) (2e ed.) Exposé 8, pp. 8.01-8.08. 51/52

(4) W. S. Massey, 'Exact couples in algebraic topology', Annals of Math. 56 (1952), pp. 363-396; Introduction, Part I and references (pp. 363-370 and 396). 51/66

(5) M. Rothenberg and N. E. Steenrod, 'The cohomology of classifying spaces of H-spaces', Bull. American Math. Soc. 71 (1965), 872-875. 74/75

(6) J.-P. Serre, 'Cohomologie modulo 2 des complexes d'Eilenberg-MacLane', Comm. Math. Helv. 27 (1953), 198-232; Introduction, §§1, 2, 4 and references (pp. 198-211, 220-225 and 232). 79/80

(7) J. F. Adams, 'On the triad connectivity theorem', from unpublished lecture notes. 101/102

(8) G. W. Whitehead, 'On the Freudenthal theorems', Annals of Math. 57 (1953), 209-228, §1 and references (pp. 209, 228). 113/114

(9) I. M. James, 'The suspension triad of a sphere', Annals of Math. 63 (1956), 407-429, extract from Introduction and references (pp. 407, 408, 428, 429). 113/116

(10) J. W. Milnor, 'On the construction FK', mimeographed lecture notes from Princeton University, 1956. 118/119

(11) J. F. Adams, 'On Chern characters and the structure of the unitary group', Proc. Camb. Phil. Soc. 57 (1961), 189-199; §2 (pp. 191-192). 137/138

Introduction

It is the object of this introduction to give a general survey of the material which faces the student of algebraic topology, and at the same time to give a guide to the sources from which this material can most conveniently be studied. It seems convenient to alternate between passages which comment on the material and passages which comment on the literature. When I have had to comment on a topic which has been treated by several authors, I have sometimes felt a responsibility to offer the student some guidance on which source to try first; I have done this by marking a recommended source with an asterisk. This does not mean that the other sources are not also good; some students may prefer them, and most will profit by seeing the same topic treated from more than one point of view. In some cases the marked source is chosen on the grounds that it gives a particularly short, simple or elementary account, while the others give longer, fuller or more advanced accounts.

In what follows, I shall refer to the following list of sources available in book form. A reference to the author's name, without further details, is a reference to this list.

J. F. Adams, 'Stable Homotopy Theory', J. Springer, 2nd ed. 1966 (Lecture Notes in Mathematics No. 3).

P. Alexandroff and H. Hopf, 'Topologie', J. Springer 1935.

M. André, 'Méthode Simpliciale en Algèbre Homologique et Algèbre Commutative', J. Springer 1967 (Lecture Notes in Mathematics No. 32).

M. F. Atiyah, 'K-Theory', Benjamin 1967.

A. Borel, 'Topics in the Homology Theory of Fibre Bundles', J. Springer 1967 (Lecture Notes in Mathematics No. 36).

H. Cartan and S. Eilenberg, 'Homological Algebra', Princeton University Press 1956 (Princeton Mathematical Series No. 19).

P. E. Conner and E. E. Floyd (1), 'Differentiable Periodic Maps', J. Springer 1964 (Ergebnisse series No. 33).

P. E. Conner and E. E. Floyd (2), 'The Relation of Cobordism to K-theories', J. Springer 1966 (Lecture notes in Mathematics No. 28).

A. Dold, 'Halbexakte Homotopie Funktoren', J. Springer 1966 (Lecture Notes in Mathematics No. 12).

J. Dugundji, 'Topology', Allyn and Bacon 1966.

B. Eckmann, 'Homotopy and Cohomology Theory', in Proceedings of the International Congress of Mathematicians 1962, Institut Mittag-Leffler 1963, pp 59-73.

S. Eilenberg and N. E. Steenrod, 'Foundations of Algebraic Topology', Princeton University Press 1952 (Princeton Mathematical Series No. 15).

P. Freyd, 'Abelian Categories', Harper and Row 1964.

P. Gabriel and M. Zisman, 'Calculus of Fractions and Homotopy Theory', J. Springer 1967 (Ergebnisse series No. 35).

R. Godement, 'Théorie des Faisceaux', Hermann 1958 (Actualités series 1252).

M. Greenberg, 'Lectures on Algebraic Topology', Benjamin 1967.

P. J. Hilton (1), 'An Introduction to Homotopy Theory', Cambridge University Press 1953 (Cambridge Tracts series No. 43).

P. J. Hilton (2), 'Homotopy Theory and Duality', Gordon and Breach, 1965.

P. J. Hilton and S. Wylie, 'Homology Theory', Cambridge University Press 1960.

F. Hirzebruch, 'Topological Methods in Algebraic Geometry (3rd ed., translated), J. Springer 1966.

J. G. Hocking and G. S. Young, 'Topology', Addison-Wesley 1961.

S. T. Hu, 'Homotopy Theory', Academic Press 1959.

W. Hurewicz and H. Wallman, 'Dimension Theory', Princeton University Press 1948 (Princeton Mathematical Series No. 4).

D. Husemoller, 'Fibre Bundles', McGraw-Hill 1966.

S. MacLane, 'Homology', J. Springer 1963 (Grundlehren series No. 114).

W. S. Massey, 'Algebraic Topology: An Introduction', Harcourt Brace and World, 1967.

J. P. May, 'Simplicial Objects in Algebraic Topology', Van Nostrand 1967 (Mathematical Studies series No. 11).

J. W. Milnor, 'Morse Theory', Princeton University Press 1963 (Annals of Mathematics Study No. 51).

B. Mitchell, 'Theory of Categories', Academic Press 1965.

R. S. Palais, 'Seminar on the Atiyah-Singer Index Theorem', Princeton University Press 1965 (Annals of Mathematics Study No. 57).

L. S. Pontryagin, 'Foundations of Combinatorial Topology', Graylock Press 1952.

H. Siefert and W. Threlfall, 'Lehrbuch der Topologie', Teubner 1934.

E. H. Spanier, 'Algebraic Topology', McGraw-Hill 1966.

N. E. Steenrod, 'The Topology of Fibre Bundles', Princeton University Press 1951 (Princeton Mathematical Series No. 14).

N. E. Steenrod and D. B. A. Epstein, 'Cohomology Operations', Princeton University Press 1962 (Annals of Mathematics Study No. 50).

R. G. Swan, 'The Theory of Sheaves', University of Chicago Press 1964).

E. Thomas, 'Seminar on Fibre Spaces', J. Springer 1966 (Lecture Notes in Mathematics No. 13).

H. Toda, 'Composition Methods in Homotopy Groups of Spheres', Princeton University Press 1962 (Annals of Mathematics Study No. 49).

A. H. Wallace, 'Algebraic Topology', Pergamon 1957.

G. W. Whitehead, 'Homotopy Theory', The M. I. T. Press 1966.

J. H. C. Whitehead, 'The Mathematical Works of J. H. C. Whitehead', Pergamon Press 1962.

In general, Spanier is the most useful single reference for the central core of the subject, followed by Husemoller for those topics which he treats.

1. A first course

I assume that most readers of this book will have had a first course in algebraic topology. This section, then, is included for completeness, and it can hardly escape a certain air of being directed at the teacher rather than the student. It is hoped that this slant diminishes in later sections.

A basic course in algebraic topology should certainly try to present a variety of phenomena typical of the subject. The author or lecturer should display a variety of spaces: cells, spheres, projective spaces, classical groups and their quotient spaces, function spaces .... Equally, one should display a variety of maps, that is, continuous functions between spaces. One must give the definition of homotopy, and one can then display a variety of phenomena or typical problems. First, we have classification problems, for example, the classification of maps $f:X \to Y$ into homotopy classes. (This can be illustrated by considering the case in which X and Y are the circle S^1; the existence and properties of the degree of a map $f:S^1 \to S^1$ can be stated as a theorem whose proof is deferred only a short time. One then has many applications to plane topology: the Brouwer fixed-point theorem for the disc E^2, the fundamental theorem of algebra, separation theorems, the topology needed for Cauchy's theorem in complex analysis, vector fields and critical-point theory in the plane.... But time presses one on.) Secondly, one has extension problems; the homotopy extension property comes in here, at least for simple pairs like the n-cell E^n and its boundary S^{n-1}. Thirdly, one has lifting problems; for this one must display and discuss fiberings, including coverings. (Some authorities prefer a separate preliminary discussion of coverings, probably in connection with the fundamental group; but personally I believe in going straight to fiberings, with coverings as an important special case.) One must also prove the homotopy lifting property, at least for simple spaces like the n-cube I^n. (At this point one can prove the theorem about the degree of a map $f:S^1 \to S^1$, by using the covering map from the real line R^1 to S^1.) By analogy with the word 'fibering', one introduces 'cofiberings' or 'cofibrations' in studying extension problems.

The basic facts about homotopy are given in Dugundji chaps. 15, 18, Greenberg part 1, *Hilton (1) chap. 1, Hocking and Young chap. 4, Hu chap. 1, and Spanier chap. 1. For classification problems, see Hu chap. 1. For extension problems, see *Hu chap. 1 or Spanier chap. 1. For lifting problems, see *Hu chap. 1 or Spanier chap. 2. For fiberings, see Dugundji chap 20, *Hilton (1) chap. 5, Hu chap. 3 or Spanier chap. 2. For cofiberings, see Spanier chap. 1.

Of course one has to face the question, what is the good category of spaces in which to do homotopy theory? Personally, I believe that one should introduce CW-complexes into even a basic course; I would advocate going as far as the theorem that every map between CW-complexes is homotopic to a cellular map. Up to this point the material belongs almost wholly to analytic topology; this theorem is usually proved by simplicial approximation, but it can be proved by an ad hoc subdivision argument, subdividing the cube by hyperplanes parallel to its faces. (Such a subdivision has already been used to prove the homotopy lifting property.)

The material on CW-complexes may be found in Hilton (1) chap. 7, Spanier chap. 7 or G. W. Whitehead chap. 2. The best source, however, is probably the original paper by *J. H. C. Whitehead, and an appropriate extract is reprinted here (see Paper no. 1).

Next, one must certainly define absolute and relative homotopy groups, and prove some of their logically elementary properties (for example, the exact sequences of a pair and a fibering). Some authorities prefer a preliminary discussion of the fundamental group $\pi_1(X)$, but personally I believe in saving time here and defining the groups $\pi_n(X, A)$ for all n at one blow. Some authors might advocate proceeding in even greater generality, defining track groups, homotopy groups of maps and so forth; but if these are needed they can quickly be obtained as homotopy groups of suitable function-spaces.

The material on homotopy groups may be found in *Hilton (1) chaps. 2, 4 and 5, Hu chaps. 4 and 5 or Spanier chap. 7. For the more general groups, see Eckmann.

At this point, or perhaps earlier, it becomes evident that one needs methods for effective calculation. This means homology theory. To give the student the feel of the subject, one should probably begin with finite simplicial homology theory. It is enough to consider only finite simplicial complexes equipped with a given ordering of the vertices; this cuts out a good deal of confusing verbiage about orientations. It is necessary to give the basic definitions and certain variations of them: relative homology, cohomology, and the use of different coefficient groups. It is not necessary to prove the topological invariance of finite simplicial homology; students at this stage usually find the proof tedious and unilluminating, and in any case the result follows from later theorems.

The material on finite simplicial homology may be found in the Sêminaire H. Cartan 1948/49 (2nd ed.) exposês 1-4, *Hilton and Wylie chaps. 2 and 5, Hocking and Young chaps. 6 and 7 or Spanier chap. 4.

Next one must introduce the Eilenberg-Steenrod axioms, set up singular homology theory, and prove that it satisfies the axioms. Here it is open to argument whether one should set up both the theory based on simplexes and the theory based on cubes, or whether one should use only simplexes. The arguments in favour of cubes are as follows. First, it may be held that the student gains from seeing that there are at least two ways of setting up a homology theory, and that any way will do providing that it works. Secondly, there are various points at which is is marginally easier or more convenient to work with cubes rather than simplexes, and at such points it is pleasant to be able to mention cubes. (Such points arise, for example, in passing from a geometrical homotopy to a chain homotopy, and in proving the Hurewicz isomorphism theorem.) Thirdly, the cubical theory is used in various classical papers which the student might want to read, such as Serre's thesis (see §5). The arguments against cubes are as follows. Against the second and third points, it appears to be true that by using extra effort, or later methods, it is possible to avoid the use of cubes at all points where they are easier or were used by classical authors. And therefore, against the first point, why spend the time and risk confusing the issue? Personally, I still like cubes. In any case, at this stage it is certainly not necessary to prove the equivalence of the two singular theories, or that the singular theories agree with the finite simplicial theory on finite simplicial complexes; both results follow from later theorems. However, one should carry the work far enough to compute the homology of a few simple spaces such as spheres.

There are many good accounts available of this material. They include Eilenberg and Steenrod chaps 1 and 7, *Greenberg part 2, Hilton and Wylie chap. 8, Spanier chap. 4 and Wallace chaps. 5, 6, 7 and 8. The original paper by S. Eilenberg ('Singular homology theory', Annals of Mathematics 45 (1944), 407-447) is as pleasant to read as any, and is recommended; but with the other sources available it would be hard to justify reprinting 40 pages. I have however found space for the original paper by Eilenberg and Steenrod (Paper no. 2) which is both elegant and lucid.

The final topic which should be included in a first course is the Hurewicz isomorphism theorem. Technically, of course, it is possible to delay the proof until further machinery is developed and one can give the painless proof due to Serre (see §8). Personally I prefer to give a fairly elementary proof at this stage. Such a proof has two main pillars: the additivity lemma, and the result that the homology of an n-connected space X can be defined in terms of singular simplexes or cubes with their n-faces at the base-point. If one uses cubes, the additivity lemma can be proved fairly easily by direct geometrical construction; alternatively, one can prove everything at once by induction over the dimension. If one uses simplexes, it is still possible to prove the additivity lemma by direct geometrical construction, but the proof is unpleasant, and in my opinion the proof by induction is preferable. The homology result is straightforward, but at this stage probably involves an irreducible amount of work, which is worse for cubes because of normalisation. (The work can be made easier if one has available the geometrical realisation of the total singular complex of X - see §3. This singular complex may be either simplicial or cubical; its 'realisation' is a CW-complex possessing a map to X, and this map can be deformed in the required way by standard theorems. However, one would not expect this 'realisation' to be available at this stage.)

Proofs of the Hurewicz isomorphism theorem are given in *Spanier chap. 7 and G. W. Whitehead chap. 2.

This completes the material appropriate to a basic course, except that some authorities would include some of the material which I have collected for convenience in §4. From this point on there is much more freedom about the order in which the material can be taken. In fact, the ordering of the sections below does not reflect the order in which I hope a student would learn the subject. Sections 6 and 7 are placed where they are because of their close relation with §5; but I would hope that a student would learn something from §§10 and 12 at an early stage.

2. Categories and functors

The student cannot escape learning about these as he goes along. Thus no special reading is necessary. If references are required, see Eilenberg and Steenrod chap. 4, Freyd, MacLane chap. I, Mitchell or *Spanier chap. 1.

3. Semi-simplicial complexes

The student should know the basic definitions; these may be found in Hilton and Wylie pp 358-359, Hu pp 140-142 or *MacLane pp 233-236. (The theory is taken rather further in the Sêminaire H. Cartan, 1956/57, exposê 1.) These complexes are useful in formalising some of the constructions and proofs about singular homology. They are also valuable in homological algebra; here they allow one to start from strictly algebraic or combinatorial foundations, and yet obtain objects to which one can apply all the techniques of algebraic topology (see for example Andrê). Personally, I am not too much impressed by the arguments that they provide a good category in which to do homotopy theory, although they have been much used in discussing Postnikov systems (see §10). The use of these complexes seems most profitable when one can consider semi-simplicial complexes with a strong algebraic structure. This subject is well represented by Milnor's paper 'On the construction FK' reprinted here as Paper no. 10. See also Bousfield, Curtis, Kan, Quillen, Rector and Schlesinger, 'The mod p lower central series and the Adams spectral sequence', Topology 5 (1966), 331-342.

If the student wishes to study semi-simplicial complexes, I would suggest the book by *J. P. May.

4. Ordinary homology and cohomology

The student will certainly require a working knowledge of ordinary homology and cohomology. Here I am taking for granted the standard 'diagram-chasing' material such as the Mayer-Vietoris sequence, the Five Lemma etc., for which all authors follow Eilenberg and Steenrod chap. 1. But the student should also cover the following: (a) direct and inverse limits, (b) the universal coefficient theorem, with enough homological algebra to understand it, (c) the Eilenberg-Zilber theorem, (d) the Kunneth theorem, (e) all relevant product operations, such as Kronecker products, cup products, cap products, cross products, Pontryagin products and slant products, (f) the Bockstein operations, (g) other primary operations such as the Steenrod squares and reduced powers, and (h) some introduction to secondary operations such as the Massey product.

One can find (a) in Eilenberg and Steenrod chap. 8 or *Spanier Introduction (although a little more generality is sometimes useful). One can find (b) and (d) in MacLane chaps. 3 and 5. One can find (b) to (f) inclusive in Hilton and Wylie chaps. 4, 5, 8 and 9 or *Spanier chaps. 5 and 6 (except that not all the products appear explicitly in both). One can find (h) in the Seminaire H. Cartan, 1958/59 exposé 13, in Spanier, 'Secondary operations on mappings and cohomology', Annals of Math. 75 (1962), 260-282, or in Adams, 'On the non-existence of elements of Hopf invariant one', Annals of Math. 72 (1960) 20-104, especially pp 21-24, 52-74. As for (g), the Steenrod operations (which link up with §§7 and 10 below) can be found in Steenrod and Epstein; however, many readers of this book will prefer to delete all references to 'regular cell complexes' and instead to define Steenrod operations for semi-simplicial complexes, using the method of acyclic models; for the Steenrod squares this may be found in Spanier chap. 5.

This of course brings one to the method of acyclic models. This method can often be bypassed by giving explicit formulae, but it is convenient, conceptual and illuminating. It may be found in Hilton and Wylie chap. 8 or in *Spanier chap. 4. Unfortunately, Hilton and Wylie seem to neglect to show that normalised cubical chains are representable - a crucial point; I understand that this omission is to be repaired in their second edition. (Spanier bypasses the difficulty by considering only unnormalised simplicial chains.) The original paper (S. Eilenberg and S. MacLane, 'Acyclic models', American Jour. Math. 79 (1953), 189-199) is not easy for a beginner to read. The best treatment is due to A. Dold, S. MacLane and U. Oberst, in 'Reports of the Midwest Category Seminar', J. Springer 1967 (Lecture Notes in Mathematics series No. 47).

The student will also need to know about spectral sequences. Here there is of course a difference of opinion; some seek to avoid the use of spectral sequences whenever possible, while others seek to introduce them as soon as possible and use them as a heavy hammer to force out subsequent theorems. Personally I incline to the latter school. For convenience, however, I have given spectral sequences a separate section of their own (see §5).

Most authors would add two further topics to this section. Some knowledge of fixed-point theorems is certainly desirable for one's general education, but surprisingly little of the rest of algebraic topology depends on it. The Brouwer fixed-point theorem may be treated as an example as soon as one has set up a homology theory; on this point all authors follow Eilenberg and Steenrod chap. 11. It is probably good to know something about the Lefschetz fixed-point formula; this may be found in Hilton and Wylie chap. 5 appendix 1 or *Spanier chap. 4.

It is even more desirable for one's general education to know something of the homology theory of manifolds (Poincaré duality, Alexander-Lefschetz duality). Indeed, for a student who proposes to work on manifolds this is essential. Treatments may be found in Greenberg parts 3 and 4, Husemoller chap. 17 and *Spanier chap. 6. The homology theory of manifolds links up with the homology theory of vector-bundles (see §11) via the tangent bundle (or microbundle) of a manifold and the Thom isomorphism. For the Thom isomorphism, see Husemoller chap. 16. Unfortunately, it is clear by now that for an idealistic treatment, one should take the whole of this subject and do it for generalised homology and cohomology theories (see §12) rather than for ordinary ones. The foundation paper in this direction is G. W. Whitehead, 'Generalised homology theories', Trans. American Math. Soc. 102 (1962), 227-283. Unfortunately, it does not do the whole of what I have just asked; the idealistic treatment parallels more closely that given in Spanier for the case of ordinary homology. A summary of some of the material in Whitehead's important paper, together with some additional material, is given here as Paper no. 13.

5. Spectral sequences

A spectral sequence is an algebraic gadget like an exact sequence, but more complicated. In particular, it contains groups $E_r^{p,q}$ for $r = 2, 3, \ldots, \infty$. Like an exact sequence, it does not provide a guarantee that one can carry out any required calculation effectively, but the experts succeed with it more often than not.

One may approach the subject as follows. Suppose given a space X filtered by a sequence (either increasing or decreasing) of subspaces X_n. Suppose given a functor (such as homology, cohomology, or homotopy) which assigns to each pair X_n, X_m an exact sequence, subject to various axioms. Then one has a whole maze of interlocking exact sequences, and from this one can distil out the algebraic gadget called a spectral sequence.

This 'distillation' can be performed in two ways, which I will call the 'explicit' and the 'implicit'. There is no essential difference. To proceed 'explicitly' one simply writes down an explicit definition for the group $E_r^{p,q}$ in terms of the given data, and similarly for anything else one wishes to define. This method is well represented by the exposé of Eilenberg, reprinted here as Paper no. 3. (Or see Cartan and Eilenberg pp 333-336.) The 'implicit' method is the method of 'exact couples'. An exact couple is an algebraic gadget abstracted from a 'maze of interlocking exact sequences', as described above. One explains how to pass from an exact couple to its 'derived exact couple', and by iteration one obtains the whole spectral sequence. The advantage of the method is its elegance. The disadvantage of the method is that at any time one may want to have explicit formulae for the groups $E_r^{p,q}$, either to bolster one's confidence or to perform particular arguments. Therefore it seems well to supplement the method of 'derived couples' with some explicit formulae. But given the explicit formulae, one can dispense with the derived couples, except perhaps for insight and guidance. The method of exact couples is due to Massey, and an extract from his work is reprinted here as Paper no. 4.

The first example and application of spectral sequences is certainly the theorem (due to Serre) that a fibering gives rise to spectral sequences in (singular) homology and cohomology. More precisely, suppose given a fibering with fibre F, total space E and base B; then there is a spectral sequence with $E_2^{p,q} \cong H^p(B; H^q(F))$ and such that the terms $E_\infty^{p,q}$ with $p+q=n$ form a series of composition-quotients for $H^n(E)$.

This theorem may be proved in two ways. On the one hand one may give a direct, explicit construction and proof, following J.-P. Serre, 'Homologie singulière des espaces fibrés', Annals of Math. 54 (1951), 425-505. This is advantageous for certain later work, such as Kudo's proof of Kudo's transgression theorem (see §7). On the other hand, it appears to be true that by using later methods one can always avoid appealing to the details of Serre's construction. A modern proof along Serre's lines is given by A. Dress, 'Zur Specktralsequenz von Fazerungen', Inventiones Math. 3 (1967), 172-178. (See also André; but readers who are only interested in this proof, rather than in homological algebra, may find that André asks them to swallow rather a lot of category-theory first.)

The second way goes back to Massey. One may suppose without loss of generality that the base B is a CW-complex with skeletons B^n. Let $p: E \to B$ be the projection. Then one can filter E by considering the subspaces $p^{-1}B^n$, and one obtains the situation considered at the beginning of this section. A treatment along these lines can be combined with a treatment of certain standard results on generalised cohomology theories (see §12, and the paper by Dold reprinted here as Paper no. 14). These results need to be known anyway, and a proof along these lines can be fairly short and conceptual. I feel that this approach can be recommended to the student.

Discussions of spectral theory without complete proofs are given in Hilton and Wylie chap. 10, *MacLane chap. 11 and G. W. Whitehead chap. 3. A proof closely following Serre is given in Hu chaps. 8 and 9. Versions of the second proof mentioned above are given in the Séminaire H. Cartan 3 (1950/51) exposé 9, in Dold and in *Spanier chap. 9 pp 446-481.

6. H*(BG)

An especially important application of spectral sequences to fiberings arises in the case when the total space E is contractible. In some cases we know the homology or cohomology groups of B and wish to infer those of F. This arises, for example, when F is the loop-space of B (see Spanier p 37). In some cases we know the homology or cohomology groups of F and wish to compute those of B. This arises when F is a Lie group G and B is its classifying space BG (see §11). It also arises when F is an Eilenberg-MacLane space $K(\pi, n)$ and B is an Eilenberg-MacLane space $K(\pi, n+1)$ (see §7).

The first theorems in this subject are due to A. Borel, 'Sur la cohomologie des espaces fibrés principaux et des espaces homogenes de groupes de Lie compacts', Annals of Math. 57 (1953), pp 115-207. (See also the book by Borel.) One version of Borel's first theorem states that if $H_*(F)$ is an exterior algebra (under the Pontryagin product), and E is contractible, as always in this section, then $H^*(B)$ is a polynomial algebra. Moreover, the suspension map

$$H^n(B) \xrightarrow{p^*} H^n(E, F) \underset{\cong}{\overset{\delta}{\leftarrow}} H^{n-1}(F)$$

yields a good correspondence between the generators in B and F. In Borel's second theorem one supposes given elements in $H^*(B)$ whose suspensions behave well in F, and one needs correspondingly less data in F.

Borel's original paper may be found hard to read, and it is not necessary to do so as there are later methods available. The foundation paper in this direction, which introduced homological algebra into the subject, is perhaps J. C. Moore, 'Algèbre homologique et homologie des espaces classifiants', Séminaire H. Cartan 12 (1959/60) exposé 7. So far as getting from H(B) to H(F) goes, the results were written up by S. Eilenberg and J. C. Moore, 'Homology and fibrations I', Commentarii Math. Helvetici 40 (1966), 201-236. For getting from H(F) to H(B), I recommend the paper by M. Rothenberg and N. E. Steenrod, reprinted here as no. 5.

7. Eilenberg-MacLane spaces and the Steenrod algebra

An Eilenberg-MacLane space is a space whose homotopy groups are all zero except for one; say $\pi_n(X) = G$, $\pi_r(X) = 0$ for $r \neq n$. We then write $X = K(G, n)$. An excellent account of Eilenberg-MacLane spaces is given by J.-P. Serre, Séminaire H. Cartan 7 (1954/55) exposé 1. Their importance is twofold. First, they are important in homotopy theory (see §10). Secondly, they are closely linked with the study of cohomology operations. To be more precise, the cohomology of an Eilenberg-MacLane space X, as above, depends on n and G, and there is a (1-1) correspondence between $H^m(X;G')$ and the set of all natural cohomology operations from $H^n(Y;G)$ to $H^m(Y;G')$. On this point, see the paper by *Serre reprinted here as no. 6. This is the paper in which Serre calculated the cohomology of Eilenberg-MacLane spaces with Z_2 coefficients. (It is important to know the cohomology of Eilenberg-MacLane spaces for both the applications mentioned above.) The cohomology with Z_p coefficients was first calculated by Cartan; see the Séminaire H. Cartan 7 (1954/55). This calculation can also be done by a version of Serre's method; see M. M. Postnikov, 'On Cartan's theorem', Russian Mathematical Surveys 21 (1966), 25-36. This method needs a lemma about the way Steenrod operations behave in the spectral sequence of a fibering. This lemma was formulated by A. Borel (see W. S. Massey, 'Some problems in algebraic topology and the theory of fibre bundles', Annals of Math. 62 (1955), 331). It was proved by T. Kudo, 'A transgression theorem', Mem. Fac. Sci. Kyusyu Univ. 9 (1956), 79-81. Fortunately, it is now possible to do the calculations a third way, following Moore, Rothenberg and Steenrod (see §6).

The mod p Steenrod algebra A is by definition the set of all natural stable operations from $H^*(Y;Z_p)$ to $H^*(Y;Z_p)$; the product is given by composition of operations. (This definition extends immediately to generalised cohomology theories - see §12.) The work of Serre and Cartan mentioned above issues in a determination of the structure of A. In order to state the results conceptually, it is convenient to begin by remarking that A is a Hopf algebra (see §11). In fact, the cup-product operation corresponds to a map

$$\mu: K(Z_p, n) \times K(Z_p, m) \to K(Z_p, n+m) ;$$

the induced map of cohomology is independent of n and m in a suitable range of dimensions, and defines a coproduct map $\mu^*: A \to A \otimes A$. (This account generalises, too.) One can now introduce the dual Hopf algebra A^* (corresponding to the homology of Eilenberg-MacLane spaces rather than their cohomology). One can introduce explicit elements in the dual A*, and so describe completely the structure of A* and A. All this is due to *J. Milnor, 'The Steenrod algebra and its dual', Annals of Math. 67 (1958), 150-171; this paper is highly recommended.

For the beginner, a sufficient introduction to the above for the case $p = 2$ is given in Adams chap. 2. There is a purely algebraic account of the Steenrod algebra in Steenrod and Epstein; but personally I prefer treatments in which the conceptual background in algebraic topology is placed first, while explicit generators and relations for the Steenrod algebra emerge as a result of calculation.

8. Serre's theory of classes of abelian groups (C-theory)

These results are of very general use, and all students of algebraic topology should certainly study them. The central theorems are the absolute Hurewicz isomorphism mod C, the relative Hurewicz isomorphism mod C, and the J. H. C. Whitehead theorem mod C. This material may be found in Hu chap. 10 or *Spanier chap. 9. The original paper by J.-P. Serre ('Groupes d'homotopie et classes de groupes abeliens', Annals of Math. 58 (1953), 258-294) is most elegant, and I recommend it highly, but with the other sources available it would be hard to justify reprinting it.

9. Obstruction theory

The idea of an obstruction is a fundamental one, and the student should get some feeling for it. It can be approached in a comparatively direct and elementary way; this is done in *Hilton and Wylie chap. 7, Hu chap. 6, Steenrod part 3 and G. W. Whitehead chap. 2. Unfortunately, this method becomes tedious if one really wishes to talk about higher obstructions. One can import a helpful geometric framework by introducing Moore-Postnikov factorisations, as is done in Spanier chap. 8. This approach would come more naturally after a student has met Postnikov systems and the method of killing homotopy groups in homotopy theory (see §10). Finally, one can achieve much the same ends by constructions involving generalised cohomology theories (see §12) and spectral sequences. See the book by Dold.

10. Homotopy theory

Here again we meet the question of the correct category in which to do homotopy theory, and at this point I give two further references: J. Milnor, 'On spaces having the homotopy type of a CW-complex', Trans. Amer. Math. Soc. 90 (1959), 272-280, and N. E. Steenrod, 'A convenient category of topological spaces', Michigan Math. Jour. 14 (1967), 133-152.

One can distinguish three strands in homotopy theory; suspension theory, explicit geometrical constructions such as Whitehead products, and the method of killing homotopy groups. The non-specialist, who wishes to apply topology to other topics such as smooth manifolds, will need to know the basic theorems of suspension-theory, so as to be able to distinguish phenomena which are 'stable' (see below) from those which are not; he should also be aware of the existence of the most usual geometrical constructions, and of the existence of the method of killing homotopy groups. The specialist will need to go further; this is one of the densest tracts of jungle we have to survey.

Let X and Y be (say) two CW-complexes with base-points, and let [X, Y] be the set of homotopy classes of maps (preserving the base-point) from X to Y. The suspension construction

(Spanier p. 41) constructs from each space X its suspension SX, and defines a function $S:[S, Y] \to [SX, SY]$. The first, 'crude' theorem in suspension-theory states that this function is a (1-1) correspondence under suitable assumptions on X and Y. For example (taking $X = S^q$, $Y = S^n$) suspension defines a homomorphism $S:\pi_q(S^n) \to \pi_{q+1}(S^{n+1})$ between the homotopy groups of spheres, and the theorem asserts that this homomorphism is an isomorphism if $q < 2n-1$. Phenomena which are independent of a dimensional parameter, in the same way that the structure of $\pi_{n+r}(S^n)$ is independent of n for $n > r + 1$, are called 'stable'. (It is unwise to attempt too rigid a definition of the word 'stable', because of the variety of the phenomena to which it is applied; for example, one says that the Steenrod squares

$$Sq^i:H^n(X;Z_2) \to H^{n+i}(X;Z_2)$$

are 'stable'.)

The most illuminating proof of the suspension theorem mentioned above proceeds by considering function-spaces. If Z is a space with base-point, let ΩZ be the loop-space of Z (Spanier p. 37). Then one has an embedding $i:Y \to \Omega SY$ and one obtains the theorem by studying this embedding. For this purpose spectral sequences are useful but not absolutely essential. See *Spanier chap. 8; the special case $Y = S^n$ is also given in the Séminaire H. Cartan 1958/59 exposés 5, 6, Hilton and Wylie chap. 10, Hu chap. 11 and G. W. Whitehead §3. 8. It is clear that the non-specialist will need to go as far as this; it is not clear that he will need to go any further.

So far we have been discussing essentially the case of absolute homotopy groups. The work can be put in a more general setting by introducing triad homotopy groups; these allow one to state the main 'crude' suspension theorem in a form which allows one to make not only deductions about absolute homotopy groups, but also deductions about relative homotopy groups, like the result given in Spanier p. 484 theorem 5. We thus reach the triad connectivity theorem. This theorem should be stated and proved in the context of Serre's C-theory (see §8). It should also be stated so as to give the structure of the first triad group which is not zero; for this purpose one needs a geometrical construction (the generalised Whitehead product). The extract from my own lecture-notes (Paper no. 7) is included to cover this topic.

Most students will not need to pursue the generalisation from triads to n-ads.

The suspension theorems which we have considered so far apply to a wide class of spaces, but give results only for a limited range of dimensions. There are further theorems in the same spirit. For example, consider the EHP sequence of G. W. Whitehead, 'On the Freudenthal theorems', Annals of Math. 57 (1953), 209-228. (The statement is reprinted here as Paper no. 8.) This sequence can be formulated for a general space X. By contrast, sophisticated suspension theorems apply only to selected spaces, but give results for all dimensions. For example, a celebrated result of James may be viewed as asserting that if n is odd, then there is a fibering $F \to E \to B$ in which F, E and B are (up to weak homotopy type) S^n, ΩS^{n+1} and ΩS^{2n+1}. If we take the exact homotopy sequence of this fibering, we obtain an exact sequence valid in all dimensions, in which each group is a homotopy group of S^n, S^{n+1} or S^{2n+1}. James' work appeared in the Annals of Math. 62 (1955), 170-197, 63 (1956), 191-247 and 63 (1956), 407-429, and a brief extract is reprinted here as Paper no. 9. There are further results of James and Toda in this direction. The exploitation of such exact sequences in computing homotopy groups of spheres usually involves geometrical constructions, such as the Toda bracket (see below); it may involve other methods as well.

For 'sophisticated' suspension-theory, see Toda chaps. 2, 13.

It will be observed that the exact sequence of 'sophisticated' suspension-theory involve not only the suspension homomorphism E, but also two other homomorphisms, say H and P, so that Im E = Ker H and Ker H = Im P. One needs to know the behaviour of both H and P with respect to suitable geometrical constructions; see Toda chaps. 2, 13. The homomorphism H is usually described as a 'Hopf invariant'; for a discussion of the different possible definitions of 'Hopf invariants' and the relations between them, see M. G. Barratt, Reports of Seminar in Topology, University of Chicago, 1957, part I §III (mimeographed notes). The homomorphism P is related to the Whitehead product (see below).

Next we come to explicit geometrical construction. The following are standard. (a) Suspension (see above). (b) Composition; also secondary (and higher) compositions, that is, 'toric constructions' or 'Toda brackets'. (c) Whitehead products, together with their generalisations, relative and generalised Whitehead products. (d) Last and perhaps least important, the Hopf construction.

For (b) see Toda chap. 1, or Spanier, 'Secondary operations on mappings and cohomology', Annals of Math. 75 (1962), 260-382. For (c) see Hilton (1) p. 92, Hu pp 138-139, Spanier pp 419-420 or the extract from my lecture-notes given as Paper no. 7. For (d) see G. W. Whitehead, 'On the homotopy groups of spheres and rotation groups', Annals of Math. 43 (1942), 634-640.

Besides the definitions of these constructions, one also has to know various identities which they satisfy. Here the method of the 'universal example' is often useful. Thus, if we have to consider operations defined on two variables $\alpha \in \pi_p(X)$, $\beta \in \pi_q(X)$ it is sufficient to consider the case in which X is the 'wedge sum' or union with one point in common, $S^p \vee S^q$, while α and β are the injections of S^p and S^q. The homotopy groups $\pi_r(S^p \vee S^q)$ were calculated (in terms of those spheres) by P. J. Hilton, 'On the homotopy groups of the union of spheres', Jour. London Math. Soc. 30 (1955), 154-172. An exposition of all this is given by Serre in the Séminaire H. Cartan 1954/55, exposé 20. Hilton's work was generalised by Milnor, whose paper 'On the construction FK' is reprinted here as Paper no. 10. The work of Hilton and Milnor sheds light on the rather complicated formula which exists for expanding the composite $(\alpha + \beta)\gamma$ ('left distributive law'). For the most illuminating work on the left distributive law, see J. M. Boardman and B. Steer, 'On Hopf invariants', Commentarii Math. Helvetici 42 (1967), 180-221.

Next we come to the method of killing homotopy groups. We observe that any space X can be approximated by an interated fibering, in which the factors are Eilenberg-MacLane spaces whose homotopy groups are those of X. The spaces and maps occurring in these fiberings are often called the 'Postnikov decomposition' of X. If we start by knowing the homotopy groups of X, and sufficiently much about the fiberings in its Postnikov decomposition, we can in principle obtain information about the homology and cohomology of X. In practice, however, we argue in the reverse direction; we begin with information about the homology and cohomology of X, and deduce information about its homotopy groups.

The construction of the Postnikov decomposition is given in Hilton (1) pp 67-68, Hu pp 155-159 and Spanier pp 437-444. Of these, Hilton is the shortest and Spanier the most general. See also the passage from one of my own papers reprinted as no. 11. To appreciate the object of the exercise and get the flavour of the homological calculations, see the original notes by Cartan and Serre reprinted here as Paper no. 12.

A reformulation of the method of killing homotopy groups, so far as it applies to stable homotopy theory, has been given by myself. It is sometimes called the 'Adams spectral sequence'. Accounts may be found in my paper 'On the structure and applications of the Steenrod algebra', Commentarii Math. Helvetici 52 (1958), 180-214, in the Séminaire H. Cartan 1958/59, exposés 18, 19, or in Adams chap. 4. An application of this method may be seen in Paper no. 23.

11. Fibre bundles and topology of groups

Under this heading I include fibre-bundles, their characteristic classes, the topology of groups and their classifying spaces, and the study of H-spaces and Hopf algebras. It will be clear that this section comes near to overlapping with the companion volume on differential topology; but most of it comes within the terms of reference of this volume. For example, even if one were a narrow-minded algebraic topologist, one might wish to work on K-theory or cobordism (considered as a generalised cohomology theory), and one would then need to know the contents of this section.

The student should certainly know the definition of a fibre-bundle with structural group G. This may be found in *Husemoller chaps. 2-6 or Steenrod part I. It is almost certainly good to avoid assuming that the action of the structural group G on the fibre F is faithful. In fact, if one contemplates the existence of fibrewise-linear maps between vector bundles with fibres of different dimensions, one sees that it is probably good to replace the structural group G by a suitable category, for example, the category of linear maps between finite-dimensional vector spaces; but this is not yet generally accepted. One has also to know something of the theory of fibre-bundles, including the classification theorem. Special attention should be given to vector-bundles.

This material may be found in *Husemoller chaps. 2-6 or Steenrod parts I, II.

The student should know the homology and cohomology (with the usual coefficient groups) of the classical groups O(n), U(n) and Sp(n) and their classifying spaces. The homology of the classical groups is computed in Steenrod and Epstein chap. 4 (although I personally would prefer to make more use of spectral sequences). Alternatively, see Borel chap. 4. The method of inferring the cohomology of the classifying space BG from that of the group G has been discussed in §6 above. The material on classifying spaces includes a knowledge of the nature and properties of various characteristic classes, namely the Stiefel-Whitney classes, Pontryagin classes, Euler classes, Chern classes and symplectic Pontryagin classes. Topologists still hope to see a book by Milnor on this subject; meanwhile, one can find much of this material in Husemoller chaps. 16, 18. For all this, see also the useful survey article by A. Borel, 'Topology of Lie groups and characteristic classes', Bull. Amer. Math. Soc. 61 (1955), 397-432.

There is a link-up between the study of vector-bundles and homotopy theory, via Thom complexes. The definition of a Thom complex may be found in Husemoller chap. 15; for further study, see M. F. Atiyah, 'Thom complexes', Proc. London Math. Soc. 11 (1961), 291-310, which is highly recommended.

An H-space is a generalisation of a topological group; the definition may be found in Dugundji chap. 19, Hocking and Young chap. 4, Hu chap. 3, Husemoller chap. 1, *Spanier chap. 1 and G. W. Whitehead part I. The concept of a Hopf algebra is abstracted from the homology or cohomology (with suitable coefficients) of an H-space; see the Séminaire H. Cartan 1959/60 exposé 2 or Spanier pp 267-269. It is useful in various places in algebraic topology. The standard reference is J. W. Milnor and J. C. Moore, 'On the structure of Hopf algebras', Annals of Math. 81 (1965), 211-264; this paper is recommended, but it is perhaps too long to justify reprinting it.

12. Generalised cohomology theories

A generalised cohomology theory is a contravariant functor satisfying the first six axioms of Eilenberg and Steenrod (see §1). Several such functors have recently become important in algebraic topology; in particular, the K-theory of Grothendieck, Atiyah and Hirzebruch, and various functors provided by cobordism theory (see below). For a survey article which says something about this, see Eckmann.

From the axioms one can deduce various elementary consequences, such as the Mayer-Vietoris sequence (see §1).

The next theorem in the subject is E. H. Brown's representability theorem; this gives necessary and sufficient conditions under which a contravariant functor of X has the form [X, Y] (see §10) for some fixed Y. A treatment may be found in Spanier pp 406-412; or see E. H. Brown, 'Cohomology theories', Annals of Math. 75 (1962), 467-484. It follows that there is a close relation between generalised cohomology theory and homotopy theory, so far as the latter studies the sets [X, Y].

There are also generalised homology theories; for example, stable homotopy provides such a theory. For the relation between homology and cohomology theories, see G. W. Whitehead, 'Generalised homology theories', Trans. Amer. Math. Soc. 102 (1962), 227-283, or Paper no. 13.

We now come to a theorem which is central to the subject. Suppose given a generalised cohomology functor K^*. (The case of homology is similar.) Suppose given also a space X having a monotone sequence (either increasing or decreasing) of subspaces X_n. By applying the functor K^*, we obtain a spectral sequence (see §5). If the spectral sequence converges, it does so towards $K^*(X)$.

If we now specialise to the case in which X is a CW-complex and X_n is its n-skeleton, then we can calculate the E_2 term of this spectral sequence; this is the ordinary cohomology of X, with coefficients in the generalised cohomology K^* of a point. The result becomes a little more explicit if X is a finite simplicial complex; in this case 'ordinary cohomology' is to be taken in the sense of finite simplicial cohomology. In particular, we obtain a proof that any 'ordinary' cohomology theory, if applied to a finite simplicial complex, gives the same result as finite simplicial cohomology (with the same coefficients). Hence, in particular, finite simplicial cohomology is a topological invariant.

For this calculation of the E_2 term, see the paper by Atiyah and Hirzebruch reprinted here as no. 19. (They only consider a particular theory K^*, but their argument is a general one.) Alternatively, see the paper by Dold reprinted here as no. 14. For the sake of history one should perhaps record that the result was a folk-theorem long before; I first heard it from G. W. Whitehead.

The theory above is often applied to calculate generalised cohomology groups. It is clear that it also has much in common with Massey's construction of the spectral sequence of a fibering (see §5 and Paper no. 14). It is also closely related to the theory which arises when we replace $K^*(X)$ by $[X, Y]$; in this case we (essentially) recover obstruction-theory for maps of X into Y (at least for suitable Y). See the lecture-notes by Dold in the main bibliography.

We have mentioned the question of the convergence of the spectral sequence when X is an infinite complex. In order to study this question, it is necessary to know about Lim^1, the derived functor of the inverse-limit functor. See the paper by Milnor reprinted here as no. 15. It is also necessary to know conditions under which $\mathrm{Lim}^1 = 0$; see extracts from Atiyah and Anderson reprinted here as Paper nos. 16, 17.

It is also advisable to know something about products in generalised homology and cohomology theories. See Paper no. 13.

An important application of generalised cohomology theories is to the theorems of 'generalised Riemann-Roch type'. On this subject I have selected the exposition by Dyer reprinted here as Paper no. 18.

We now turn to particular cohomology functors which have been found useful. The first is K-theory. The standard reference is the paper by Atiyah and Hirzebruch reprinted here as no. 19; but Husemoller part 2 is also useful and perhaps more inclusive. It may be felt that now that a book by *Atiyah has appeared, it takes precedence over other sources. The lectures by Hirzebruch reprinted here as Paper no. 20 are also recommended. In order to set up K-theory, one needs the Bott isomorphism theorem.

Various proofs have been given. For the complex case the standard reference is probably M. F. Atiyah and R. Bott, 'On the periodicity theorem for complex vector bundles', Acta Mathematica 112 (1964), 229-247; this proof may be found in Husemoller chap. 10. For the real case I recommend M. F. Atiyah, 'K-theory and reality', Quart. Jour. Math. 17 (1966), 367-386; this paper is reprinted in his book.

In K-theory, one can introduce cohomology operations Ψ^k; see J. F. Adams, 'Vector fields on spheres', Annals of Math. 75 (1962), 603-632, Atiyah p. 135 or Husemoller chap. 12. A summary of my paper is reprinted here as no. 21. For some sample applications of K-theory, see J. F. Adams (loc. cit.) or Husemoller chap. 15; J. F. Adams and M. F. Atiyah, 'K-theory and the Hopf invariant', Quart. Jour. Math. 17 (1966), 31-38, Atiyah p. 137 or Husemoller chap. 14; or J. F. Adams, 'On the groups J(X). IV', Topology 5 (1966), 21-71. Extracts from the last are reprinted here as Paper no. 22.

The theory of cobordism gives rise to various cohomology functors of interest. The most accessible account is in the two books by Conner and Floyd, which give the references of historic importance. A summary on complex cobordism is given as Paper no. 23. Good information about the algebra of cohomology operations on complex cobordism has recently been obtained by S. P. Novikov, 'The methods of algebraic topology from the viewpoint of cobordism theories', Izvestiya Akademii Nauk SSSR (Ser. Mat.) 31 (1967), 855-951.

13. Final touches

Finally, a student's course should include some survey of the current state of the art, and some reading of one or two pieces of current work which are either elegant, or stimulating, or serve in some way to give the feel of research. The narrower purpose might perhaps be served by some of the references in the last two paragraphs of §12. For the wider purpose, I have no hesitation in recommending S. P. Novikov's survey, reprinted here as Paper no. 24. See also the list of problems published by R. Lashof, 'Problems in differential and algebraic topology', Annals of Math. 81 (1965), 565-591 and the comments on it by S. P. Novikov, Uspeki Mat. Nauk 20 (1965), 147-170 (= Russian Mathematical Surveys 20 (1965), 145-167).

Before the student writes anything himself, he should soak himself in papers which are well written. For this purpose I would recommend practically anything written by J. -P. Serre or J. W. Milnor.

1

The first extract is from J. H. C. Whitehead's fundamental paper on CW-complexes, which are a most useful class of spaces in which to do homotopy theory. There is always an analogy between what we can do topologically with a space, and what we can do algebraically with its chain groups, etc.; in this class of spaces the analogy reaches its maximum strength. The main prerequisite for reading this extract is a sound knowledge of general topology. On p. 40 Whitehead also uses the homotopy extension property for the pair E^n, S^{n-1}. Whitehead also makes two references to his earlier papers; the first, on p. 40, is to a geometrical construction which the reader can supply for himself; the second, on p. 42, is to the subdivision argument referred to in §1 above. At the foot of p. 41 Whitehead uses the word 'cellular'; a map $f:X \to Y$ between CW-complexes is said to be cellular if $f(X^n) \subset Y^n$ for each n, where X^n is as defined on p. 33.

1

COMBINATORIAL HOMOTOPY

J. H. C. Whitehead

4. **Cell complexes.**[16] By a *cell complex*, K, or simply a *complex*, we mean a Hausdorff space, which is the union of disjoint (open) cells, to be denoted by e, e^n, e_i^n, etc., subject to the following condition. The closure, $\bar{e}^n$, of each n-cell, $e^n \in K$, shall be the image of a fixed n-simplex, σ^n, in a map, $f:\sigma^n \to \bar{e}^n$, such that

(4.1) (a) *$f|\sigma^n - \partial\sigma^n$ is a homeomorphism onto e^n,*

(b) *$\partial e^n \subset K^{n-1}$, where $\partial e^n = f\partial\sigma^n = \bar{e}^n - e^n$ and K^{n-1} is the $(n-1)$-section*[17] *of K, consisting of all the cells whose dimensionalities do not exceed $n-1$.*

Such a map will be called a *characteristic map* for the cell e^n. If $f:\sigma^n \to \bar{e}^n$ is a characteristic map for e^n, so obviously is $fh:\sigma^n \to \bar{e}^n$, where $h:(\sigma^n, \partial\sigma^n) \to (\sigma^n, \partial\sigma^n)$ is any map such that $h|\sigma^n - \partial\sigma^n$ is a homeomorphism of $\sigma^n - \partial\sigma^n$ onto itself. No restriction other than $\partial e^n \subset K^{n-1}$ is placed on $f|\partial\sigma^n$. Therefore $\bar{e}^n$ need not coincide, as a point set, with a subcomplex of K. Since K, and hence $\bar{e}^n$, is a Hausdorff space and since σ^n is compact it follows that $\bar{e}^n$ has the identification topology determined[18] by f. A complex is defined as a topological space with a certain cell structure. Therefore we shall not need a separate letter to denote a complex and the space on which it lies.[19] Notice that, in the absence of further restrictions, any (Hausdorff) space may be re-

[16] The use of these complexes was suggested in [**3**, p. 1235]. They are now called cell complexes, rather than membrane complexes, in conformity with [**14**].

[17] K^n is defined for every value of n. If there are no m-cells in K for $m>n$ then $K^n = K$.

[18] I.e., $Y \subset \bar{e}^n$ is closed if, and only if, $f^{-1}Y$ is closed. In other words the closed sets in $\bar{e}^n$ are precisely the sets fX for every closed set, $X \subset \sigma^n$, which is saturated with respect to f, meaning that $f^{-1}fX = X$ (cf. [**23**, pp. 61, 95] and [**24**, p. 52]).

[19] N.B. $e \in K$ will mean that e is a cell of the complex K and $e \subset K$, $\bar{e} \subset K$, etc., will mean that the sets of points e, $\bar{e}$, etc., are subsets of the space K.

garded as a complex. For example, we may take it to be the complex $K=K^0$, which consists entirely of 0-cells, each point in K being a 0-cell.

A subcomplex, $L\subset K$, is the union of a subset of the cells of K, which are the cells of L, such that, if $e\subset L$ then $\bar{e}\subset L$. Clearly L is a subcomplex if it is the union of a subset of the cells in K, which is a closed set of points in K. However the above example shows that a subcomplex need not be a closed set of points. Clearly K^n is a subcomplex, for each $n\geqq 0$, and we admit the empty set as the subcomplex K^{-1}. Also the union and intersection of any set of subcomplexes, finite or infinite, are obviously subcomplexes. If $X\subset K$ is an arbitrary set of points we shall use $K(X)$ to stand for the intersection of all the subcomplexes of K, which contain X. Obviously $K(p) = K(e) = K(\bar{e})$, where p is any point in K and $e\in K$ is the cell which contains p. A finite subcomplex, L (i.e. one which contains but a finite number of cells) is a closed, and indeed a compact subset of K. For it is the union of the finite aggregate of compact sets, $\bar{e}$, for each cell $e\in L$.

The topological product, $K_1\times K_2$, of complexes K_1, K_2 is a complex, whose cells are the products, $e^{n_1+n_2}=e_1^{n_1}\times e_2^{n_2}$, of all pairs of cells $e_1^{n_1}\in K_1$, $e_2^{n_2}\in K_2$. For let $f_i:\sigma^{n_i}\rightarrow\bar{e}^{n_i}$ $(i=1, 2)$ be a characteristic map for e^{n_i}, let $g:\sigma^{n_1}\times\sigma^{n_2}\rightarrow\bar{e}^{n_1+n_2}$ be given by $g(p_1, p_2)=(f_1p_1, f_2p_2)$ and let $h:\sigma^{n_1+n_2}\rightarrow\sigma^{n_1}\times\sigma^{n_2}$ be a homeomorphism (onto). Then $gh:\sigma^{n_1+n_2}\rightarrow\bar{e}^{n_1+n_2}$ obviously satisfies the conditions (4.1). Therefore $K_1\times K_2$ is a complex, with this cell structure. In particular $K\times I$ is a complex, which consists of the cells $e\times 0$, $e\times 1$, $e\times(0, 1)$, for each cell $e\in K$, where $(0, 1)$ is the open interval $0<t<1$.

Let K be a locally connected complex, let $\tilde{K}$ be a (locally connected) covering space of K and let $p:\tilde{K}\rightarrow K$ be the *covering map*. That is to say there is a basis, $\{U\}$, for the open sets in K such that, if $U\in\{U\}$ then p maps each component of $p^{-1}U$ homeomorphically onto U (cf. [**20**, p. 40]). Let $\tilde{x}\in\tilde{K}$ be a given point and let $e^n\in K$ be the cell which contains $x=p\tilde{x}$. Then a characteristic map, $f:\sigma^n\rightarrow\bar{e}^n$, can be "lifted" into a unique map,[20] $\tilde{f}:\sigma^n\rightarrow\tilde{K}$, such that $f=p\tilde{f}$ and $\tilde{f}(f^{-1}x)=\tilde{x}$. Let $\tilde{e}^n=\tilde{f}(\sigma^n-\partial\sigma^n)$ and let $p_0=p\,|\,\tilde{e}^n$. Then $f\,|\,\sigma^n-\partial\sigma^n=p_0(\tilde{f}\,|\,\sigma^n-\partial\sigma^n)$ and since $f\,|\,\sigma^n-\partial\sigma^n$ is a (1-1) map onto e^n it follows that p_0 is (1-1) and is onto e_n. Since p, and hence p_0, is an open mapping it follows that p_0 is a homeomorphism. Since

$$\tilde{f}\,|\,\sigma^n-\partial\sigma^n = p_0^{-1}(f\,|\,\sigma^n-\partial\sigma^n)$$

[20] See [**21**, Theorem 2, p. 40] or [**22**]. We shall sometimes use the same symbol, f or g, to denote two maps, $f:A\rightarrow B$, $g:A\rightarrow C\subset B$, such that $fa=ga$ for each point $a\in A$, even though $B\neq C$.

it follows that $\tilde{f}|\sigma^n - \partial\sigma^n$ is a homeomorphism, which, according to the definition of $\tilde{e}^n$, is onto $\tilde{e}^n$. Also $\tilde{f}\partial\sigma^n \subset \tilde{K}^{n-1} = p^{-1}K^{n-1}$. Therefore $\tilde{f}$ satisfies the conditions (4.1). It follows that $\tilde{K}$ is a complex, each of whose cells is mapped by p homeomorphically onto a cell of K.

Let Q be a subcomplex of $\tilde{K}$ and let e be a given cell in Q. Then $p\bar{e}$ is closed, since $\bar{e}$ is compact, and $p\bar{e} \subset pQ$. Therefore $\overline{pe} = p\bar{e} \subset pQ$. Therefore pQ is a subcomplex of K, which consists of the cells pe for each cell $e \in Q$.

5. **CW-complexes.** We shall describe a complex, K, as *closure finite* if, and only if, $K(e)$ is a finite subcomplex, for every cell $e \in K$. Since $K(p) = K(e)$ if $p \in e$ this is equivalent to the condition that $K(p)$ is finite for each point $p \in K$. If $L \subset K$ is a subcomplex and $e \in L$ then obviously $L(e) = K(e)$. Therefore any subcomplex of a closure finite complex is closure finite.

We shall say that K has the *weak topology* (cf. [1, pp. 316, 317]) if, and only if, a subset $X \subset K$ is closed (open) provided $X \cap \bar{e}$ is closed (relatively open) for each cell $e \in K$. If K is closure finite this is equivalent to the condition that X is closed provided $X \cap L$ is closed for every finite subcomplex $L \subset K$. For $X \cap L$ is the union of the finite number of sets $X \cap \bar{e}$ $(e \in L)$. Therefore $X \cap L$ is closed if each set $X \cap \bar{e}$ is closed. Conversely, if $X \cap L$ is closed for each finite subcomplex, L, and if $K(\bar{e})$ is finite, then $X \cap \bar{e}$ is closed, since $X \cap \bar{e} = X \cap K(\bar{e}) \cap \bar{e}$.

By a CW-*complex* we mean one which is closure finite and has the weak topology. Any finite complex, K, is obviously closure finite and it has the weak topology since $X \subset K$ is the union of the finite number of sets $X \cap \bar{e}$ $(e \in K)$. Therefore any finite complex is a CW-complex. Also a complex, K, is a CW-complex if it is *locally finite*, meaning that each point $p \in K$ is an inner point of some finite subcomplex of K. For let K be locally finite. Then $K(p)$ is finite, for each point $p \in K$. Therefore K is closure finite. Let $X \subset K$ be such that $X \cap L$ is closed for each finite subcomplex $L \subset K$. Let L be a finite subcomplex of which a given point $p \in K - X$ is an inner point. Since $X \cap L$ is closed, p is an inner point of $L - X = L - (X \cap L)$. Therefore X is closed and K has the weak topology. It may be verified that the number of cells, and hence the number of finite subcomplexes of a connected, locally finite complex, K, is countable. Hence, and from (G) below, it may be proved that K is a separable metric space.

If the cells in a CW-complex, K, have a maximum dimensionality we call this the dimensionality, dim K, of K. If there is no such maximum we write dim $K = \infty$.

Examples of complexes which are not CW-complexes are:

(1) $\partial\sigma^n$ $(n>1)$ regarded as a "0-dimensional" complex, K^0, whose cells are the points of $\partial\sigma^n$. This is closure finite but does not have the weak topology.

(2) σ^n $(n>1)$, regarded as a complex $K^n = K^0 \cup e^n$, where $e^n = \sigma^n - \partial\sigma^n$ and $K^0 = \partial\sigma^n$, as in (1). This has the weak topology, since $\bar{e}^n = K^n$, but is not closure finite.

(3) a simplicial complex, which has a metric topology but which is not locally finite (e.g. a complex covering the coordinate axes in Hilbert space). The weak topology in such a complex cannot be metricized (cf. [1, pp. 316, 317]).

Let K be a CW-complex. We establish some properties of K.

(A) *A map, $f: X \to Y$, of a closed (open) subset, $X \subset K$, in any space, Y, is continuous provided $f|X \cap \bar{e}$ is continuous for each cell $e \in K$.*

Let $f_e = f|X \cap \bar{e}$ be continuous, for each cell $e \in K$. Let Y_0 be any closed (open) subset of Y. Obviously $\bar{e} \cap f^{-1}Y_0 = f_e^{-1}Y_0$ and it follows from the continuity of f_e that $\bar{e} \cap f^{-1}Y_0$ is a relatively closed (open) subset of $X \cap \bar{e}$. But $X \cap \bar{e}$ is a closed (relatively open) subset of $\bar{e}$, whence $\bar{e} \cap f^{-1}Y_0$ is closed (relatively open) in $\bar{e}$. Therefore $f^{-1}Y_0$ is closed (open) in K, and a fortiori in X. Therefore f is continuous.

(B) *A subcomplex, $L \subset K$, is a closed subspace of K and the topology induced by K is the weak topology in L.*

Let $Y \subset L$ be such that $Y \cap L_0$ is closed, and hence compact, for each finite subcomplex $L_0 \subset L$. Since $Y \cap L_0$ is compact it is a closed subset of K. Let K_0 be any finite subcomplex of K. Then $L_0 = L \cap K_0$ is a finite subcomplex of L and

$$Y \cap K_0 = Y \cap L \cap K_0 = Y \cap L_0.$$

Therefore $Y \cap K_0$ is closed, whence Y is closed in K, and a fortiori in L. Therefore L has the weak topology. Also, taking $Y = L$, it follows that L is closed, which establishes (B).

(C) *If K is connected so is K^n for each $n > 0$.*

Let $n > 0$ and let K^n be the union of disjoint, nonvacuous closed sets K_1^n, K_2^n. Since the closure of a cell $e \in K$ is connected it follows that $\bar{e} \subset K_i^n$ if $e \cap K_i^n \neq 0$ $(i = 1, 2)$. Therefore K_i^n is a subcomplex of K. Clearly ∂e^{n+1} is connected $(e^{n+1} \in K)$, whence it lies either in K_1^n or in K_2^n. Therefore K^{n+1} is the union of disjoint subcomplexes, K_1^{n+1}, K_2^{n+1}, where $K_i^n \subset K_i^{n+1}$ and $e^{n+1} \in K_i^{n+1}$ if $\partial e^{n+1} \subset K_i^n$. A similar (inductive) argument shows that K^m is the union of disjoint subcomplexes, K_1^m, K_2^m, such that $K_i^m \subset K_i^{m+1}$ $(m = n, n+1, \cdots)$. Let K_i be the union of the K_i^m for $m = n, n+1, \cdots$. Then $K_i \cap K^m = K_i^m$ and

$$K_1 \cap K_2 = \bigcup_m (K_1 \cap K_2) \cap K^m = \bigcup_m K_1^m \cap K_2^m = 0.$$

Also $K_i \neq 0$, since $K_i^n \subset K_i$, and K_i is a closed subset of K, according to (B). Therefore K is not connected, which establishes (C).

(D) *If $X \subset K$ is compact, then $K(X)$ is a finite complex.*

If X meets but a finite number of cells, $e_1, \cdots, e_k \subset K$, it is contained in the finite union of the (finite) subcomplexes $K(e_1), \cdots, K(e_k)$. Assume that there is an infinite set of cells, $\{e_i\}$, each of which meets X and let $p_i \in X \cap e_i$. Then a finite subcomplex, $L \subset K$, contains but a finite set of the cells in $\{e_i\}$ and $e_i \cap L = 0$ unless $e_i \in L$. Therefore L contains but a finite number of points in the set $P = \{p_i\}$, whence P is closed. Similarly any subset of P is closed, whence P is discrete. But this is absurd, since P is compact, being a closed subset of X. Therefore (D) is established.

(E) *If a complex L, and also L^n for each $n \geqq 0$, all have the weak topology, then L is a CW-complex.*

Certainly L^0 is closure finite. Assume that L^{n-1} is closure finite, and hence a CW-complex, for some $n > 0$. Let e^n be a given n-cell in L^n. Since ∂e^n is compact it follows from (D) that $L(\partial e^n)$ is finite. But obviously $L(e^n) = L(\partial e^n) \cup e^n$ and it follows from induction on n that L is closure finite, which establishes (E).

Let $f: K \rightarrow L$ be a map of K onto a closure finite complex L, which has the indentification topology[18] determined by f. Further let the subcomplex $L(f\bar{e})$ be finite for each cell $e \in K$.

(F) *Subject to these conditions L is a CW-complex.*

Let $Y \subset L$ be such that $Y \cap L_0$ is closed for each finite subcomplex $L_0 \subset L$. Let $L_0 \bar{e} = L(f\bar{e})$ for a given cell $e \in K$. Then $\bar{e} \subset f^{-1}L_0$ and

$$f^{-1}Y \cap \bar{e} = f^{-1}Y \cap \bar{e} \cap \bar{e} \subset (f^{-1}Y \cap f^{-1}L_0) \cap \bar{e} \subset f^{-1}(Y \cap L_0) \cap \bar{e},$$

since $f^{-1}A \cap f^{-1}B \subset f^{-1}(A \cap B)$ for any sets $A, B \subset L$. But $f^{-1}(Y \cap L_0) \subset f^{-1}Y$. Therefore

$$f^{-1}Y \cap \bar{e} = f^{-1}(Y \cap L_0) \cap \bar{e}.$$

Since $Y \cap L_0$ is closed it follows that $f^{-1}Y \cap \bar{e}$ is closed. Therefore $f^{-1}Y$ is closed, since K has the weak topology. Since L has the identification topology determined by f it follows that Y is closed. Therefore L has the weak topology. Since L is closure finite by hypothesis this proves (F).

(G) *K is a normal space.*

Let $X_1, X_2 \subset K$ be disjoint, closed subsets and let $X_i^r = X_i \cap K^r$ ($i = 1, 2$; $r \geqq 0$). Clearly K^0 is a discrete set, and hence normal. Let $n > 0$ and assume that there are disjoint, relatively open subsets, U_1^{n-1}, $U_2^{n-1} \subset K^{n-1}$, such that $X_i^{n-1} \subset U_i^{n-1}$. Then $X_i \cap \bar{U}_j^{n-1} = 0$ ($i, j = 1, 2$; $i \neq j$). If $K^n = K^{n-1}$ we define $U_i^n = U_i^{n-1}$. Otherwise let

$f:\sigma^n \to \bar{e}^n$ be a characteristic map for a given n-cell $e^n \in K$ and let

$$V_i = f^{-1} U_i^{n-1} \subset \partial\sigma^n, \qquad Y_i = f^{-1} X_i \subset \sigma^n.$$

Since $X_1 \cap X_2 = 0$, $X_i \cap \bar{U}_j^{n-1} = 0$ we have $Y_1 \cap Y_2 = 0$, $Y_i \cap \bar{V}_j = 0$. Let p_0 be the centroid of σ^n and let r, p be polar coordinates for σ^n ($r \in I$, $p \in \partial\sigma^n$) such that (r, p) is the point which divides the rectilinear segment $p_0 p$ in the ratio $r:1-r$. Let $V_i' \subset \sigma^n$ be the (open) subset, which consists of all points (r, p) with $p \in V_i$ and $1-\epsilon < r \leqq 1$, where $0 < \epsilon < 1$. Since $Y_i \cap \bar{V}_j = 0$ it follows that, if ϵ is sufficiently small, then $Y_i \cap \bar{V}_j' = 0$, which we assume to be the case. Since $fY_i \subset X_i$, $f\partial\sigma^n \subset K^{n-1}$ and $X_i \cap K^{n-1} \subset U_i^{n-1}$ it follows that $Y_i \cap \partial\sigma^n \subset V_i$. Let V_i'' be an η-neighborhood of Y_i, defined in terms of a metric for σ^n, where η is so small that $V_1'' \cap V_2'' = 0$, $V_i'' \cap \bar{V}_j' = 0$ and $V_i'' \cap (\partial\sigma^n - V_i) = 0$. Then $V_i'' \cap \partial\sigma^n \subset V_i$. Let

$$W_i = V_i' \cup V_i''.$$

Then $Y_i \subset W_i$ and $W_1 \cap W_2 = 0$. Obviously $V_i' \cap \partial\sigma^n = V_i$, whence

$$W_i \cap \partial\sigma^n = V_i = f^{-1} U_i^{n-1}. \tag{5.1}$$

Since $f:\sigma^n - \partial\sigma^n$ is a (1-1) map onto e^n and $f\partial\sigma^n \cap e^n = 0$ it follows that W_i is saturated[18] with respect to f. Therefore fW_i is a relatively open subset of $\bar{e}^n$. From (5.1) we have

$$fW_i \cap K^{n-1} = U_i^{n-1} \cap \partial e^n \tag{5.2}$$

and it follows that $fW_1 \cap fW_2 = 0$.

Let us write $W_i = W_i(e^n)$ and let

$$U_i^n = U_i^{n-1} \cup \bigcup_{e^n \in K} fW_i(e^n).$$

Then it follows from (5.2) that $U_i^n \cap K^{n-1} = U_i^{n-1}$ and that

$$U_i^n \cap \partial e^n = U_i^{n-1} \cap \partial e^n = fW_i(e^n) \cap K^{n-1}.$$

Also $fW_i(e^n) \subset K^{n-1} \cup e^n$ and $U_i^n \cap e^n = fW_i(e^n) \cap e^n$. Therefore

$$\begin{aligned} U_i^n \cap \bar{e}^n &= (U_i^n \cap \partial e^n) \cup (U_i^n \cap e^n) \\ &= fW_i(e^n) \cap (K^{n-1} \cup e^n) \\ &= fW_i(e^n). \end{aligned}$$

Therefore U_i^n is a relatively open subset of K^n. Obviously $X_i^n \subset U_i^n$ and $U_1^n \cap U_2^n = 0$. Therefore such sets, U_i^n, may be defined inductively for every value of n. Let them be so defined and let

$$(1) \qquad U_i = \bigcup_n U_i^n.$$

Since $U_i^{n+1} \cap K^n = U_i^n$ it follows by induction on $m > n$ that

$$U_i^m \cap K^n = U_i^m \cap K^{m-1} \cap K^n = U_i^{m-1} \cap K^n = U_i^n$$

and hence that $U_i \cap K^n = U_i^n$. Therefore it follows, first that U_i is an open subset of K and second that $U_1 \cap U_2 = 0$. Obviously $X_i \subset U_i$, which completes the proof of (G).

(H) *If L is a locally finite*[21] *complex then $K \times L$ is a CW-complex.*

If $e \in K$, $e' \in L$ are cells in K and L respectively, then the cell $e \times e' \in K \times L$ is contained in the finite subcomplex $K(e) \times L(e') \subset K \times L$. Therefore $K \times L$ is closure finite.

Let the cells in K be indexed and with each m-cell, $e_i^m \in K$, $(m = 0, 1, \cdots)$ let us associate an m-element, E_i^m, as follows. The points in E_i^m shall be the pairs (x, e_i^m), for every point $x \in \sigma^m$, and E_i^m shall have the topology which makes the map $x \to (x, e_i^m)$ a homeomorphism. No two of these elements have a point in common and we unite them into a topological space,

$$P = \bigcup_{m,i} E_i^m,$$

in which each E_i^m, with its own topology, is both open and closed. Let $f_i^m : \sigma^m \to \bar{e}_i^m$ be a characteristic map for e_i^m and let $\phi : P \to K$ be the map which is given by $\phi(x, e_i^m) = f_i^m x$, for each point $(x, e_i^m) \in P$. Since $\bar{e}_i^m$ has the identification topology determined by f_i^m it follows that the weak topology in K is the identification topology determined by ϕ.

Let a space,

$$Q = \bigcup_{n,j} E_j^n,$$

and a map, $\psi : Q \to L$, be similarly associated with L. Then $K \times L = \theta(P \times Q)$, where $\theta : P \times Q \to K \times L$ is given by $\theta(p, q) = (\phi p, \psi q)$ $(p \in P, q \in Q)$. Also $P \times Q$ is the union of the $(m+n)$-elements $E_i^m \times E_j^n$, and $\theta(E_i^m \times E_j^m) = \bar{e}_{ij}^{m+n}$, where $e_{ij}^{m+n} = e_i^m \times e_j^m$. Therefore the weak topology in $K \times L$ is obviously the same as the identification topology determined by θ.

Let $V \subset L$ be an open subset and $y \in V$ an arbitrary point in V. Since y is an inner point of a finite subcomplex, $L_0 \subset L$, it is contained in a subset, $V_0 \subset V \cap L_0$, which is open in L. Since L is normal there is a neighborhood, W, of y such that $\overline{W} \subset V_0$. Since $\overline{W} \subset V_0 \subset L_0$ and

[21] I do not know if this restriction on L is necessary.

since V_0 is open in L, it follows that $\overline{W} \cap \bar{e} = 0$ for any cell $e \in L - L_0$. Therefore there are only a finite number of cells in L, whose closures meet $\overline{W}$. Therefore $\psi^{-1}\overline{W}$ is contained in the union of a finite subset of the components $E_i^n \subset Q$. Therefore $\psi^{-1}\overline{W}$ is compact and (H) follows from Lemma 4 in [7].

(I) *A homotopy, $f_t: X \to Y$, of a closed (open) subset, $X \subset K$, in an arbitrary space, Y, is continuous provided $f_t | X \cap \bar{e}$ is continuous for each cell $e \in K$.*

This follows from (H), with $L = I$, and (A), applied to the subset $X \times I \subset K \times I$ and the map $f: X \times I \to Y$, which is given by $f(x, t) = f_t x$.

(J) (*Homotopy extension.*) *Let $f_0: K \to X$ be a given map of K in an arbitrary space X. Let $g_t: L \to X$ be a homotopy of $g_0 = f_0 | L$, where L is a subcomplex of K. Then there is a homotopy, $f_t: K \to X$, such that $f_t | L = g_t$.*

Let $K_r = L \cup K^r$ ($r \geqq -1$; $K_{-1} = L$) and assume that g_t has been extended to a homotopy, $f_t^{n-1}: K_{n-1} \to X$, such that $f_0^{n-1} = f_0 | K_{n-1}$, $f_t^{n-1} | L = g_t$ ($n \geqq 0$). The homotopy f_t^{n-1} can be extended throughout $K_{n-1} \cup e^n$, for each n-cell[22] $e^n \in K_n - L$, and hence, by (I), to a (continuous) homotopy $f_t^n: K_n \to X$. Starting with $f_t^{-1} = g_t$ it follows by induction on n that there is a sequence of homotopies, $f_t^n: K_n \to X$ ($n = 0, 1, \cdots$), such that $f_0^n = f_0 | K_n$, $f_t^n | K_{n-1} = f_t^{n-1}$. It follows from (I) that a homotopy, $f_t: K \to X$, which satisfies the requirements of (J), is given by $f_t | K_n = f_t^n$.

Let $X_0 \subset X_1 \subset \cdots$ be a sequence of subspaces of a given space, X, such that any map, $(\sigma^n, \partial\sigma^n) \to (X, X_{n-1})$, is homotopic, rel. $\partial\sigma^n$, to a map[23] $\sigma^n \to X_n$ ($n = 0, 1, \cdots$). Let $L \subset K$ be a given subcomplex, which may be empty, and let $f_0: K \to X$ be a map such that $f_0 L^n \subset X_n$, for each $n = 0, 1, \cdots$.

(K) *There is a homotopy, $f_t: K \to X$, rel. L, such that $f_1 K^n \subset X_n$ for each $n = 0, 1, \cdots$.*

Since each point in X is joined by an arc to some point in X_0 there is a homotopy, $f_t^0: K^0 \to X$, rel. L^0, such that $f_0^0 = f_0 | K^0$ and $f_1^0 K^0 \subset X_0$. Let $n > 0$ and assume that there is a homotopy f_t^{n-1}: $K^{n-1} \to X$, rel. L^{n-1}, such that $f_0^{n-1} = f_0 | K^{n-1}$, $f_1^{n-1} K^{n-1} \subset X_{n-1}$. It follows from (J) that f_t^{n-1} can be extended, first throughout L^n by writing $f_t^{n-1} | L^n = f_0 | L^n$, and then to a homotopy, $\xi_t: K^n \to X$, rel. L^n ($\xi_0 = f_0 | K^n$). Since $\xi_1 K^{n-1} \subset X_{n-1}$ it follows from a standard argument (see [6, §8]), and the condition on $X_0, X_1, \cdots$, that there is a homotopy, $\eta_t: K^n \to X$, rel. $(K^{n-1} \cup L^n)$, such that $\eta_0 = \xi_1$, $\eta_1 K^n \subset X_n$. If dim $K < \infty$ we

[22] See [5, Lemma 10 in §16].

[23] If $n = 0$ this simply means that each point in X is joined by an arc to some point in X_0.

define $f_t^n : K^n \to X$ as the resultant of ξ_t followed by η_t. Then f_t^n may be defined inductively for every $n \geqq 0$ and we take $f_t = f_t^m$, where $m = \dim K$. But if $\dim K = \infty$ this method fails and we shall define f_t^n as an extension of f_t^{n-1}, not as the resultant of ξ_t followed by η_t.

If $K^n = K^{n-1}$ we define $f_t^n = f_t^{n-1}$. Otherwise let $g : \sigma^n \to \bar{e}^n$ be a characteristic map for a given n-cell $e^n \in K$. Let r, p be polar coordinates for σ^n, defined as in (G), and let $\rho_t : \bar{e}^n \to X$ be defined by

$$(5.3)\qquad \begin{aligned} \rho_t g(r, p) &= \xi_{2t/(1+r)} g(r, p) && (\text{if } 0 \leqq 2t \leqq 1 + r) \\ &= \eta_{(2t-1-r)/(1-r)} g(r, p) && (\text{if } 1 + r < 2t \leqq 2). \end{aligned}$$

Since $\eta_0 = \xi_1$ and $g^{-1} | e^n$ is a homeomorphism onto $\sigma^n - \partial\sigma^n$ it follows that $\rho_t | e^n$ is single-valued and continuous. Since $\rho_t x = \xi_t x$ for any point $x = g(1, p) \in \partial e^n$ it follows that ρ_t is single-valued. Also ρ_t is continuous at $\{g(r, p), t\}$ if $r < 1$ and, obviously, if $t < 1$. I say that it is continuous at $\{g(1, p), 1\} = (gp, 1)$. For $gp \in K^{n-1}$ and $\eta_t | K^{n-1} = \eta_0 | K^{n-1} = \xi_1 | K^{n-1}$. Therefore, given a neighborhood, $U \subset X$, of $\xi_1 gp = \eta_t gp$, it follows from the compactness of I that there is a neighborhood, $V \subset \bar{e}^n$, of gp such that $\eta_t x \in U$ for every $t \in I$, provided $x \in V$. There is also a neighborhood, $V' \subset \bar{e}^n$, of gp, and a $\delta > 0$ such that $\xi_t x \in U$ if $x \in V'$, $1 - 2\delta < t \leqq 1$. Since $(2 - 2\delta)/(1 + r) > 1 - 2\delta$ it follows that $\rho_t x \in U$ if $x \in V \cap V'$, $1 - \delta < t \leqq 1$. Therefore ρ_t is continuous. Also

$$\begin{aligned} \rho_t g(1, p) &= \xi_t g(1, p) = f_t^{n-1} g(1, p), \\ \rho_0 g(r, p) &= \xi_0 g(r, p) = f_0 g(r, p), \\ \rho_1 g(r, p) &= \eta_1 g(r, p) \in X_n. \end{aligned}$$

Therefore a homotopy, $f_t^n : K^n \to X$, rel. L^n, such that

$$f_t^n | K^{n-1} = f_t^{n-1}, \qquad f_0^n = f_0 | K^n,\ f_1 K^n \subset X_n,$$

is defined by $f_t^n | K^{n-1} = f_t^{n-1}$, $f_t^n | \bar{e}^n = \rho_t$, for each n-cell $e^n \in K^n$. It follows from induction on n that such a homotopy is defined for each $n \geqq 0$ and a homotopy, $f_t : K \to X$, which satisfies the requirements of (K), is defined by $f_t | K^n = f_t^n$.

Let $f_0 : K \to P$ be a map of K into a CW-complex, P, such that $f_0 | L$ is cellular, where $L \subset K$ is a subcomplex. Also let $g_t : K \to P$ be a homotopy such that the maps g_0, g_1 and the homotopy $g_t | L$ are cellular.

(L) *There is a homotopy, $f_t | K \to P$, rel. L, of f_0 into a cellular map f_1. There is a cellular homotopy, $g_t' : K \to P$, such that $g_0' = g_0$, $g_1' = g_1$, $g_t' | L = g_t | L$.*

Since any continuous image of σ^n in P is compact it is contained in a finite subcomplex $Q \subset P$, according to (D). Any map $(\sigma^n, \partial\sigma^n) \to (Q, Q^{n-1})$ is homotopic,[24] rel. $\partial\sigma^n$, in Q to a map $\sigma^n \to Q^n$. Therefore the first part follows from (K). The second part follows from the first part with K, L, f_0 replaced by $K \times I$, $(K\times 0) \cup (L \times I) \cup (K \times 1)$, $g: K\times I \to P$, where $g(p, t) = g_t p$.

(M) *K is locally contractible.*

Let $a_0 \in K$ be a given point, let $U \subset K$ be a given neighborhood of a_0 and let $e^r \in K$ be the cell which contains a_0. Let $E^r \subset U \cap e^r$ be an r-element, which contains a_0 in its interior, $V^r = E^r - \partial E^r$, and let $f_t^r: V^r \to V^r$ be a homotopy such that $f_0^r = 1$, $f_1^r V^r = a_0$. Using induction on n we shall define sequences of relatively open subsets, $V^n \subset K^n$ $(n = r, r+1, \cdots)$, such that $V^{n+1} \cap K^n = V^n$, $\overline{V}^n \subset U$, and of homotopies $f_t^n: V^n \to V^n$, such that $f_t^{n+1} | V^n = f_t^n$, $f_0^n = 1$, $f_1^n V^n = a_0$. Assuming that this has been done, let

$$V = \bigcup_n V^n$$

and let $f_t: V \to V$ be defined by $f_t | V^n = f_t^n$. Then it follows from the definition of the weak topology and from (I) that V is open in K and f_t continuous. Obviously $V \subset U$, $f_0 = 1$, $f_1 V = a_0$ and (M) follows.

Assume that V^{n-1} and f_t^{n-1} satisfy the above conditions for some $n > r$. Let $g: \sigma^n \to \bar{e}^n$ be a characteristic map for a given n-cell, $e^n \in K$, and let polar coordinates, r, p, for σ^n be defined as in (G). If $g\partial\sigma^n \cap V^{n-1} = 0$, let $W \subset \sigma^n$ be the empty set. Otherwise let $W \subset \sigma^n$ be the (open) subset, which consists of all points, (r, p), such that

$$1 - \epsilon < r \leqq 1, \qquad p \in g^{-1}V^{n-1},$$

where $0 < \epsilon < 1$. Since $\overline{V}^{n-1} \subset U$, whence $g^{-1}\overline{V}^{n-1} \subset g^{-1}U$, it follows that $\overline{W} \subset g^{-1}U$ if ϵ is sufficiently small, which we assume to be the case. Let $\xi_t: W \to W$ be the "radial projection," which is defined by

$$\xi_t(r, p) = (r + t - rt, p),$$

and let $\theta_t gW \to V^{n-1} \cup gW$ be given by

$$\begin{aligned} \theta_t g(r, p) &= g\xi_{2t/(1-r)}(r, p) && (\text{if } 0 \leqq 2t < 1 - r) \\ &= f^{n-1}_{(2t-1+r)/(1+r)} g(1, p) && (\text{if } 1 - r \leqq 2t \leqq 2). \end{aligned}$$

Since $\xi_t(1, p) = (1, p)$ and $g\xi_1(r, p) = g(1, p) = f_0^{n-1} g(1, p)$ it follows from an argument similar to the one which comes after (5.3) that θ_t is single-valued and continuous. Alsc

---

[24] [5, §16, Theorem 6]. It follows from Theorem 6 in [5] that the condition $f_t x \in P(f_0 x)$ $(x \in K)$ may be imposed on the homotopy f_t in (L).

(1)

$$
(5.4)\qquad
\begin{aligned}
\theta_t g(1, p) &= f_t^{n-1} g(1, p),\\
\theta_0 g(r, p) &= g\xi_0(r, p) = g(r, p),\\
\theta_1 g(r, p) &= f_1^{n-1} g(1, p) = a_0.
\end{aligned}
$$

Let V^n be the union of V^{n-1} and the sets gW, which are thus defined for all the n-cells in K. Arguments used in (G) show that V^n is a relatively open subset of K^n and that $V^n \cap K^{n-1} = V^{n-1}$. Also it follows from the definition of W that $\partial\sigma^n \cap \overline{W} \subset g^{-1}\overline{V}^{n-1}$, whence

$$K^{n-1} \cap g\overline{W} \subset \overline{V}^{n-1}.$$

Hence it follows from the definition of the weak topology that $\overline{V}^n$ is the union of $\overline{V}^{n-1}$ and the sets $g\overline{W}$, which are closed since $\overline{W}$ is compact. Since $\overline{V}^{n-1} \subset U$, $\overline{W} \subset g^{-1}U$ it follows that $\overline{V}^n \subset U$. Finally define $f_t^n: V^n \to V^n$ by $f_t^n \mid V^{n-1} = f_t^{n-1}$, $f_t^n \mid gW = \theta_t$. It follows from (5.4) and from (I) that f_t^n is single-valued and continuous and that $f_0^n = 1$, $f_1^n V^n = a_0$. Therefore (M) follows by induction on n.

(N) *Any covering complex, $\tilde{K}$, of K is a* CW-*complex.*

Since $\tilde{K}$ is locally connected, by the definition of a covering space, each of its components is both open and closed and is a covering complex of a component of K. A locally connected complex is obviously a CW-complex if, and only if, each of its components is a CW-complex. Therefore (N) will follow when we have proved it in case K and $\tilde{K}$ are connected. We assume that this is so and also, to begin with, that $\tilde{K}$ is a regular covering complex of K. That is to say the group, G, of covering transformations[25] in $\tilde{K}$ operates transitively on the set $\mathbf{p}^{-1}q$, for any point $q \in K$, where $\mathbf{p}: \tilde{K} \to K$ is the covering map. We shall describe an open set, $U \subset K$, as an *elementary neighborhood* if, and only if, each component of $\mathbf{p}^{-1}U$ is mapped by $\mathbf{p}$ topologically onto U. We shall describe an elementary neighborhood in K as a *basic neighborhood* if, and only if, its closure is contained in an elementary neighborhood. We shall describe a subset of $\tilde{K}$ as a *basic neighborhood* if and only if it is a component of $\mathbf{p}^{-1}U$, where U is a basic neighborhood in K. If $\tilde{U} \subset \tilde{K}$ is a basic neighborhood the component of $\mathbf{p}^{-1}(\mathbf{p}\tilde{U})$ are the sets $T\tilde{U}$ for every $T \in G$. It follows from the definition of $\tilde{K}$ and the normality of K that the basic neighborhoods constitute a basis for the open sets, both in K and in $\tilde{K}$.

Let $U \subset K$ be a basic neighborhood and let V be an elementary neighborhood such that $\overline{U} \subset V$. Then the components of $\mathbf{p}^{-1}V$ are disjoint open sets in $\tilde{K}$, each of which contains exactly one component of $\mathbf{p}^{-1}\overline{U}$. Let $Q \subset \mathbf{p}^{-1}\overline{U}$ be a set of points, of which at most

[25] I.e., the group of homeomorphisms, $T: \tilde{K} \to \tilde{K}$, such that $\mathbf{p}T = \mathbf{p}$.

one lies in each component of $p^{-1}\overline{U}$. Then Q is a closed discrete set. For if Q has a limit point, $\tilde{q}$, then $p\tilde{q}\in\overline{U}\subset V$, whence $\tilde{q}$ lies in one of the components, $\tilde{V}$, of $p^{-1}V$. But this is absurd, since $\tilde{V}$ contains at most one point of Q. Therefore Q is closed and discrete.

Let $\tilde{U}\subset\tilde{K}$ be a basic neighborhood, let U^* be its closure[26] and let $C\subset\tilde{K}$ be compact. I say that only a finite number of the sets TC meet U^*, where $T\in G$. For if TC meets U^* then C meets $T^{-1}U^*$. Let $q_T\in C\cap T^{-1}U^*$. Since $T'U^*\cap T''U^*=0$ if $T'\neq T''$ it follows from the preceding paragraph that the aggregate of points q_T, for every T such that $U^*\cap TC\neq 0$, is a discrete, closed subset of C. Since C is compact the set $\{q_T\}$ is finite, which proves our assertion.

We now prove that $\tilde{K}$ has the weak topology. Let $\tilde{X}\subset\tilde{K}$ be a subset such that $\tilde{X}\cap e^*$ is closed, for every cell $\tilde{e}\in\tilde{K}$. In order to prove that $\tilde{X}$ is closed it is enough to prove that $\tilde{X}\cap U^*$ is closed, where U^* is the closure of an arbitrary basic neighborhood $\tilde{U}\subset\tilde{K}$. For this implies that $\tilde{U}-\tilde{X}=\tilde{U}-(\tilde{X}\cap U^*)$ is open, whence it follows that $\tilde{K}-\tilde{X}$ is open. Therefore, to simplify the notation, we assume that $\tilde{X}\subset U^*$, where $\tilde{U}$ is a basic neighborhood in $\tilde{K}$. Let $X=p\tilde{X}$ and let e be a given cell in K. Then[27]

$$X\cap\bar{e}=p(\tilde{X}\cap p^{-1}\bar{e}).$$

Let $\tilde{e}\in\tilde{K}$ be a cell which covers e. Then $p^{-1}\bar{e}$ consists of the sets Te^* for every $T\in G$, and Te^* is the closure of the cell $T\tilde{e}\in\tilde{K}$. Since e^* is compact it follows from the preceding paragraph that only a finite number of the sets Te^*, say $T_1e^*,\cdots,T_ke^*$, meet U^*. Let $P_i=\tilde{X}\cap T_ie^*$ $(i=1,\cdots,k)$. Then

$$X\cap\bar{e}=p(\tilde{X}\cap p^{-1}\bar{e})=p(P_1\cup\cdots\cup P_k).$$

But P_i is closed, by the hypothesis concerning $\tilde{X}$, and hence compact, since T_ie^* is compact. Therefore, $P_1\cup\cdots\cup P_k$ and hence $X\cap\bar{e}$ are compact. Since the cell $e\in K$ is arbitrary it follows that X is closed. Therefore $p^{-1}X$ is closed. Since $U^*\cap TU^*=0$ if $T\neq 1$ it follows that

$$\tilde{X}=U^*\cap\bigcup_T T\tilde{X}=U^*\cap p^{-1}X.$$

Therefore $\tilde{X}$ is closed and it follows that $\tilde{K}$ has the weak topology.

Since K^0 is discrete it follows that $\tilde{K}^0=p^{-1}K^0$ is a discrete set of points. That is to say, $\tilde{K}^0$ has the weak topology. If $n>0$ then K^n is connected, according to (C), and $\tilde{K}^n$ is obviously a covering complex of K^n. It follows from (L) that the injection homomorphism, $\pi_1(K^n)$

[26] We shall denote the closure of a set $\tilde{P}\subset\tilde{K}$ by P^*.

[27] If $f:P\to Q$ is any map and $A\subset P$, $B\subset Q$, then $f(A\cap f^{-1}B)=(fA)\cap B$.

$\to\pi_1(K)$, is onto, whence $\tilde{K}^n$ is connected. Obviously $T\tilde{K}^n=\tilde{K}^n$ for any $T\in G$ and it follows that $\tilde{K}^n$ is a regular covering complex of K^n. Therefore $\tilde{K}^n$ has the weak topology, according to what we have just proved. It follows from (E) that $\tilde{K}$ is a CW-complex.

Now let $\tilde{K}$ be a (connected) covering complex of K, which is not regular. Then a universal covering complex, $\hat{K}$ of $\tilde{K}$ is a universal covering complex of K. Therefore $\hat{K}$ is a CW-complex. Let $p:\hat{K}\to\tilde{K}$ be the covering map. Since p is an open map it follows that $\tilde{K}$ has the identification topology determined by p. It follows from the final paragraph in §4 that $\tilde{K}$ is closure finite and that the remaining condition of (F) is satisfied. Therefore it follows from (F) that $\tilde{K}$ is a CW-complex, which completes the proof of (N).

References

1. J. H. C. Whitehead, Proc. London Math. Soc. vol. 45 (1939) pp. 243–327.
2. ———, Ann. of Math. vol. 42 (1941) pp. 409–428.
3. ———, Ann. of Math. vol. 42 (1941) pp. 1197–1239.
4. ———, Proc. London Math. Soc. vol. 48 (1945) pp. 243–291.
5. ———, *On simply connected, 4-dimensional polyhedra*, Comment. Math. Helv. vol. 22 (1949) pp. 48–92.
6. ———, *On the homotopy type of ANR's*, Bull. Amer. Math. Soc. vol. 54 (1948) pp. 1133–1145.
7. ———, *Note on a theorem due to Borsuk*, Bull. Amer. Math. Soc. vol. 54 (1948) pp. 1125–1132.
8. K. Reidemeister, *Topologie der Polyeder*, Leipzig, 1938.
9. R. H. Fox, Ann. of Math. vol. 42 (1941) pp. 333–370.
10. ———, Ann. of Math. vol. 44 (1943) pp. 40–50.
11. N. E. Steenrod, Ann. of Math. vol. 48 (1947) pp. 290–320.
12. S. Eilenberg, Fund. Math. vol. 32 (1939) pp. 167–175.
13. ———, Trans. Amer. Math. Soc. vol. 61 (1947) pp. 378–417.
14. S. Eilenberg and N. E. Steenrod, *Foundations of Algebraic topology*, not yet published.
15. W. Hurewicz, K. Akademie van Wetenschappen, Amsterdam, Proceedings vol. 38 (1935) pp. 521–528.
16. ———, K. Akademie van Wetenschappen, Amsterdam, Proceedings vol. 39 (1936) pp. 215–224.
17. H. Hopf, Comment. Math. Helv. vol. 14 (1942) pp. 257–309.
18. ———, Comment. Math. Helv. vol. 15 (1942) pp. 27–32.
19. ———, Comment. Math. Helv. vol. 17 (1945) pp. 307–326.
20. Claude Chevalley, *Theory of Lie groups*, Princeton, 1946.
21. Charles Ehresmann, Bull. Soc. Math. France, vol. 72 (1944) pp. 27–54.
22. W. Hurewicz and N. E. Steenrod, Proc. Nat. Acad. Sci. U.S.A. vol. 27 (1941) pp. 60–64.
23. P. Alexandroff and H. Hopf, *Topologie*, Berlin, 1935; Also Ann Arbor, 1945.
24. N. Bourbaki, *Topologie générale*, vol. 3, Paris, 1940.
25. A. L. Blakers, Ann. of Math. vol. 49 (1948) pp. 428–461.

2

The following note by Eilenberg and Steenrod is the first announcement of their axiomatic approach to homology theory. Of course, this work has influenced later developments very strongly. It should perhaps be pointed out that in 1945, when this note appeared, we had essentially no examples of what are now called 'generalised homology theories'; nor did we have the techniques, such as spectral sequences, which are now used for dealing with them. The only prerequisite is a minimal acquaintance with homology groups (see §1 of the introduction).

2 *AXIOMATIC APPROACH TO HOMOLOGY THEORY*

By Samuel Eilenberg and Norman E. Steenrod

Department of Mathematics, University of Michigan

Communicated February 21, 1945

1. Introduction.—The present paper provides a brief outline of an axiomatic approach to the concept. homology group. It is intended that a full development should appear in book form.

The usual approach to homology theory is by way of the somewhat complicated idea of a complex. In order to arrive at a purely topological concept, the student of the subject is required to wade patiently through a large amount of analytic geometry. Many of the ideas used in the constructions, such as orientation, chain and algebraic boundary, seem artificial. The motivation for their use appears only in retrospect.

Since, in the case of homology groups, the definition by construction is so unwieldy, it is to be expected that an axiomatic approach or definition by properties should result in greater logical simplicity and in a broadened point of view. Naturally enough, the definition by construction is not eliminated by the axiomatic approach. It constitutes an existence proof or proof of consistency.

2. Preliminaries.—The concepts of a topological space and of a group are assumed to be known. The symbol (X, A) stands for a pair consisting of a topological space X and a closed subset A. A map $f:(X, A) \rightarrow (Y, B)$ of one such pair into another is a continuous map of X into Y which maps A into B. In case A is the vacuous set (X, A) is written as (X). If f_0, f_1 are two maps of (X, A) into (Y, B), they are homotopic if there exists a homotopy $f(x, t)$ connecting the two maps of X into Y such that $f(x, t) \in B$ for any $x \in A$ and all t.

3. Basic Concepts.—The fundamental concept to be axiomatized is a function $H_q(X, A)$ (called the *q-dimensional, relative homology group of X mod A*) defined for all triples consisting of an integer $q \geqq 0$ and a pair (X, A). The value of the function is an abelian group.

The first subsidiary concept is that of boundary. For each $q \geqq 1$ and each (X, A), there is a homomorphism

$$\partial : H_q(X, A) \rightarrow H_{q-1}(A)$$

called the *boundary operator.*

The second subsidiary concept is that of the induced homomorphism. If f is a map of (X, A) into (Y, B) and $q \geqq 0$, there is an attached homomorphism

(2)

$$f_*:H_q(X, A) \to H_q(Y, B)$$

called the *homomorphism induced by f*.

4. *Axioms.*—These three concepts have the following properties.

AXIOM 1. *If* f = *identity, then* f_* = *identity.*

That is to say, if f is the identity map of (X, A) on itself, then f_* is the identity map of $H_q(X, A)$ on itself.

AXIOM 2. $(gf)_* = g_*f_*$.

Explicitly, if $f:(X, A) \to (Y, B)$ and $g:(Y, B) \to (Z, C)$, then the combination of the induced homomorphisms $f_*:H_q(X, A) \to H_q(Y, B)$ and $g_*:H_q(Y, B) \to H_q(Z, C)$ is the induced homomorphism $(gf)_*:H_q(X, A) \to H_q(Z, C)$.

An immediate consequence of Axioms 1 and 2 is that homeomorphic pairs (X, A) and (Y, B) have isomorphic homology groups.

AXIOM 3. $\partial f_* = f_*\partial$.

Explicitly, if $f:(X, A) \to (Y, B)$ and $q \geqq 1$, the axiom demands that two homomorphisms of $H_q(X, A)$ into $H_{q-1}(B)$ shall coincide. The first is the combination of $\partial:H_q(X, A) \to H_{q-1}(A)$ followed by $(f|A)_*: H_{q-1}(A) \to H_{q-1}(B)$. The second is the combination of $f_*:H_q(X, A) \to H_q(Y, B)$ followed by $\partial:H_q(Y, B) \to H_{q-1}(B)$.

AXIOM 4. *If f is homotopic to g, then* $f_* = g_*$.

Definition: The *natural system* of the pair (X, A) is the sequence of groups and homomorphisms

$$\ldots \to H_q(X) \to H_q(X, A) \to H_{q-1}(A) \to H_{q-1}(X) \to \ldots \to H_0(X, A)$$

where $H_q(X) \to H_q(X, A)$ is induced by the identity map $(X) \to (X, A)$. $H_q(X, A) \to H_{q-1}(A)$ is the boundary operation, and $H_{q-1}(A) \to H_{q-1}(X)$ is induced by the identity map $(A) \to (X)$.

AXIOM 5. *In the natural system of (X, A) the last group, $H_0(X, A)$, is the image of $H_0(X)$. In any other group of the sequence, the image of the preceding group coincides with the kernel of the succeeding homomorphism.*

At first glance, this axiom may seem strange even to one familiar with homology theory. It is equivalent to three propositions usually stated as follows: (1) the boundary of a cycle of X mod A bounds in A if and only if the cycle is homologous mod A to a cycle of X; (2) a cycle of A is homologous to zero in X if and only if it is the boundary of a cycle of X mod A; (3) a cycle of X is homologous to a cycle of A if and only if it is homologous to zero mod A.

Definition: An open set U of X is strongly contained in A, written $U \Subset A$, if the closure $\overline{U}$ is contained in an open set $V \subset A$.

AXIOM 6. *If $U \Subset A$, then the identity map: $(X - U, A - U) \to (X, A)$ induces isomorphisms $H_q(X - U, A - U) \cong H_q(X, A)$ for each $q \geqq 0$.*

This axiom expresses the intuitive idea that $H_q(X, A)$ is pretty much independent of the internal structure of A.

AXIOM 7. *If P is a point, then $H_q(P) = 0$ for $q \geqq 1$.*

A particular reference point P_0 is selected, and $H_0(P_0)$ is called the *coefficient group* of the homology theory.

5. *Uniqueness.*—On the basis of these seven axioms, one can deduce the entire homology theory of a complex in the usual sense. Some highlights of the procedure are the following.

If σ is an n-simplex, and $\dot{\sigma}$ its point-set boundary, then $H_n(\sigma, \dot{\sigma})$ is isomorphic to the coefficient group. Further, $H_q(\sigma, \dot{\sigma}) = 0$ for $q \neq n$, and the boundary operator $\partial : H_n(\sigma, \dot{\sigma}) \rightarrow H_{n-1}(\dot{\sigma})$ is an isomorphism onto for $n > 1$, and into for $n = 1$.

Let f be the simplicial map of σ on itself which interchanges two vertices and leaves all others fixed. Then, for any $z \in H_n(\sigma, \dot{\sigma})$, we have $f_*(z) = -z$. This permits the usual division of permutations into the classes of even and odd, and leads naturally to a definition of orientation—a concept which is quite troublesome in the usual approach.

Definition: Let H, H' be two homology theories satisfying Axioms 1 through 7. A *homomorphism*

$$h : H \rightarrow H'$$

is defined to be a system of homomorphisms

$$h(q, X, A) : H_q(X, A) \rightarrow H_q'(X, A)$$

defined for all q, (X, A), which commute properly with the boundary operator and induced homomorphisms:

$$h(q-1, A)\partial = \partial' h(q, X, A), \qquad h(q, Y, B)f_* = f_*' \, h(q, X, A). \qquad \text{(I)}$$

If h gives an isomorphism of the coefficient groups $h(0, P_0) : H_0(P_0) \cong H_0'(P_0)$, then h is called a *strong homomorphism*. If each $h(q, X, A)$ is an isomorphism, then h is called an *equivalence* and H and H' are called *equivalent*.

Since the usual homology theory of complexes is deducible from the axioms, there follows the

UNIQUENESS THEOREM: *Any two homology theories having the same coefficient group coincide on complexes.*

Explicitly, if $i : H_0(P_0) \cong H_0'(P_0)$ is an isomorphism between the coefficient groups of H and H', then isomorphisms

$$h(q, X, A) : H_q(X, A) \cong H_q'(X, A)$$

can be defined for X a complex, A a subcomplex such that $h(0, P_0)$ coincides with i, and the relations (I) hold in so far as they are defined (f need not be simplicial). Indeed, there is just one way of constructing $h(q, X, A)$. The uniqueness theorem implies that any strong homomorphism $h : H \rightarrow$

H' is an equivalence as far as complexes are concerned. In view of Axiom 4, the uniqueness theorem holds for spaces having the same homotopy type as complexes. These include the absolute neighborhood retracts.

6. *Existence.*—As is to be expected, homology theories exist which satisfy the axioms. Both the Čech homology theory H^1 and the singular homology theory H^0 satisfy the axioms. This is fairly well known, although the proofs of some of the axioms are only implicitly contained in the literature. It is well known that the two homology theories differ for some pairs (X, A). Thus, the axioms do not provide uniqueness for all spaces.

The surprising feature of H^0 and H^1 that appears in this development is that they play extreme roles in the family of all homology theories, and have parallel definitions. They can be defined as follows: The homology groups of the simplicial structure of a complex (using chains, etc.) are defined as usual. (As a first step of an existence proof, this is quite natural since the definition has been deduced from the axioms.) Using maps $K \to X$ of complexes into the space X, the singular homology groups $H^0_q(X, A)$ can be defined using a suitable limiting process. Similarly using maps $X \to K$ of the space into complexes, the Čech homology groups $H^1_q(X, A)$ are obtained. It is then established that H^0 and H^1 are minimal and maximal in the family of all homology theories with a prescribed coefficient group in the sense that, if H is any homology theory, there exist strong homomorphisms $H^0 \to H \to H^1$. This is an indication of how it is possible to characterize H^0 or H^1 by the addition of a suitable Axiom 8.

7. *Generalizations.*—A suitable refinement of the axioms will permit the introduction of topologized homology groups.

Cohomology can be axiomatized in the same way as homology. It is only necessary to reverse the directions of the operators ∂ and f_* in the above axioms and make such modifications in the statements as these reversals entail. The analogous uniqueness theorems can be proved.

The products of elements of two cohomology groups with values in a third (in the usual sense) may also be axiomatized and characterized uniquely.

3&4

The next two pieces constitute an introduction to spectral sequences, which today form an almost indispensable part of the topologist's tool kit. For applications of spectral sequences, see §§5-8, 10, 12 of the introduction. The only prerequisite for reading the exposé by Eilenberg is a familiarity with the axiomatic approach to homology theory (see §1 of the introduction). In the extract by Massey the possible applications are more varied; some familiarity with elementary homotopy theory would be useful.

3

Séminaire H. Cartan, E. N. S., 1950/51. Topologie algébrique

LA SUITE SPECTRALE. I: CONSTRUCTION GÉNÉRALE

Exposé de Samuel Eilenberg, le 22. 1. 1951

1. Fondations

Nous considérerons un ensemble muni d'une relation d'ordre (partielle), notée $A < B$, qui est réflexive ($A < A$) et transitive ($A < B$ et $B < C$ entraînent $A < C$). On supposera que l'ensemble contient un plus petit élément, noté 0, et un plus grand élément, noté 1; on a donc $0 < A < 1$ pour tout A. Nous considérerons des paires (A, B) où $B < A$, et nous écrirons $(A, B) < (A', B')$ lorsque $A < A'$ et $B < B'$. De même, nous considérerons des triples (A, B, C) où $C < B < A$, et nous écrirons $(A, B, C) < (A', B', C')$ lorsque $A < A'$, $B < B'$, et $C < C'$. Le triple $(A, B, 0)$ sera identifié à la paire (A, B), et la paire $(A, 0)$ sera identifiée à l'élément A.

Nous supposerons qu'à toute paire (A, B) l'on ait associé un groupe abélien (ou un module sur un anneau), noté $H(A, B)$, qu'à toute inégalité $(A, B) < (A', B')$ l'on ait associé un homomorphisme $H(A, B) \to H(A', B')$, et qu'à tout triple (A, B, C) l'on ait associé un homomorphisme $d: H(A, B) \to H(B, C)$.

Ces trois termes primitifs seront assujettis aux axiomes suivants:

(H. 1) $H(A, B) \to H(A, B)$ est l'identité.

(H. 2) Si $(A, B) < (A', B') < (A'', B'')$, le diagramme suivant est commutatif:

$$\begin{array}{ccc} H(A, B) & \rightarrow & H(A'', B'') \\ & \searrow \quad \nearrow & \\ & H(A', B') & \end{array}$$

(H. 3) Si $(A, B, C) < (A', B', C')$, le diagramme suivant est commutatif:

$$\begin{array}{ccc} H(A, B) & \xrightarrow{d} & H(B, C) \\ \downarrow & & \downarrow \\ H(A', B') & \xrightarrow{d} & H(B', C') \end{array}$$

(H. 4) Pour tout triple (A, B, C), les inégalités $(B, C) < (A, C) < (A, B)$ donnent lieu à la suite exacte:

$$\ldots \rightarrow H(A, B) \xrightarrow{d} H(B, C) \rightarrow H(A, C) \rightarrow H(A, B) \xrightarrow{d} H(B, C) \rightarrow \ldots$$

Le fait que $H(A) \rightarrow H(A)$ est un isomorphisme, joint à l'exactitude de la suite:

$$H(A) \rightarrow H(A) \rightarrow H(A, A) \rightarrow H(A) \rightarrow H(A)$$

entraîne:

$$H(A, A) = 0 \tag{1.1}$$

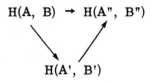

Remarque: Il suffit de supposer que l'homomorphisme d: $H(A, B) \rightarrow H(B, C)$ est défini seulement pour les paires, c'est-à-dire quand $C = 0$. Les axiomes (H. 3) et (H. 4) devront être énoncés avec C (et C') égal à 0. On pourra alors définir $d: H(A, B) \to H(B, C)$ pour un triple (A, B, C) par la composition des homomorphismes:

$$H(A, B) \xrightarrow{d} H(B) \to H(B, C) ,$$

et l'on pourra démontrer les axiomes (H. 3) et (H. 4) sous la forme énoncée plus haut.

2. Les suite fondamentales

Supposons maintenant que, dans notre ensemble partiellement ordonné, nous nous soyons donné un élément A, et une suite $A_p(-\infty < p < +\infty)$ telle que:

$$\ldots < A_p < A_{p+1} < \ldots < A .$$

Les groupes ci-dessous sont définis comme noyau-image du groupe du millieu de la suite exacte a trois termes écrite à droite:

$$B_p: \quad H(A_p) \to H(A) \to H(A, A_p)$$

$$C_p: \quad H(A_p) \to H(A_p, A_{p-1}) \to H(A_{p-1})$$

$$D_p: \quad H(A, A_p) \xrightarrow{d} H(A_p, A_{p-1}) \to H(A, A_{p-1})$$

$$C_p^k: \quad H(A_p, A_{p-k}) \to H(A_p, A_{p-1}) \xrightarrow{d} H(A_{p-1}, A_{p-k})$$

$$D_p^k: \quad H(A_{p+k-1}, A_p) \xrightarrow{d} H(A_p, A_{p-1}) \to H(A_{p+k-1}, A_{p-1})$$

(avec $k = 1, 2, \ldots$). Si nous convenons de poser $A_{-\infty} = 0$ et $A_{+\infty} = A$, alors $C_p = C_p^{\infty}$ et $D_p = D_p^{\infty}$. Nous avons les relations suivantes:

$$\ldots \subset B_{p-1} \subsetneq B_p \subset \ldots \subset H(A) \tag{2.1}$$

$$0 = D_p^1 \subset \ldots \subset D_p^k \subset D_p^{k+1} \subset \ldots \subset D_p \subset C_p \subset \ldots \subset$$
$$C_p^{k+1} \subset C_p^k \subset \ldots \subset C_p^1 = H(A_p, A_{p-1}) \tag{2.2}$$

Ces inclusions proviennent des relations de commutation dans les diagrammes suivants:

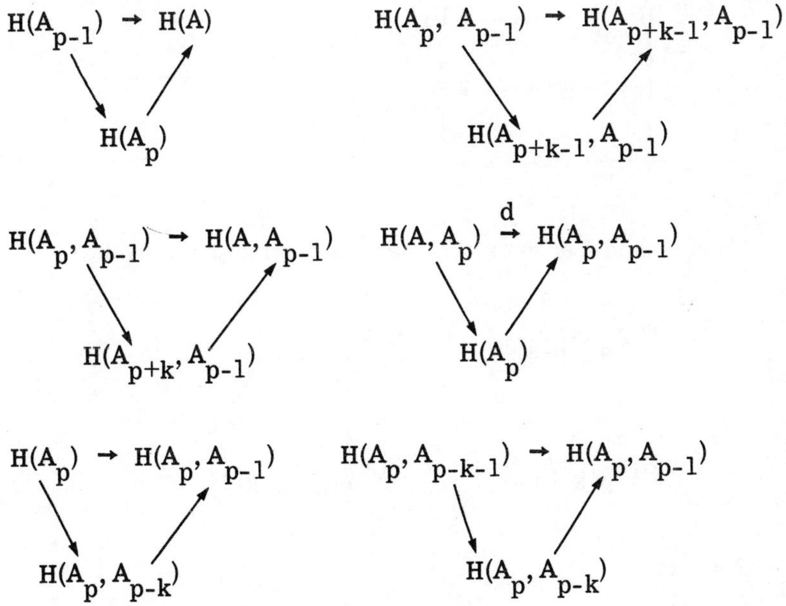

Posons

$$E_p = C_p/D_p \quad \text{et} \quad E_p^k = C_p^k/D_p^k. \tag{2.3}$$

(3)

En particulier, on a:

$$E^1_p = H(A_p, A_{p-1}) . \tag{2.4}$$

Demontrons maintenant:

$$E_p \approx B_p/B_{p-1} . \tag{2.5}$$

Cet isomorphisme résulte directement du diagramme ci-dessous, dont toutes les lignes et toutes colonnes sont exactes:

$$\begin{array}{ccccccc}
H(A_p, A_{p-1}) & \longrightarrow & H(A, A_p) & \xrightarrow{d} & D_p & \longrightarrow & 0 \\
\downarrow & & \downarrow & & \downarrow & & \downarrow \\
H(A_{p-1}) & \longrightarrow & H(A_p) & \longrightarrow & C_p & \longrightarrow & 0 \\
\downarrow & & \downarrow & & & & \\
B_{p-1} & \longrightarrow & B_p & & & & \\
\downarrow & & \downarrow & & & & \\
0 & \longrightarrow & 0 & & & &
\end{array}$$

Nous considérerons maintenant le diagramme suivant:

$$\begin{array}{ccc}
H(A_p, A_{p-k}) & \xrightarrow{\alpha} & H(A_p, A_{p-1}) \\
\Big\downarrow \gamma & & \Big\downarrow \delta \\
H(A_{p-k}, A_{p-k-1}) & \xrightarrow{\beta} & H(A_{p-1}, A_{p-k-1})
\end{array} \tag{2.6}$$

Nous avons:

$$\text{Image de } \alpha = C^k_p \supset C^{k+1}_p = \text{Noyau de } \delta$$

$$\text{Image de } \gamma = D^{k+1}_{p-k} \supset D^k_{p-k} = \text{Noyau de } \beta$$

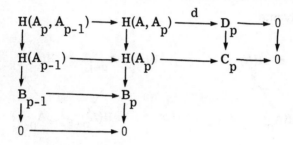

(3)

Pour tout $x \in C_p^k$, choisissons un $y \in H(A_p, A_{p-k})$ tel que $\alpha y = x$. Alors $\gamma y \in H(A_{p-k}, A_{p-k-1})$; ceci définit un homomorphisme:

$$\Delta : C_p^k \to H(A_{p-k}, A_{p-k-1}), \text{ avec}$$

$$\text{Noyau de } \Delta = C_p^{k+1}$$

$$\text{Image de } \Delta = D_{p-k}^{k+1} / D_{p-k}^k .$$

Puisque $D_p^k \subset C_p^{k+1}$ et $D_{p-k}^{k+1} \subset C_{p-k}^k$, il s'enseuit que D_p^k est contenu dans le noyau de Δ, et que C_{p-k}^k / D_{p-k}^k contient l'image de Δ. Ainsi, Δ définit un homomorphisme:

$$d_p^k : C_p^k / D_p^k \to C_{p-k}^k / D_{p-k}^k , \text{ ou:}$$

$$d_p^k : E_p^k \to E_{p-k}^k ,$$

avec:

$$\text{Noyau de } d_p^k = C_p^{k+1} / D_p^k \text{ et Image de } d_p^k = D_{p-k}^{k+1} / D_{p-k}^k$$

Si nous considérons la suite:

$$\cdots \to E_{p+k}^k \xrightarrow{d_{p+k}^k} E_p^k \xrightarrow{d_p^k} E_{p-k}^k \to \cdots \tag{2.7}$$

alors:

$$\text{Noyau de } d_p^k = C_p^{k+1} / D_p^k \text{ et Image de } d_{p+k}^k = D_p^{k+1} / D_p^k . \tag{2.8}$$

Ainsi $d_p^k d_{p+k}^k = o$, et (2. 7) est un complexe de chaînes, dont le groupe d'homologie, calculé en E_p^k n'est autre que:

$$C_p^{k+1} / D_p^{k+1} = E_p^{k+1} .$$

Exemple: $k = 1$.

Si $k = 1$ les homomorphismes α et β du diagramme (2. 6) sont les applications identiques, et $\gamma = \delta = \Delta$ est l'opérateur

$$d{:}H(A_p, A_{p-1}) \to H(A_{p-1}, A_{p-2})$$

du triple (A_p, A_{p-1}, A_{p-2}). Puisque $D_p^1 = D_{p-1}^1 = 0$, il s'ensuit que l'opérateur

$$d_p^1 : E_p^1 \to E_{p-1}^1$$

coincide avec l'opérateur d:

$$d{:}\ H(A_p, A_{p-1}) \to H(A_{p-1}, A_{p-2}) .$$

Résumé: Les résultats précédents peuvent être résumés comme suit. Posons:

$$E^k = \sum_p E_p^k \qquad k \geq 1 .$$

Les opérateurs d_p^k définissent alors une dérivation $d^k{:}E^k \to E^k$, de degré $-k$; et par rapport à cette dérivation, on a:

$$E^{k+1} = H(E^k) .$$

3. Le cas gradué

Dans les applications, le groupe H(A, B) sera fréquemment donné comme un groupe gradué $\sum_n H_n(A, B)$, n entier; l'homomorphisme de H(A, B) dans H(A', B') préservera le degré, tandis que l'opérateur d augmentera le degré de (les cas les plus fréquents sont = ± 1). La suite exacte de l'axiome (H. 4) prend alors la forme:

$$\ldots \to H_{n-}(A, B) \to H_n(B, C) \to H_n(A, C) \to H_n(A, B) \to$$

$$H_{n+}(B, C) \to \ldots$$

Tous les groupes qui ont été définis ci-dessus se décomposent en sommes directes:

$$B_p = \sum_n B_{n,p}, \ \ldots, \ E_p^k = \sum_n E_{n,p}^k .$$

L'isomorphisme (2. 5) préserve les degrés, et prend la forme:

$$E_{n,p} = B_{n,p}/B_{n,p-1} \qquad (3.1)$$

Le diagramme (2. 6) devient:

$$\begin{array}{ccc}
H_n(A_p, A_{p-k}) & \xrightarrow{\ \alpha\ } & H_n(A_p, A_{p-1}) \\
\downarrow \gamma & & \downarrow \delta \\
H_{n+}(A_{p-k}, A_{p-k-1}) & \xrightarrow{\ \beta\ } & H_{n+}(A_{p-1}, A_{p-k-1})
\end{array}$$

avec:

$$\text{Image de } \alpha = C^k_{n,p} \subset C^{k+1}_{n,p} = \text{Noyau de } \delta$$

$$\text{Image de } \gamma = D^{k+1}_{n+\ ,p-k} \subset D^k_{n+\ ,p-k} = \text{Noyau de } \beta\ .$$

Il s'ensuit que l'homomorphisme d^k_p donne:

$$d^k_{n,p} : E^k_{n,p} \to E^k_{n+\ ,p-k}\ ,$$

avec $C^{k+1}_{n,p}/D^k_{n,p}$ pour noyau, et $D^{k+1}_{n+\ ,p-k}/D^k_{n+\ ,p-k}$ pour image. La suite (2. 7) devient:

$$\ldots \to E^k_{n-\ ,p+k} \to E^k_{n,p} \to E^k_{n+\ ,p-k} \to \ldots \qquad (3.\ 2)$$

et son groupe d'homologie, calculé en $E^k_{n,p}$ est $E^{k+1}_{n,p}$.

Nous supposerons maintenant que, pour tout n fixé, l'on a:

$$H_n(A_p) = 0 \text{ pour } p \text{ assez petit} \qquad (3.\ 3)$$

$$H_n(A,\ A_p) = 0 \text{ pour } p \text{ assez grand.} \qquad (3.\ 4)$$

Il suit de (3. 3) que:

$$B_{n,p} = 0 \text{ pour } p \text{ assez petit,} \qquad (3.\ 5)$$

tandis que (3. 4) entraîne que $H_n(A_p) \to H_n(A)$ est sur. On a donc

$$B_{n,p} = H_n(A) \text{ pour } p \text{ assez grand .} \qquad (3.\ 6)$$

(3)

Fixons maintenant les indices n et p. Puisque $H_{n+}(A_{p-k}) = 0$ pour k assez grand, il en résulte que $H_n(A_p) \to H_n(A_p, A_{p-k})$ est sur, ce qui donne:

$$C_{n,p} = C^k_{n,p} \text{ pour } k \text{ assez grand}. \tag{3.7}$$

Puisque $H_{n+}(A, A_{p+k-1}) = 0$ pour k assez grand, il en résulte que $H_{n-}(A_{p+k-1}, A_p) \to H_{n-}(A, A_p)$ est sur. Ceci donne:

$$D_{n,p} = D^k_{n,p} \text{ pour } k \text{ assez grand}. \tag{3.8}$$

En combinant (3. 7) et (3. 8), on obtient:

$$E_{n,p} = E^k_{n,p} \text{ pour } k \text{ assez grand}. \tag{3.9}$$

Puisque le noyau de $d^k_{n,p}: E^k_{n,p} \to E^k_{n+\ ,p-k}$ est $C^{k+1}_{n,p}/D^k_{n,p}$, on a:

$$d^k_{n,p} = 0 \text{ pour } k \text{ assez grand}. \tag{3.10}$$

Dans tout ce qui prócède, nous avons supposé que la suite $\{A_p\}$ était croissante, c'est-à-dire que $A_p < A_{p+1}$. On aurait pu aussi bien considérer des suites descendantes vérifiant $A_{p+1} < A_p$. Le passage des suites ascendantes aux suites descendantes peut se faire par changement du signe des indices. Dans (2. 1) et (2. 3), $p - 1$ devra être remplacé par $p + 1$. Dans le diagramme due paragraphe 2, $p - 1$, $p - k$, $p - k - 1$ devront être remplacés par $p + 1$, $p + k$, $p + k + 1$. Ainsi:

$$d_p^k : E_p^k \to E_{p+k}^k .$$

Dans les formules (3. 3) a (3. 6), les mots 'grand' et 'petit' devront être échangés.

4. Le cas contravariant

Le foncteur H considéré dans ce qui précède se comporte comme une théorie de l'homologie, et, en particulier, est covariant. En théorie de la cohomologie, on a à considérer des foncteurs H(A, B) pour lesquels une inégalité $(A, B) < (A', B')$ donne naissance à un homomorphisme $H(A', B') \to H(A, B)$, et un triple (A, B, C) donne naissance à un homomorphisme $d: H(B, C) \to H(A, B)$. On énonce les axiomes (H. 1), ..., (H. 4) en renversant le sens des flèches. Dans les définitions du début du paragraphe 2, les flèches sont renversées, et les relations (2. 1) et (2. 2) deviennent:

$$\ldots \subset B_p \subset B_{p+1} \subset \ldots \subset H(A) \tag{4.1}$$

$$0 = C_p^1 \subset \ldots \subset C_p^k \subset C_p^{k+1} \subset \ldots \subset C_p \subset D_p \subset \ldots \subset$$

$$D_p^{k+1} \subset D_p^k \subset \ldots \subset D_p^1 = H(A_p, A_{p-1}) . \tag{4.2}$$

Les définitions (2. 3) deviennent:

$$E_p = D_p/C_p \qquad E_p^k = D_p^k/C_p^k . \tag{4.3}$$

L'isomorphisme (2. 5) devient:

$$E_p \approx B_{p-1}/B_p \tag{4. 4}$$

et se démontre en utilisant le diagramme:

$$\begin{array}{ccccccc}
H(A_p) & \longrightarrow & H(A_{p-1}) & \longrightarrow & C_p & \longrightarrow & 0 \\
\downarrow & & \downarrow & & \downarrow & & \\
H(A, A_p) & \longrightarrow & H(A, A_{p-1}) & \longrightarrow & D_p & \longrightarrow & 0 \\
\downarrow & & \downarrow & & & & \\
B_p & \longrightarrow & B_{p-1} & & & & \\
\downarrow & & \downarrow & & & & \\
0 & \longrightarrow & 0 & & & &
\end{array}$$

Le diagramme (2. 6) devient, après avoir élevé tous les indices de k:

$$\begin{array}{ccc}
H(A_{p+k}, A_p) & \xleftarrow{\ \alpha\ } & H(A_{p+k}, A_{p+k-1}) \\
\uparrow \gamma & & \uparrow \delta \\
H(A_p, A_{p-1}) & \xleftarrow{\ \beta\ } & H(A_{p+k-1}, A_{p-1})
\end{array}$$

avec:

$$\text{Image de } \beta = D_p^k \supset D_p^{k+1} = \text{Noyau de } \gamma$$

$$\text{Image de } \delta = C_{p+k}^{k+1} \supset C_{p+k}^k = \text{Noyau de } \alpha \, .$$

Ainsi $\delta\beta^{-1}$ donne:

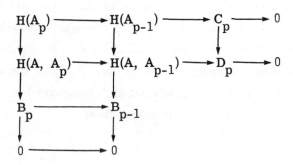

(3)

$$\Delta : D_p^k \to H(A_{p+k}, A_{p+k-1})/C_{p+k}^k\,, \quad \text{qui définit les}$$

$$d_p^k : E_p^k \to E_{p+k}^k\,,$$

avec les mêmes propriétés que ci-dessus.

La discussion séparée du cas contravariant peut être évitée en utilisant l'astuce suivante: on considère l'ensemble partiellement ordonné S^* obtenu en renversant l'ordre de l'ensemble original S. La théorie contravariante H sur S donne alors une théorie covariante sur S^*. Ceci permet de transporter les résultats par un simple changement de notations.

5. Le cas algébrique

Suit A un module (sur un certain anneau $\wedge$) muni d'une dérivation d. Nous considérerons l'ensemble de tous les sous-modules de A, stable vis-à-vis de d, avec la relation d'ordre définie par l'inclusion. Si (A', B') est une paire dans cet ensemble, alors A'/B' est un module à dérivation, et nous pouvons poser:

$$H(A', B') = H(A'/B')\,.$$

Une inclusion $(A', B') < (A'', B'')$ définit un homomorphisme permis de A'/B' dans A''/B'', et induit donc un homomorphisme $H(A', B') \to H(A'', B'')$. Un triple (A', B', C') donne une suite exacte:

$$0 \to B'/C' \to A'/C' \to A'/B' \to 0\,,$$

(3)

qui donne un opérateur d: H(A', B') → H(B', C'). Les axiomes (H. 1), . . . , (H. 4) se vérifient tout de suite.

Supposons maintenant que nous nous soyons donné une suite:

$$\dots \subset A_p \subset A_{p+1} \subset \dots \subset A .$$

La théorie précédente donne des groupes gradués E^k, avec dérivations d^k telles que $E^{k+1} = H(E^k)$, $k = 1, 2, \dots$.

Posons:

$$E_p^0 = A_p/A_{p-1}, \quad E^0 = \sum_p A_p/A_{p-1},$$

soit d_p^0 l'opérateur de dérivation de E_p^0 induit par celui de A, et soit d^0 la dérivation que en résulte dans E^0. Puisque $E_p^1 = H(A_p/A_{p-1})$, on a:

$$E^1 = H(E^0) .$$

Ainsi, dans le cas algébrique, on peut étendre la suite des $\{E^k\}$, en acceptant la valeur $k = o$.

4

EXACT COUPLES IN ALGEBRAIC TOPOLOGY

(Parts I and II)

By W. S. Massey

(Received December 11, 1951)

Introduction[1]

The main purpose of this paper is to introduce a new algebraic object into topology. This new algebraic structure is called an *exact couple of groups* (or of modules, or of vector spaces, etc.). It apparently has many applications to problems of current interest in topology. In the present paper it is shown how exact couples apply to the following three problems: (a) To determine relations between the homology groups of a space X, the Hurewicz homotopy groups of X, and certain additional topological invariants of X; (b) To determine relations between the cohomology groups of a space X, the cohomotopy groups of X, and certain additional topological invariants of X; (c) To determine relations between the homology (or cohomology) groups of the base space, the bundle space, and the fibre in a fibre bundle.

In each of these problems, the final result is expressed by means of a Leray-Koszul sequence. The notion of a Leray-Koszul sequence (also called a spectral homology sequence or spectral cohomology sequence) has been introduced and exploited by topologists of the French school. It is already apparent as a result of their work that the solution to many important problems of topology is best expressed by means of such a sequence. With the introduction of exact couples, it seems that the list of problems, for which the final answer is expressed by means of a Leray-Koszul sequence, is extended still further.

This paper is divided into five parts. The first part gives the purely algebraic aspects of the idea of an exact couple. The remaining four parts give applications of the algebraic machinery developed in part one to topological problems. Part two shows how exact couples may be applied to express relations between the homology and homotopy groups of a space, and certain new groups which are topological invariants of the space. Part three treats what may be called the dual situation. Exact couples are applied to obtain relations between the cohomology and cohomotopy groups of a space. In part four the duality which exists between the applications in parts two and three is given a precise formulation, and connection is made with the usual duality between the homology and cohomology groups of a space. In both parts two and three the relations involved are expressed by means of a Leray-Koszul sequence, and it seems rather unlikely that these Leray-Koszul sequences could be obtained from a differential-filtered group.

[1] An abstract of the principal results of this paper was submitted to the American Mathematical Society in March, 1951; cf. Bull. Amer. Math. Soc., 57 (1951), pp. 281–282.

The fifth and last part applies the theory of exact couples to the study of the cohomology structure of a fibre bundle for which the base space is a cell complex. A Leray-Koszul sequence is obtained which gives relations between the cohomology groups of the base space, bundle space, and fibre. The results obtained in this case are included among those obtained by Leray [11] under more general hypotheses. Furthermore, we have not included the multiplicative structure of the cohomology ring into the Leray-Koszul sequence obtained, and thus our results are not as complete as those of Leray. This application of exact couples to obtain the Leray-Koszul sequence of a fibre bundle is published in spite of these shortcomings mainly because it is our belief that the methods we use are closer to the usual methods of algebraic topology, and hence can be understood by most topologists with less effort than the methods of Leray. They should serve as an introduction to the important papers of Leray.

The notations, definitions, and conventions used for homology, cohomology, homotopy, and cohomotopy groups are collected together for ready reference in the appendix. Here also are contained the explicit statements of some lemmas from homotopy theory which are needed.

Parts I and II are published in the present issue of these ANNALS; parts III, IV, and V will appear in these ANNALS in the near future.

PART I. GENERAL ALGEBRAIC THEORY

1. Differential Groups

The principle algebraic objects with which we shall be concerned in this paper are abelian groups with certain additional elements of structure, together with certain homomorphisms of these abelian groups.

Let A be an abelian group. An endomorphism $d: A \to A$ is called a *differential operator* if $d^2 = 0$ (i.e., $d[d(a)] = 0$ for any $a \in A$). A *differential group* is a pair (A, d) consisting of an abelian group A and a differential operator d. If (A, d) is a differential group, we will denote by $\mathcal{Z}(A)$ the kernel of d, and by $\mathcal{B}(A)$ the image, $d(A)$. Both are subgroups of A, and from $d^2 = 0$, it follows that $\mathcal{B}(A) \subset \mathcal{Z}(A)$. The factor group $\mathcal{Z}(A)/\mathcal{B}(A)$ will be denoted by $\mathcal{H}(A)$, and called the *derived group*.

Let (A, d) and (A', d') be differential groups; a homomorphism $f: A \to A'$ is called *allowable* if the commutativity relation $d' \circ f = f \circ d$ holds. Such an allowable homomorphism f has the property that $f[\mathcal{Z}(A)] \subset \mathcal{Z}(A')$ and $f[\mathcal{B}(A)] \subset \mathcal{B}(A')$, and hence f *induces* a homomorphism $f^*: \mathcal{H}(A) \to \mathcal{H}(A')$. This operation of assigning to each allowable homomorphism the induced homomorphism of the derived groups has the following two obvious, but important, properties: (1) The identity homomorphism $i: A \to A$ is allowable, and the induced homomorphism $i^*: \mathcal{H}(A) \to \mathcal{H}(A)$ is also the identity. (2) Let (A, d), (A', d'), and (A'', d'') be differential groups, and let $f: A \to A'$, $g: A' \to A''$ be allowable homomorphisms. Then the composition $g \circ f: A \to A''$ is also allowable, and $(g \circ f)^* = g^* \circ f^*$. In the language of Eilenberg and MacLane [3], these two facts may be conveniently expressed by saying that the set of all differential groups and allowable homomorphisms constitutes a category, and the operation of assigning to each differential group its derived group and to each homomorphism its induced homomorphism is a covariant functor.

Let (A, d) and (A', d') be differential groups, and $f, g: A \to A'$ allowable homomorphisms. Then f and g are said to be *algebraically homotopic* (notation: $f \simeq g$) if there exists a homomorphism $\xi: A \to A'$ which satisfies the following condition:

$$f - g = d' \circ \xi + \xi \circ d.$$

It is readily verified that this relation is an equivalence relation. Furthermore, if f and g are algebraically homotopic, then the induced homomorphisms $f^*, g^*: \mathcal{H}(A) \to \mathcal{H}(A')$ are the same.

A subgroup B of a group A with differential operator d is said to be *allowable* if $d(B) \subset B$; if this is the case, then d defines differential operators on B and on the factor group A/B in an obvious fashion; furthermore, the inclusion homomorphism $B \to A$ and the natural projection of A onto the factor group A/B are both allowable homomorphisms in the above mentioned sense.

2. Graded and Bigraded Groups

An abelian group A is said to be *graded*, or to have a *graded structure*, if there is prescribed a sequence of subgroups, A_n, $n = 0, \pm 1, \pm 2, \cdots$, such that A can be expressed as a direct sum,

$$A = \sum_{-\infty}^{+\infty} A_n .$$

An abelian group A is said to be *bigraded*, or to have a *bigraded structure*, if there is prescribed a double sequence of subgroups, A_{mn}, $m, n = 0, \pm 1, \pm 2, \cdots$, such that $A = \sum_{m,n} A_{m,n}$. In case of a graded group, $A = \sum_m A_m$, the elements of the subgroup A_p are said to be *homogeneous of degree* p; in the case of a bigraded group, $A = \sum_{m,n} A_{m,n}$, the elements of the subgroup $A_{p,q}$ are said to be *homogeneous of degree* (p, q).

When dealing with graded or bigraded groups, only a certain limited class of homormorphisms are of interest, the so-called homogeneous homomorphisms. If $A = \sum A_m$ and $B = \sum B_m$ are graded groups, then a homomorphism $f: A \to B$ is said to be *homogeneous of degree* p provided $f(A_m) \subset B_{m+p}$ for all values of m. If $A = \sum A_{m,n}$ and $B = \sum B_{m,n}$ are bigraded, then a homomorphism $f: A \to B$ is *homogeneous of degree* (p, q) provided $f(A_{m,n}) \subset B_{m+p,n+q}$ for all pairs (m, n). Note that the identity homomorphism of a graded or bigraded group onto itself is homogeneous, and that the composition of two homogeneous homomorphisms is again homogeneous.

Let $A = \sum A_{m,n}$ be a bigraded group; a subgroup $B \subset A$ is said to be *allowable* (with respect to the bigraded structure) in case $B = \sum (B \cap A_{m,n})$. If this is true, then B has a bigraded structure defined by $B = \sum B_{m,n}$, where $B_{m,n} = B \cap A_{m,n}$. We will express this fact by saying that the allowable subgroup B *inherits* a bigraded structure from A. It is readily verified that the factor group A/B is isomorphic to the direct sum $\sum A_{m,n}/B_{m,n}$; moreover, this isomorphism is natural in the sense of Eilenberg and MacLane [3]. We will agree to identify these naturally isomorphic groups. This representation of A/B as a direct sum defines a bigraded structure on A/B. This bigraded structure will also be referred to as the bigraded structure that A/B *inherits* from A.

It is clear how to define in an analogous way the concept of an allowable subgroup of a graded group, etc. Note that the kernel and image of a homogeneous homomorphism are always allowable subgroups.

Often in algebraic topology we have to deal with differential groups which also have a graded (or bigraded) structure, and for which the differential operator is homogeneous. In this case the derived group inherits a graded (or bigraded) structure from the given group.

If A and G are abelian groups, then we will use the notation $A \otimes G$ to denote their tensor product (see Whitney, [24]). If (A, d) is a differential group, and G is an arbitrary abelian group, then we define a differential operator $d': A \otimes G \rightarrow A \otimes G$ in the obvious way:

$$d'(a \otimes g) = (da) \otimes g$$

for any $a \in A$ and $g \in G$. If $A = \sum A^p$ is a graded group, then the direct sum decomposition

$$A \otimes G = \sum (A^p \otimes G),$$

defines a graded structure on the tensor product $A \otimes G$. An analogous definition is applicable in case A is a bigraded group. If (A, d) is a differential group, and G is an abelian group, then for the sake of convenience we will use the notation $\mathcal{H}(A, G)$ for the derived group of $(A \otimes G, d')$; i.e., $\mathcal{H}(A, G) = \mathcal{H}(A \otimes G)$.

3. Definition of a Leray-Koszul Sequence

A sequence of differential groups, (A^n, d^n), where the index n ranges over all integers larger than some given integer N, is called a *Leray-Koszul sequence* in case each group in the sequence is the derived group of the preceding:

$$A^{n+1} = \mathcal{H}(A^n).$$

In a Leray-Koszul sequence there exist natural homomorphisms

$$\kappa_n : \mathcal{Z}(A^n) \rightarrow A^{n+1}.$$

κ_n is defined by assigning to each element of the subroup $\mathcal{Z}(A^n)$ its coset modulo $\mathcal{B}(A^n)$. Thus κ_n is a homomorphism of a subgroup of A^n onto A^{n+1}. We will define a homomorphism κ_n^p of a subgroup of A^n onto A^{n+p} by the formula

$$\kappa_n^p = \kappa_{n+p-1} \circ \kappa_{n+p-2} \circ \cdots \circ \kappa_n .$$

Then $\kappa_n^1 = \kappa_n$. The precise definition of the domain of definition of κ_n^p is left to the reader. Let $\bar{A}^n$ denote the subgroup of A^n consisting of these elements $a \in A^n$ such that $\kappa_n^p(a)$ is defined for all values of p. Define $\bar{\kappa}_n : \bar{A}^n \rightarrow \bar{A}^{n+1}$ to be the restriction of κ_n to the subgroup $\bar{A}^n$. Then the sequence of groups $\{\bar{A}^n\}$ and homomorphisms $\{\bar{\kappa}_n\}$ constitutes a direct sequence of groups in the usual sense (see, for example, [10, ch. VIII, definition VIII 12]). The limit group of this direct sequence of groups will be called the *limit group* of the given Leray-Koszul sequence.

In most cases we shall have to deal with Leray-Koszul sequences (A^n, d^n) for which each of the groups A^n is bigraded, $A^n = \sum A^n_{p,q}$, each of the differential operators d^n is homogeneous, and A^{n+1} inherits its bigraded structure from A^n. In this case each of the homomorphisms κ_n is homogeneous of degree (0, 0), and there is determined a bigraded structure on the limit group in a natural way. Also, it will usually be true that for each pair of integers (p, q) there exists an integer N such that if $n > N$, then κ_n maps $A^n_{p,q}$ isomorphically onto $A^{n+1}_{p,q}$. This makes it possible to determine any homogeneous component of the limit group by an essentially finite process.

4. Definition of an Exact Couple; The Derived Couple

An *exact couple* of abelian groups consists of a pair of abelian groups, A and C, and three homomorphisms:

$$f: A \to A,$$

$$g: A \to C,$$

$$h: C \to A.$$

These homomorphisms are required to satisfy the following "exactness" conditions:

$$\text{image } f = \text{kernel } g,$$

$$\text{image } g = \text{kernel } h,$$

$$\text{image } h = \text{kernel } f.$$

These three conditions can be easily remembered if one makes the following triangular diagram,

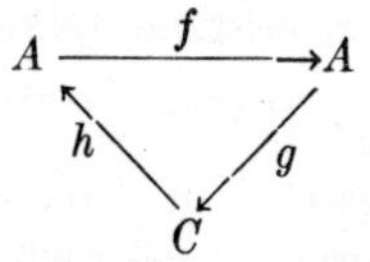

and observes that the kernel of each homomorphism is required to be the image of the preceding homomorphism. We shall denote such an exact couple by the notation $\langle A, C; f, g, h\rangle$. When there is no danger of confusion, we shall often abbreviate this to $\langle A, C\rangle$.

There is an important operation which assigns to an exact couple $\langle A, C; f, g, h\rangle$ another exact couple, $\langle A', C'; f', g', h'\rangle$, called the *derived* exact couple. This derived exact couple is defined as follows.

Define an endomorphism $d:C \to C$ by $d = g \circ h$. Then $d^2 = d \circ d = g \circ h \circ g \circ h = 0$, since $h \circ g = 0$ by exactness. Therefore d is a differential operator. Let $C' = \mathcal{H}(C)$, the derived group of the differential group (C, d). Let $A' = f(A) =$ image $f =$ kernel g. Define $f':A' \to A'$ by $f' = f \mid A'$, the restriction of f to the subgroup A'. The homomorphism $h':C' \to A'$ is induced by h: it is readily verified that $h[\mathcal{Z}(C)] \subset A'$, and $h[\mathcal{B}(C)] = 0$, hence h induces a homomorphism of the factor group $C' = \mathcal{Z}(C)/\mathcal{B}(C)$ into A'. The definition of $g':A' \to C'$ is more complicated. Let $a \in A'$; choose an element $b \in A$ such that $f(b) = a$. Then $g(b) \in \mathcal{Z}(C)$, and $g'(a)$ is defined to be the coset of $g(b)$ modulo $\mathcal{B}(C)$. It is easily verified that this definition is independent of the choice made of the element $b \in A$, and that g' is actually a homomorphism.

Of course, it is necessary to verify that the homomorphisms f', g', and h' satisfy the exactness condition of an exact couple. This verification is straightforward, and is left to the reader.

It is clear that this process of derivation can be applied to the derived exact couple $\langle A', C'; f', g', h' \rangle$ to obtain another exact couple $\langle A'', C''; f'', g'', h'' \rangle$, called the *second derived couple*, and so on. In general, we shall denote the nth derived couple by $\langle A^{(n)}, C^{(n)}; f^{(n)}, g^{(n)}, h^{(n)} \rangle$.

5. Maps of Exact Couples

Let $\langle A, C; f, g, h \rangle$ and $\langle A_0, C_0; f_0, g_0, h_0 \rangle$ be two exact couples; a *map*,

$$(\phi, \psi): \langle A, C; f, g, h \rangle \to \langle A_0, C_0; f_0, g_0, h_0 \rangle.$$

consists of a pair of homomorphisms,

$$\phi : A \to A_0,$$

$$\psi : C \to C_0,$$

which satisfy the following three commutativity conditions:

$$\phi \circ f = f_0 \circ \phi$$
$$\psi \circ g = g_0 \circ \phi$$
$$\phi \circ h = h_0 \circ \psi.$$

If $d = g \circ h: C \to C$ and $d_0 = g_0 \circ h_0: C_0 \to C_0$ denote the differential operators on C and C_0 respectively, then our definitions imply the following commutativity relation:

$$\psi \circ d = d_0 \circ \psi.$$

Therefore ψ is an allowable homomorphism, in the sense defined in the preceding section, and hence induces a homomorphism

$$\psi': C' \to C_0'$$

of the corresponding derived groups. Also, it is clear that $\phi(A') \subset A_0'$; therefore ϕ defines a homomorphism

$$\phi': A' \to A_0'.$$

It can now be verified without difficulty that the pair of homomorphisms (ϕ', ψ') constitute a map of the first derived exact couples,

$$(\phi', \psi'):\langle A', C'\rangle \to \langle A_0', C_0'\rangle$$

in the sense just defined. We will say that the map (ϕ', ψ') is *induced* by (ϕ, ψ). By iterating this process, one obtains a map $(\phi^{(n)}, \psi^{(n)}):\langle A^{(n)}, C^{(n)}\rangle \to \langle A_0^{(n)}, C_0^{(n)}\rangle$ which is induced by the given map (ϕ, ψ).

The set of all exact couples and maps of exact couples constitutes a category in the sense of Eilenberg and MacLane [3], and the operation of derivation is a covariant functor.

Let (ϕ_0, ψ_0) and $(\phi_1, \psi_1):\langle A, C; f, g, h\rangle \to \langle A_0, C_0; f_0, g_0, h_0\rangle$ be two maps of exact couples in the sense we have just defined. The maps (ϕ_0, ψ_0) and (ϕ_1, ψ_1) are said to be *algebraically homotopic*[2] (notation: $(\phi_0, \psi_0) \simeq (\phi_1, \psi_1)$) if there exists a homomorphism $\xi: C \to C_0$ such that for any element $c \in C$,

$$\psi_1(c) - \psi_0(c) = \xi[d(c)] + d_0[\xi(c)],$$

and for any $a \in A$,

$$\phi_1(a) - \phi_0(a) = h_0 \xi g(a).$$

It is readily verified that the relation so defined is reflexive, transitive, and symmetric, and hence is an equivalence relation. The main reason for the importance of this concept is the following proposition:

THEROEM 5.1. *If the maps*

$$(\phi_0, \psi_0), (\phi_1, \psi_1):\langle A, C; f, g, h\rangle \to \langle A_0 C_0; f_0, g_0, h_0\rangle$$

are algebraically homotopic, then the induced maps (ϕ_0', ψ_0') *and* (ϕ_1', ψ_1') *of the derived couples are the same.*

The proof is entirely trivial. It follows that the induced maps of the nth derived couples, $(\phi_0^{(n)}, \psi_0^{(n)})$ and $(\phi_1^{(n)}, \psi_1^{(n)})$ are also the same.

6. Bigraded Exact Couples; The Associated Leray-Koszul Sequence

In the applications later on it will usually be true that groups occurring in the exact couples with which we are concerned will be bigraded groups, and that all the homomorphisms involved will be homogeneous homomorphisms. Then the groups of the successive derived couples will inherit a bigraded structure from the original groups, and the homomorphisms in the successive derived couples will also be homogeneous. To be precise, if $\langle A, C; f, g, h\rangle$ is a bigraded exact couple, and $\langle A', C'; f', g', h'\rangle$ denotes the first derived couple, then f' and f have the same degree of homogeneity, as do h' and h; however, the degree of homogeneity of g' is that of g minus that of f.

Let $\langle A, C; f, g, h\rangle$ be an exact couple, and let $\langle A^{(n)}, C^{(n)}; f^{(n)}, g^{(n)}, h^{(n)}\rangle$, $n = 1, 2, \cdots$ denote the successive derived couples. Let $d^{(n)} = g^{(n)} \circ h^{(n)}: C^{(n)} \to C^{(n)}$ denote the differential operator of $C^{(n)}$. Then the sequence of differential groups

[2] This definition is patterned after a similar one given by J. H. C. Whitehead, [20].

$(C^{(n)}, d^{(n)})$ is a Leray-Koszul sequence. It will be referred to as *the Leray-Koszul sequence associated with the exact couple* $\langle A, C\rangle$.

Bibliography

1. P. Alexandroff and H. Hopf, Topologie I. Berlin, J. Springer, 1935.
2. A. L. Blakers and W. S. Massey, *Homotopy Groups of a Triad I*. Ann. of Math., 53 (1951), 161–205.
3. S. Eilenberg and S. MacLane, *Natural Isomorphisms in Group Theory*. Proc. Nat. Acad. Sci. U.S.A., 28 (1942), 537–543.
4. S. Eilenberg and S. MacLane, *Relations between Homology and Homotopy Groups of Spaces*. Ann. of Math., 46 (1945), 480–509.
5. S. Eilenberg and N. E. Steenrod, Foundations of Algebraic Topology. Princeton University Press, 1952.
6. H. Freudenthal, *Über die Klassen der Sphärenabbildungen* I. Compositio Math., 5 (1937), 299–314.
7. S. T. Hu, *An Exposition of the Relative Homotopy Theory*. Duke Math. J., 14 (1937), 991–1033.
8. S. T. Hu, *Mappings of a Normal Space in an Absolute Neighborhood Retract*. Trans. Amer. Math. Soc., 64 (1948), 336–358.
9. S. T. Hu, *A Group Multiplication for Relative Homotopy Groups*. J. London Math. Soc., 22 (1947), 61–67.
10. W. Hurewicz and H. Wallman, Dimension Theory. Princeton University Press, 1941.
11. J. Leray, *L'Homologie d'un Espace Fibré dont la Fibre Est Connexe*. Jour. Math. Pures Appl., 29 (1950), 169–213.
12. S. MacLane and J. H. C. Whitehead, *On the 3-type of a Complex*. Proc. Nat. Acad. Sci. U.S.A., 36 (1950), 41–48.
13. W. S. Massey, *Homotopy Groups of Triads*. Proc. Int. Congress of Math. Vol. II (1950), 371–382.
14. E. Spanier, *Borsuk's Cohomotopy Groups*. Ann. of Math., 50 (1949), 203–245.
15. N. E. Steenrod, *Homology with Local Coefficients*. Ann. of Math., 44 (1943), 610–627.
16. N. E. Steenrod, *Products of Cocycles and Extensions of Mappings*. Ann. of Math., 48 (1947), 290–320.
17. N. E. Steenrod, The Topology of Fibre Bundles. Princeton University Press, 1951.
18. G. W. Whitehead, *A Generalization of the Hopf Invariant*. Ann. of Math., 51 (1950), 192–237.
19. J. H. C. Whitehead, *A Note on Suspension*. Quart. J. Math. (2nd series) 1 (1950), 9–22.
20. J. H. C. Whitehead, *A Certain Exact Sequence*. Ann. of Math., 52 (1950), 51–110.
21. J. H. C. Whitehead, *Combinatorial Homotopy* I. Bull. Amer. Math. Soc., 55 (1949), 213–245.
22. J. H. C. Whitehead, *Combinatorial Homotopy* II. Bull. Amer. Math. Soc., 55 (1949), 453–496.
23. J. H. C. Whitehead, *On the Realizability of Homotopy Groups*. Ann. of Math., 50 (1949), 261–263.
24. H. Whitney, *Tensor Products of Abelian Groups*. Duke Math. J., 4 (1938), 495–528.

5

The scope of the next paper has been explained in §6. As one of the later papers, it assumes a fair familiarity with the machinery of algebraic topology.

5 THE COHOMOLOGY OF CLASSIFYING SPACES OF *H*-SPACES

BY M. ROTHENBERG AND N. E. STEENROD[1]

Communicated by N. E. Steenrod, June 17, 1965

Let G denote an associative H-space with unit (e.g. a topological group). We will show that the relations between G and a classifying space B_G are more readily displayed using a geometric analog of the resolutions of homological algebra. The analogy is quite sharp, the stages of the resolution, whose base is B_G, determine a filtration of B_G. The resulting spectral sequence for cohomology is independent of the choice of the resolution, it converges to $H^*(B_G)$, and its E_2-term is $\operatorname{Ext}_{H(G)}(R, R)$ (R=ground ring). We thus obtain spectral sequences of the Eilenberg-Moore type [**5**] in a simpler and more geometric manner.

1. **Geometric resolutions.** We shall restrict ourselves to the category of compactly generated spaces. Such a space is Hausdorff and each subset which meets every compact set in a closed set is itself closed (a k-space in the terminology of Kelley [**3**, p. 230]). Subspaces are usually required to be closed, and to be deformation retracts of neighborhoods.

Let G be an associative H-space with unit e. A right G-action on a space X will be a continuous map $X \times G \to X$ with $xe = x$, $x(g_1g_2) = (xg_1)g_2$ for all $x \in X$, $g_1, g_2 \in G$. A space X with a right G-action will be called a G-space. A G-space X and a sequence of G-invariant closed subspaces $X_0 \subset X_1 \subset \cdots \subset X_n \subset \cdots$ such that $X_0 \neq \emptyset$, $X = \bigcup_{i=0}^{\infty} X_i$, and X has the weak topology induced by $\{X_i\}$ will be called a *filtered* G-space.

1.1. DEFINITION. (a) A filtered G-space X is called *acyclic* if for some point $x_0 \in X_0$, X_n is contractible to x_0 in X_{n+1} for every n.

(b) A filtered G-space X is called *free* if, for each n, there exists a closed subspace D_n ($X_{n-1} \subset D_n \subset X_n$) such that the action mapping $(D_n, X_{n-1}) \times G \to (X_n, X_{n-1})$ is a relative homeomorphism.

(c) A filtered G-space X is called a *G-resolution* if X is both free and acyclic.

Under the restrictions we have imposed on subspaces, the acyclicity condition implies that X is contractible.

---

[1] This work was partially supported by the National Science Foundation under NSF grants GP 3936 and GP 2425.

1.2. THEOREM. *If G is a topological group, any G-resolution X is a principal G-bundle over $B_G = X/G$ with action $X \times G \to X$ as principal map.*

When G is a topological group, Milnor's construction [4], where X_n is the join of $n+1$ copies of G, is a G-resolution. In the general case, the existence of a G-resolution is given by the Dold-Lashof construction [2].

There is also a comparison theorem. Let G, G' be H-spaces, $\Phi: G \to G'$ a morphism, X, X' filtered G, G'-spaces. An *extension* Φ' of Φ is a map $\Phi': X \to X'$ with $\Phi'(X_n) \subset X_n'$ and $\Phi'(xg) = \Phi'(x)\Phi(g)$. If Φ', Φ'' are two extensions of Φ, a *homotopy* h will be a map $h: X \times I \to X'$ with $h_0 = \Phi'$, $h_1 = \Phi''$, $h(X_n \times I) \subset X_{n+1}'$, and $h(xg, t) = h(x, t)\Phi(g)$.

1.3. MAPPING THEOREM. *If $\Phi: G \to G'$ is a morphism, X a free filtered G-space, X' an acyclic filtered G'-space, then Φ has an extension Φ': $X \to X'$. Furthermore, any two such extensions are homotopic.*

Thus in particular, for any two resolutions X, X' of G there exists an equivariant $\mu: X \to X'$, unique up to equivariant homotopy.

We define the product of two filtered spaces X, X' to be the product space $X \times X'$ filtered by $(X \times X')_n = \bigcup_{i=0}^{n} X_i \times X_{n-i}$.

1.4. THEOREM. *If X is a G-resolution and X' a G'-resolution, then $X \times X'$ is a $G \times G'$-resolution.*

2. **The spectral sequence.** When X is a G-resolution, let $B = X/G$ denote the decomposition space by maximal orbits, let $p: X \to B$ be the projection and $B_n = p(X_n)$. If R is a coefficient ring, the filtration $\{B_n\}$ of B determines two spectral sequences, the homology spectral sequence $E_*(B, R) = \{E^r, d_r\}$ and the cohomology spectral sequence $E^*(B, R) = \{E_r, d^r\}$.

2.1. THEOREM. (a) *The spectral sequences E_*, E^* are functors from the category of H-spaces and continuous morphisms to the category of bigraded spectral sequences. (We regard all spectral sequences as beginning with E^2, E_2.)*

(b) *If the homology algebra $H(G) = H(G; R)$ is R-free, then as a bigraded R-module*

$$E^2 \cong \mathrm{Tor}^{H(G)}(R, R), \qquad E_2 \cong \mathrm{Ext}_{H(G)}(R, R).$$

(c) $E_* \Rightarrow H(B; R)$. *If R is compact or $H(G)$ is free then $E^* \Rightarrow H^*(B; R)$.*

Proposition (a) follows from 1.3, (c) is true in any filtered space, and (b) is proved using the Milnor-Dold-Lashof construction, in fact the E^1-term in this case is precisely the bar resolution of R over the algebra $H(G)$.

In order to deepen these results to include products, we develop the theory of $\times$-products for the spectral sequences of filtered spaces X, Y. These are natural transformations μ: $E^r(X) \otimes E^r(Y) \to E^r(X \times Y)$, ν: $E_r(X) \otimes E_r(Y) \to E_r(X \times Y)$ which behave nicely with respect to differentials. They are isomorphisms when R is a field and $E_1(X)$ is of finite type.

The diagonal morphism $\Delta: G \to G \times G$ induces, by 2.1(a), a mapping of the cohomology spectral sequences $\Delta^*: E_r(B_G \times B_G) \to E_r(B_G)$. Composing Δ^* with ν (where $X = Y = B_G$) gives the multiplication in E_r.

2.2. Theorem. *With respect to this multiplication, $E_r(B_G)$ is a commutative, associative, bigraded, differential algebra with unit. The multiplication on E_{r+1} is induced by that on E_r. The multiplications commute with the convergence* 2.1(c). *When $H(G)$ is R-free, the second isomorphism of* 2.1(b) *preserves products.*

When R is a field, the composition $\mu^{-1}\Delta_*$ defines a co-algebra structure in the homology spectral sequence having dual properties.

3. Co-algebra structure. We assume in this section that R is a field and $H(G)$ is of finite type. When G is commutative the multiplication $m: G \times G \to G$ is also a morphism. Then the composition $m_*\mu$ gives an algebra structure on E_*, and $\nu^{-1}m^*$ a co-algebra structure in E^*. Actually the same is true if G is the loop space of an H-space. This yields

3.1. Theorem. *If G is commutative or the loop space of an H-space, then E_r, E^r are bicommutative, biassociative, differential, bigraded Hopf algebras with (E^r, d_r) the dual algebra to (E_r, d^r). The Hopf algebra structure on $E_2 = \mathrm{Ext}_{H(G)}(R, R)$ is the natural one arising from the Hopf algebra structure on $H(G)$. Moreover if G is connected and R is perfect, then E_r is primitively generated on elements of bi-degree $(1, q)$, $(2, q')$, and $d^r = 0$ except for $r = p^k - 1$ or $2p^k - 1$ where $p = \mathrm{Char}\, R$. If $G = \Omega(H)$, H homotopy associative, then $E_\infty \approx H^*(B; R)$ as an algebra.*

Actually one can give an explicit description of E_{r+1} in terms of E_r and $d^r(x^{1,q})$, $d^r(x^{2,q'})$, where $x^{1,q}$, $x^{2,q'}$ are primitive generators.

4. Applications. Moore pointed out [5] that his spectral sequence gives an easy proof of the theorem of Borel which states: *If $H(G)$ is an exterior algebra with generators of odd dimensions and is R-free, then $H^*(B_G)$ is a polynomial algebra on corresponding generators of one higher dimension.* Moore argues that a brief computation shows that the E_2-term, $\operatorname{Ext}_{H(G)}(R, R)$, is just such a polynomial algebra. Then all terms of E_2 of odd total degree are zero. Hence every $d^r=0$, so $E_2=E_\infty$. Since E_∞ is a polynomial algebra, it is algebraically free; and therefore $H^*(B_G)\approx E_\infty$ as an algebra.

An Eilenberg-MacLane space of type (π, n) can be realized by a commutative topological group G, and its B_G is of type $(\pi, n+1)$. Consequently $H(\pi, n)$ and $H^*(\pi, n+1)$ are connected by a spectral sequence of Hopf algebras $E_r(B_G)$.

4.1. THEOREM. *If G is of type (π, n), π is a finitely generated abelian group, and $R=Z_p$ where p is a prime, then the spectral sequence collapses*

$$\operatorname{Ext}_{H(G)}(Z_p, Z_p) \approx E_2 = E_\infty \approx H^*(B_G).$$

This implies that $H^*(\pi, n; Z_p)$ is a free commutative algebra for every n. In fact an algorithm is obtained for computing $H^*(\pi, n; Z_p)$ as a primitively generated Hopf algebra over the algebra of reduced pth powers. These results confirm and amplify results of H. Cartan.

For another application, let K be a compact, simply-connected Lie group, and let G be the loop space of K. Using Bott's result [1] that $H(G; Z)$ is torsion free, we obtain

4.2. THEOREM. (a) *If $p>5$, the spectral sequence collapses*

$$\operatorname{Ext}_{H(G)}(Z_p, Z_p) \approx E_2= E_\infty \approx H^*(K; Z_p) \approx \Lambda(x_1, \cdots, x_r)$$

where $x_1, \cdots, x_r$ are generators of the dimensions of the primitive invariants of K. In particular K has no p-torsion, and $H^*(K; Z_p) \approx H^*(K; Z)\otimes Z_p$.

(b) *If $p=3$ or 5, there is at most one nonzero differential, namely, d^{2p-1}. Moreover $H^*(K; Z_p)$ and $H_*(G; Z_p)$ can be constructed explicitly from the Betti numbers of K and the dimensions of the kernels of the maps $x\to x^p$ and $x\to x^{p^2}$ where $x\in H^2(G; Z_p)$.*

(c) *For any $p>2$, we have $u^p=0$ for all $u\in \tilde{H}^*(K; Z_p)$.*

BIBLIOGRAPHY

1. R. Bott, *The space of loops of a Lie group*, Michigan Math. J. **5** (1958), 35–61.

2. A. Dold and R. Lashof, *Principal quasifibrations and fibre homotopy equivalences*, Illinois J. Math **3** (1959), 285–305.

3. J. L. Kelley, *General topology*, Van Nostrand, Princeton, N. J., 1955.

4. J. Milnor, *Construction of universal bundles*. II, Ann. of Math. **63** (1956), 430–436.

5. J. C. Moore, *Algèbre homologique et cohomologie des espaces classifants*, Séminaire H. Cartan, 1959–1960, Exposé 7.

UNIVERSITY OF CHICAGO AND
PRINCETON UNIVERSITY

6

The next extract contains the major part of Serre's elegant and important paper on Eilenberg-MacLane spaces. The scope and usefulness of this paper has been explained in §7. This work is perhaps one of the main sources for the idea of 'universal example' or 'representablefunctor' (see also the application of the same idea to homotopy summarised in paper 7). The paper is fairly self-contained but it leans heavily on a theorem of A. Borel. For this, see the remarks and references in §6 plus Adams chap. 2, p. 11. The student might find the paper easier if he is familiar with homology theory before reading it; but equally, reading this paper would be a good way to acquire familiarity with homology theory.

6

Cohomologie modulo 2 des complexes d'Eilenberg-MacLane

Par Jean-Pierre Serre, Paris

Introduction

On sait que les complexes $K(\Pi, q)$ introduits par Eilenberg-MacLane dans [4] jouent un rôle essentiel dans un grand nombre de questions de topologie algébrique. Le présent article est une contribution à leur étude.

En nous appuyant sur un théorème démontré par A. Borel dans sa thèse [2], nous déterminons les algèbres de cohomologie modulo 2 de ces complexes, tout au moins lorsque le groupe Π possède un nombre fini de générateurs. Ceci fait l'objet du § 2. Dans le § 3 nous étudions le comportement asymptotique des séries de Poincaré des algèbres de cohomologie précédentes ; nous en déduisons que, lorsqu'un espace X vérifie des conditions très larges (par exemple, lorsque X est un polyèdre fini, simplement connexe, d'homologie modulo 2 non triviale), il existe une infinité d'entiers i tels que le groupe d'homotopie $\pi_i(X)$ contienne un sous-groupe isomorphe à Z ou à Z_2. Dans le § 4 nous précisons les relations qui lient les complexes $K(\Pi, q)$ et les diverses «opérations cohomologiques»; ceci nous fournit notamment une méthode permettant d'étudier les relations entre i-carrés itérés. Le § 5 contient le calcul des groupes $\pi_{n+3}(S_n)$ et $\pi_{n+4}(S_n)$; ce calcul est effectué en combinant les résultats des §§ 2 et 4 avec ceux d'une Note de H. Cartan et l'auteur ([3], voir aussi [14]). Les §§ 4 et 5 sont indépendants du § 3.

Les principaux résultats de cet article ont été résumés dans une Note aux Comptes Rendus [9].

§ 1. Préliminaires

1. Notations

Si X est un espace topologique et G un groupe abélien, nous notons $H_i(X, G)$ le i-ème groupe d'homologie singulière de X à coefficients dans G ; nous posons $H_*(X, G) = \sum_{i=0}^{\infty} H_i(X, G)$, le signe $\sum$ représentant une somme directe.

De façon analogue, nous notons $H^i(X, G)$ les groupes de cohomologie de X, et nous posons $H^*(X, G) = \sum_{i=0}^{\infty} H^i(X, G)$.

Les groupes d'homologie et de cohomologie relatifs d'un couple (X, Y) sont notés $H_i(X, Y; G)$ et $H^i(X, Y; G)$.

Nous notons Z le groupe additif des entiers et Z_n le groupe additif des entiers modulo n.

2. Les *i*-carrés de Steenrod

N. E. Steenrod a défini dans [12] (voir aussi [13]) des homomorphismes :

$$Sq^i: \ H^n(X, Y; Z_2) \to H^{n+i}(X, Y; Z_2) \qquad (i \text{ entier } \geqslant 0) ,$$

où (X, Y) désigne un couple d'espaces topologiques, avec $Y \subset X$. Ces opérations ont les propriétés suivantes[1]) :

2.1. $Sq^i \circ f^* = f^* \circ Sq^i$, lorsque f est une application continue d'un couple (X, Y) dans un couple (X', Y').

2.2. $Sq^i \circ \delta = \delta \circ Sq^i$, δ désignant le cobord de la suite exacte de cohomologie.

2.3. $Sq^i(x \cdot y) = \sum_{j+k=i} Sq^j(x) \cdot Sq^k(y)$, $x \cdot y$ désignant le cup-produit.

2.4. $Sq^i(x) = x^2$ si $\dim. x = i$, $Sq^i(x) = 0$ si $\dim. x < i$.

2.5. $Sq^0(x) = x$.

On sait que toute suite exacte $0 \to A \to B \to C \to 0$ définit un opérateur cobord δ: $H^n(X, Y; C) \to H^{n+1}(X, Y; A)$. En particulier :

2.6. Sq^1 coïncide avec l'opérateur cobord attaché à la suite exacte

$$0 \to Z_2 \to Z_4 \to Z_2 \to 0 .$$

On a donc une suite exacte :

2.7. $$\ldots \to H^n(X, Y; Z_4) \to H^n(X, Y; Z_2) \overset{Sq^1}{\to} H^{n+1}(X, Y; Z_2) \to H^{n+1}(X, Y; Z_4) \to \ldots$$

3. Les *i*-carrés itérés

On peut composer entre elles les opérations Sq^i. On obtient ainsi les *i-carrés itérés* $Sq^{i_1} \circ Sq^{i_2} \circ \ldots \circ Sq^{i_r}$ qui appliquent $H^n(X, Z_2)$ dans le groupe $H^{n+i_1+\ldots+i_r}(X, Z_2)$. Une telle opération sera notée Sq^I, I désignant la suite d'entiers $\{i_1, \ldots, i_r\}$. Nous supposerons toujours que

---

[1]) Ces propriétés sont démontrées dans [12], à l'exception de 2.3, dont on trouvera la démonstration dans une Note de H. Cartan aux Comptes Rendus **230**, 1950, p. 425.

les entiers $i_1, \ldots, i_r$ sont > 0 (ceci ne restreint pas la généralité, à cause de 2.5).

Les définitions suivantes joueront un rôle essentiel par la suite :

3.1. L'entier $n(I) = i_1 + \cdots + i_r$ est appelé le *degré* de I.

3.2. Une suite I est dite *admissible* si l'on a :

$$i_1 \geqslant 2i_2, i_2 \geqslant 2i_3, \ldots, i_{r-1} \geqslant 2i_r \ .$$

3.3. Si une suite I est admissible, on définit son *excès* $e(I)$ par :

$$\begin{aligned} e(I) &= (i_1 - 2i_2) + (i_2 - 2i_3) + \cdots + (i_{r-1} - 2i_r) + i_r \\ &= i_1 - i_2 - \ldots - i_r = 2i_1 - n(I) \ . \end{aligned}$$

Par définition, $e(I)$ est un entier $\geqslant 0$, et si $e(I) = 0$ la suite I est vide (l'opération Sq^I correspondante est donc l'identité).

4. Les complexes d'Eilenberg-MacLane

Soient q un entier, Π un groupe (abélien si $q \geqslant 2$). Nous dirons qu'un espace X est *un espace* $\mathsf{K}(\Pi, q)$ si $\pi_i(X) = 0$ pour $i \neq q$, et si $\pi_q(X) = \Pi$. On sait (cf. [4]) que les groupes d'homologie et de cohomologie de X sont isomorphes à ceux du complexe $K(\Pi, q)$ défini de façon purement algébrique par Eilenberg-MacLane. Nous noterons ces groupes $H_i(\Pi; q, G)$ et $H^i(\Pi; q, G)$, G étant le groupe de coefficients.

Pour tout couple (Π, q) il existe un espace X qui est un espace $\mathsf{K}(\Pi, q)$ (cf. J. H. C. Whitehead, Ann. Math. **50**, 1949, p. 261—263). Soit X' le complexe cellulaire obtenu en «réalisant géométriquement» le complexe singulier de X [2]); on sait que $\pi_i(X') = \pi_i(X)$ pour tout $i \geqslant 0$, donc X' est un espace $\mathsf{K}(\Pi, q)$. Comme d'autre part on peut subdiviser simplicialement X', on obtient finalement :

4.1. *Pour tout couple* (Π, q) *il existe un espace* $\mathsf{K}(\Pi, q)$ *qui est un complexe simplicial.*

(Ici, comme dans toute la suite, nous entendons par *complexe simplicial* un complexe K qui peut avoir une infinité de simplexes et qui est muni de la topologie *faible* : une partie de K est fermée si ses intersections avec les sous-complexes finis de K sont fermées.)

---

[2]) L'espace X' est défini et étudié dans les articles suivants: 1) *J. B. Giever*, On the equivalence of two singular homology theory, Ann. Math. 51, 1950, p. 178—191; 2) *J. H. C. Whitehead*, A certain exact sequence, Ann. Math. 52, 1950, p. 51—110 (voir notamment les nos 19, 20, 21).

5. Propriétés élémentaires des espaces $K(\Pi, q)$

5.1. *Pour tout couple* (Π, q) *il existe un espace fibré contractile dont la base est un espace* $K(\Pi, q)$ *et dont la fibre est un espace* $K(\Pi, q-1)$.

Rappelons ([8], p. 499) que l'on obtient un tel espace fibré en prenant l'espace des chemins d'origine fixée sur un espace $K(\Pi, q)$.
L'énoncé suivant est évident:

5.2. *Si X est un espace* $K(\Pi, q)$ *et si X' est un espace* $K(\Pi', q)$, *le produit direct* $X \times X'$ *est un espace* $K(\Pi + \Pi', q)$.

Soit maintenant X un espace $K(\Pi, q)$, le groupe Π étant abélien (ce n'est une restriction que si $q = 1$). On a alors $H_q(X, Z) = \Pi$, d'où $H^q(X, \Pi) = \mathrm{Hom}\,(\Pi, \Pi)$. Le groupe $H^q(X, \Pi)$ contient donc une «classe fondamentale» u qui correspond dans $\mathrm{Hom}\,(\Pi, \Pi)$ à l'application identique de Π sur Π. Soit alors $f: Y \to X$ une application continue d'un espace Y dans l'espace X; l'élément $f^*(u)$ est un élément bien défini de $H^q(Y, \Pi)$ et il résulte de la théorie classique des obstructions (cf. S. Eilenberg, *Lectures in Topology*, Michigan 1941, p. 57—100) que l'on a :

5.3. *Si Y est un complexe simplicial,* $f \to f^*(u)$ *met en correspondance biunivoque les classes d'homotopie des applications de Y dans X et les éléments de* $H^q(Y, \Pi)$.

(On trouvera dans [5], IV un résultat très proche du précédent.)
Si Y est un espace $K(\Pi', q)$, on a $H^q(Y, \Pi) = \mathrm{Hom}\,(\Pi', \Pi)$, d'où :

5.4. *Si un complexe simplicial Y est un espace* $K(\Pi', q)$, *les classes d'homotopie des applications de Y dans un espace* $K(\Pi, q)$ *correspondent biunivoquement aux homomorphismes de Π' dans Π.*

6. Fibrations des espaces $K(\Pi, q)$ [3])

Donnons-nous un entier q, et une suite exacte de groupes abéliens :

$$0 \to A \to B \to C \to 0 \ .$$

6.1. *Il existe un espace fibré E, de fibre F et base X, où F est un espace* $K(A, q)$, *E un espace* $K(B, q)$, *X un espace* $K(C, q)$, *et dont la suite exacte d'homotopie (en dimension q) est la suite exacte donnée.*

Soient Y un complexe simplicial qui soit un espace $K(B, q)$, X un espace $K(C, q)$ et $f: Y \to X$ une application continue telle que

$$f_0: \ \pi_q(Y) \to \pi_q(X)$$

soit l'homomorphisme donné de B sur C (cf. 5.4).

---

[3]) Ces fibrations m'ont été signalées par H. Cartan.

On prend pour espace E l'espace des couples $(y, \alpha(t))$, où $y \in Y$, et où $\alpha(t)$ est un chemin de X tel que $\alpha(0) = f(y)$. L'espace E est rétractile sur Y, c'est donc un espace $K(B, q)$. L'application $(y, \alpha(t)) \to \alpha(1)$ fait de E un espace fibré de base X (c'est une généralisation immédiate de la Proposition 6 de [8], Chapitre IV). La suite exacte d'homotopie montre alors que la fibre F de cette fibration est un espace $K(C, q)$; plus précisément, la suite :

$$\pi_{q+1}(X) \to \pi_q(F) \to \pi_q(E) \to \pi_q(X) \to \pi_{q-1}(F) ,$$

est identique à la suite exacte $0 \to A \to B \to C \to 0$ donnée.

On montre de façon tout analogue l'existence d'un espace fibré où :

6.2. *L'espace fibré est un espace* $K(A, q)$, *la fibre est un espace* $K(C, q-1)$ *et la base est un espace* $K(B, q)$.

De même, il existe un espace fibré où :

6.3. *L'espace fibré est un espace* $K(C, q-1)$, *la fibre est un espace* $K(B, q-1)$ *et la base est un espace* $K(A, q)$.

§ 2. Détermination de l'algèbre $H^*(\Pi; q, Z_2)$

7. Un théorème de A. Borel

Soient X un espace et $A = H^*(X, Z_2)$ l'algèbre de cohomologie de X à coefficients dans Z_2. On dit ([2], Définition 6.3) qu'une famille (x_i) $(i = 1, \ldots)$, d'éléments de A est un *système simple de générateurs* de A si :

7.1. Les x_i sont des éléments homogènes de A,

7.2. Les produits $x_{i_1} . x_{i_2} \ldots x_{i_r}$ $(i_1 < i_2 < \cdots < i_r, r \geqslant 0$ quelconque) forment une base de A, considéré comme espace vectoriel sur Z_2.

Nous pouvons maintenant rappeler le théorème de A. Borel ([2], Proposition 16.1) qui est à la base des résultats de ce paragraphe :

Théorème 1. *Soit E un espace fibré de fibre F et base B connexes par arcs, vérifiant les hypothèses suivantes :*

α) *Le terme E_2 de la suite spectrale de cohomologie de E (à coefficients dans Z_2) est $H^*(B, Z_2) \otimes H^*(F, Z_2)$ (c'est le cas, comme on sait, si $\pi_1(B) = 0$ et si les groupes d'homologie de B ou de F sont de type fini).*

β) *$H^i(E, Z_2) = 0$ pour tout $i > 0$.*

γ) *$H^*(F, Z_2)$ possède un système simple de générateurs (x_i) qui sont transgressifs.*

Alors, si les y_i sont des éléments homogènes de $H^*(B, Z_2)$ qui correspondent aux x_i par transgression, $H^*(B, Z_2)$ est l'algèbre de polynômes ayant les y_i pour générateurs.

(En d'autres termes, les y_i engendrent $H^*(B, Z_2)$ et ne vérifient aucune relation non triviale.)

Nous utiliserons ce théorème principalement dans le cas particulier où $H^*(F, Z_2)$ est elle-même une algèbre de polynômes ayant pour générateurs des éléments transgressifs z_i, de degrés n_i. Il est immédiat que $H^*(F, Z_2)$ admet alors pour système simple de générateurs les puissances (2^r)-èmes des z_i $(i = 1, \ldots, ;\ r = 0, 1, \ldots)$. Si a et r sont deux entiers, désignons par $L(a, r)$ la suite $\{2^{r-1} a, \ldots, 2a, a\}$; d'après 2.4 on a $z_i^{(2^r)} = Sq^{L(n_i, r)}(z_i)$, les notations étant celles du n° 3. Soient alors $t_i \in H^{n_i+1}(B, Z_2)$ des éléments qui correspondent par transgression aux z_i ; puisque les Sq^i commutent à la transgression ([8], p. 457), les éléments $z_i^{(2^r)}$ sont transgressifs et leurs images par transgression sont les $Sq^{L(n_i, r)}(t_i)$. Appliquant le Théorème 1, on obtient donc :

7.3. *Sous les hypothèses précédentes, $H^*(B, Z_2)$ est l'algèbre de polynômes ayant pour générateurs les $Sq^{L(n_i, r)}(t_i)$ $(i = 1, \ldots ;\ r = 0, 1, \ldots)$.*

8. Détermination de l'algèbre $H^*(Z_2 ; q, Z_2)$

On a $H^i(Z_2 ; q, Z_2) = 0$ pour $0 < i < q$, et $H^q(Z_2 ; q, Z_2) = Z_2$. Nous désignerons par u_q l'unique générateur de ce dernier groupe.

Théorème 2. *L'algèbre $H^*(Z_2 ; q, Z_2)$ est l'algèbre de polynômes ayant pour générateurs les éléments $Sq^I(u_q)$, où I parcourt l'ensemble des suites admissibles d'excès $< q$ (au sens du n° 3).*

On sait que l'espace projectif réel à une infinité de dimensions est un espace $K(Z_2, 1)$; $H^*(Z_2 ; 1, Z_2)$ est donc l'algèbre de polynômes ayant u_1 pour unique générateur ; comme d'autre part $e(I) < 1$ entraîne que I soit vide, le théorème est vérifié pour $q = 1$.

Supposons-le vérifié pour $q - 1$ et démontrons-le pour q. Considérons la fibration 5.1. Par hypothèse, $H^*(Z_2 ; q - 1, Z_2)$ est l'algèbre de polynômes ayant pour générateurs les éléments $z_J = Sq^J(u_{q-1})$, où J parcourt l'ensemble des suites admissibles d'excès $e(J) < q - 1$. Nous noterons s_J le degré de l'élément z_J ; on a $s_J = q - 1 + n(J)$. Il est clair que u_{q-1} est transgressif et que son image par la transgression τ

est u_q. D'après [8], loc. cit., z_J est donc aussi transgressif et $\tau(z_J) = Sq^J(u_q)$. Il s'ensuit que l'on peut appliquer 7.3 à la fibration 5.1, ce qui montre que $H^*(Z_2\,;q\,,Z_2)$ est l'algèbre de polynômes ayant pour générateurs les éléments $Sq^{L(s_J,r)} \circ Sq^J(u_q)$, où r parcourt l'ensemble des entiers $\geqslant 0$, et J l'ensemble des suites admissibles d'excès $<q-1$. La démonstration du Théorème 2 sera donc achevée si nous prouvons le Lemme suivant :

Lemme 1. *Si à tout entier* $r \geqslant 0$, *et à toute suite admissible* $J = \{j_1, \ldots, j_k\}$ *d'excès* $<q-1$, *on fait correspondre la suite :*

$$I = \{2^{r-1}\cdot s_J, \ldots, 2s_J, s_J, j_1, \ldots, j_k\}\ , \qquad \textit{où} \qquad s_J = q-1+n(J)\ ,$$

on obtient toutes les suites admissibles d'excès $<q$ *une fois et une seule.*

Notons d'abord que $s_J - 2j_1 = n(J) - 2j_1 + q - 1 = q-1-e(J) > 0$, donc I est une suite *admissible.* Si $r = 0$, on a $I = J$, d'où $e(I) = e(J) < q-1$; si $r > 0$, on a $e(I) = e(J) + s_J - 2j_1 = q-1$. Ainsi, en prenant $r = 0$ on trouve toutes les suites admissibles d'excès $e(I) < q-1$, et en prenant $r > 0$ on trouve des suites admissibles d'excès $q-1$.

Inversement, si l'on se donne une suite admissible $I = \{i_1, \ldots, i_p\}$ d'excès $q-1$, r et J sont déterminés sans ambiguïté :

$$\begin{cases} r \text{ est le plus grand entier tel que } i_1 = 2i_2, \ldots, i_{r-1} = 2i_r\ , \\ J = \{i_{r+1}, \ldots, i_p\}\ . \end{cases}$$

La suite associée au couple (r, J) est bien I car l'on a :

$$q - 1 = e(I) = i_r - 2i_{r+1} + e(J) = i_r - 2i_{r+1} + 2i_{r+1} - n(J)\ ,$$

d'où $i_r = n(J) + q - 1 = s_J$, et $i_{r-1} = 2s_J, \ldots, i_1 = 2^{r-1}\cdot s_J$.

Le Lemme 1 est donc démontré.

9. Exemples

$H^*(Z_2\,;1\,,Z_2)$ est l'algèbre de polynômes engendrée par u_1.

$H^*(Z_2\,;2\,,Z_2)$ est l'algèbre de polynômes engendrée par :

$u_2, Sq^1u_2, Sq^2Sq^1u_2, \ldots, Sq^{2^k}Sq^{2^{k-1}} \ldots Sq^2Sq^1u_2, \ldots$.

$H^*(Z_2\,;3\,,Z_2)$ est l'algèbre de polynômes engendrée par :

$u_3, Sq^2u_3, Sq^4Sq^2u_3, \ldots, Sq^{2^r}Sq^{2^{r-1}} \ldots Sq^2u_3, \ldots$

$Sq^1u_3, Sq^3Sq^1u_3, Sq^6Sq^3Sq^1u_3, \ldots, Sq^{3\cdot 2^r}Sq^{3\cdot 2^{r-1}} \ldots Sq^3Sq^1u_3, \ldots$

. . . . . .

$Sq^{2^{k-1}} \ldots Sq^2Sq^1u_3, \ldots, Sq^{(2^k+1)2^r} \ldots Sq^{2^k+1}Sq^{2^{k-1}} \ldots Sq^2Sq^1u_3, \ldots$

. . . .

10. Détermination de l'algèbre $H^*(Z\,;q,Z_2)$

Le cercle S_1 est un espace $K(Z,1)$; ceci détermine $H^*(Z\,;1,Z_2)$. Nous pouvons donc nous borner au cas où $q \geqslant 2$. Nous désignerons encore par u_q l'unique générateur de $H^q(Z\,;q,Z_2)$.

Théorème 3. *Si* $q \geqslant 2$, *l'algèbre* $H^*(Z\,;q,Z_2)$ *est l'algèbre de polynômes ayant pour générateurs les éléments* $Sq^I(u_q)$ *où* I *parcourt l'ensemble des suites admissibles* $\{i_1,\ldots,i_r\}$, *d'excès* $<q$, *et telles que* $i_r > 1$.

On sait que l'espace projectif complexe à une infinité de dimensions est un espace $K(Z,2)$; $H^*(Z\,;2,Z_2)$ est donc l'algèbre de polynômes ayant u_2 pour unique générateur ; comme d'autre part $e(I)<2$ et $i_r>1$ entraînent que I soit vide, le théorème est vérifié pour $q=2$.

A partir de là on raisonne par récurrence sur q, exactement comme dans la démonstration du Théorème 2. Il faut simplement observer que, si $q \geqslant 3$, les suites I dont le dernier terme est >1 correspondent, par la correspondance du Lemme 1, aux couples (r, J) où le dernier terme de J est >1.

Corollaire. *Si* $q \geqslant 2$, *l'algèbre* $H^*(Z\,;q,Z_2)$ *est isomorphe au quotient de l'algèbre* $H^*(Z_2\,;\,q,Z_2)$ *par l'idéal engendré par les* $Sq^I(u_q)$ *où* I *est admissible, d'excès* $<q$, *et de dernier élément égal à* 1.

De façon plus précise, l'homomorphisme canonique $Z \to Z_2$ définit (grâce à 5.4) un homomorphisme de $H^*(Z_2\,;q,Z_2)$ dans $H^*(Z\,;q,Z_2)$, et les théorèmes 2 et 3 montrent que cet homomorphisme applique la première algèbre sur la seconde, le noyau étant l'idéal défini dans l'énoncé du corollaire.

11. Détermination de l'algèbre $H^*(Z_m\,;q,Z_2)$ lorsque $m=2^h$, $h \geqslant 2$

L'algèbre $H^*(Z_m\,;1,Z_2)$ n'est pas autre chose que l'algèbre de cohomologie modulo 2 du groupe Z_m, au sens de Hopf. Sa structure est bien connue (on peut la déterminer soit algébriquement, soit en utilisant les espaces lenticulaires) :

C'est le produit tensoriel d'une algèbre extérieure de générateur u_1 et d'une algèbre de polynômes de générateur un élément v_2 de degré 2. L'élément v_2 peut être défini ainsi :

Soit δ_h l'opérateur cobord attaché à la suite exacte de coefficients $0 \to Z_2 \to Z_{2^{h+1}} \to Z_{2^h} \to 0$. Soit u'_1 le générateur canonique de $H^1(Z_m\,;1,Z_m)$; on a alors $v_2 = \delta_h(u'_1)$.

Si h était égal à 1, on aurait $\delta_h = Sq^1$, d'après 2.6; mais comme nous avons supposé $h \geqslant 2$, δ_h diffère de Sq^1 (on a d'ailleurs $Sq^1(u_1) = u_1^2 = 0$), Nous écrirons : $v_2 = Sq_h^1(u_1)$, lorsque cette écriture ne pourra pas prêter à confusion.

Le raisonnement de [8], p. 457, montrant que les Sq^i commutent à la transgression, se laisse adapter sans difficulté à l'opération δ_h, et montre ainsi que v_2 est un élément transgressif de $H^2(Z_m; 1, Z_2)$ dans la fibration qui a $K(Z_m, 1)$ pour fibre et $K(Z_m, 2)$ pour base. Comme $H^*(Z_m; 1, Z_2)$ a pour système simple de générateurs le système :

$$u_1, v_2 = Sq_h^1(u_1), Sq^2 Sq_h^1(u_1), \ldots, Sq^{2^k} \ldots Sq^2 Sq_h^1(u_1), \ldots ,$$

le théorème 1 montre que $H^*(Z_m; 2, Z_2)$ est l'algèbre de polynômes ayant pour générateurs les éléments :

$$u_2, Sq_h^1(u_2), \ldots, Sq^{2^k} \ldots Sq^2 Sq_h^1(u_2), \ldots .$$

Ceci nous conduit à la notation suivante : si $I = \{i_1, \ldots, i_r\}$ est une suite admissible, on définit $Sq_h^I(u_q)$ comme étant égal à $Sq^I(u_q)$ si $i_r > 1$, et à $Sq^{i_1} \ldots Sq^{i_{r-1}} Sq_h^1(u_q)$ si $i_r = 1$ ($Sq_h^1(u_q)$ a le même sens que plus haut, autrement dit $Sq_h^1(u_q) = \delta_h(u_q')$, u_q' désignant le générateur canonique de $H^q(Z_m; q, Z_m)$).

La détermination de $H^*(Z_m; q, Z_2)$ se poursuit alors par récurrence sur q, exactement comme celle de $H^*(Z_2; q, Z_2)$, à cela près que les Sq_h^I remplacent les Sq^I. On obtient finalement :

Théorème 4. *Si $q \geqslant 2$, l'algèbre $H^*(Z_m; q, Z_2)$, où $m = 2^h$ avec $h \geqslant 2$, est l'algèbre de polynômes ayant pour générateurs les éléments $Sq_h^I(u_q)$ où I parcourt l'ensemble des suites admissibles d'excès $< q$.*

Comme les Sq_h^I correspondent biunivoquement aux Sq^I, on a :

Corollaire. *$H^*(Z_m; q, Z_2)$ et $H^*(Z_2; q, Z_2)$ sont isomorphes en tant qu'espaces vectoriels sur le corps Z_2.*

Le résultat précédent est valable même si $q = 1$.

12. Détermination de l'algèbre $H^*(\Pi; q, Z_2)$ lorsque Π est un groupe abélien de type fini

Le résultat suivant peut être considéré comme classique :

Théorème 5. *Soient Π et Π' deux groupes abéliens, Π étant de type fini, et soit k un corps commutatif. L'algèbre $H^*(\Pi + \Pi'; q, k)$ est isomorphe au produit tensoriel sur k des algèbres $H^*(\Pi; q, k)$ et $H^*(\Pi'; q, k)$.*

Rappelons la démonstration : Soient X un espace $K(\Pi, q)$ et X' un espace $K(\Pi', q)$. L'espace $X \times X'$ est un espace $K(\Pi + \Pi', q)$, comme nous l'avons déjà signalé (5.2). Puisque Π est de type fini, les groupes d'homologie de X sont de type fini en toute dimension d'après [8], p. 500 (voir aussi [11], Chapitre II, Proposition 8). Appliquant alors un cas particulier de la formule de Künneth[4]), on a :

$$H^*(X \times X', k) = H^*(X, k) \otimes H^*(X', k) \ ,$$

ce qui démontre le Théorème 5.

Comme tout groupe abélien de type fini est somme directe de groupes isomorphes à Z et de groupes cycliques d'ordre une puissance d'un nombre premier, le Théorème 5 ramène le calcul de $H^*(\Pi\,; q, Z_2)$ aux trois cas particuliers : $\Pi = Z$, $\Pi = Z_{2^h}$, $\Pi = Z_{p^h}$ avec p premier $\neq 2$. Les deux premiers cas ont été traités dans les n^{os} précédents et l'on sait par ailleurs (cf. [8] et [11], loc. cit.) que $H^n(Z_m\,; q, Z_2) = 0$ pour $n > 0$, si m est un entier impair ; le troisième cas conduit donc à une algèbre de cohomologie triviale, et la détermination de $H^*(\Pi\,; q, Z_2)$ est ainsi achevée, pour tout groupe Π de type fini.

13. Relations entre les diverses algèbres $H^*(\Pi\,; q, Z_2)$

Dans ce qui précède nous avons traité indépendamment les cas $\Pi = Z$, $\Pi = Z_2$, $\Pi = Z_{2^h}$. Il y a cependant des relations entre ces trois cas, qui proviennent des fibrations du n° 6. Nous allons en donner un exemple :

Posons $m = 2^h$, avec $h \geqslant 1$. Considérons la suite exacte

$$0 \to Z \to Z \to Z_m \to 0 \ ,$$

où le premier homomorphisme est la multiplication par m. En appliquant 6.3 on en déduit l'existence d'une fibration où l'espace fibré est un espace $K(Z_m, q-1)$, où la fibre est un espace $K(Z, q-1)$ et la base un espace $K(Z, q)$. Soit u_{q-1} l'unique générateur du groupe $H^{q-1}(Z\,; q, Z_2)$; l'image de u_{q-1} par la transgression τ est nulle, car sinon $H^{q-1}(Z_m\,; q-1, Z_2)$ serait nul, ce qui n'est pas ; puisque les Sq^I commutent à la transgression, on a $\tau(Sq^I u_{q-1}) = 0$ pour toute suite I, et comme $H^*(Z\,; q-1, Z_2)$ est engendré par les $Sq^I u_{q-1}$, il s'ensuit que toutes les différentielles d_r $(r \geqslant 2)$ de la suite spectrale de cohomologie modulo 2 de la fibration précédente sont identiquement nulles. Le terme E_∞ de cette suite spectrale est donc isomorphe au terme E_2, ce qui donne :

---

[4]) Ce cas particulier est démontré dans [8], p. 473.

13.1. *L'algèbre graduée associée à* $H^*(Z_m; q-1, Z_2)$, *convenablement filtrée, est isomorphe à* $H^*(Z; q, Z_2) \otimes H^*(Z; q-1, Z_2)$.

En particulier :

13.2. $H^*(Z_m; q-1, Z_2)$ *et* $H^*(Z; q, Z_2) \otimes H^*(Z; q-1, Z_2)$ *sont isomorphes en tant qu'espaces vectoriels sur le corps* Z_2.

On notera que 13.2 fournit une nouvelle démonstration du Corollaire au Théorème 4. D'un autre côté, il serait facile de tirer 13.2 des Théorèmes 2, 3, 4.

14. Les groupes stables; cas de la cohomologie

Π et G étant deux groupes abéliens, nous poserons[5]) :

14.1. $A_n(\Pi, G) = H_{n+q}(\Pi; q, G)$, avec $q > n$.

On sait (cf. [5] ainsi que [8], p. 500) que ces groupes ne dépendent pas de la valeur de q choisie, mais seulement de Π, G et n. Ce sont les «groupes stables».

Le raisonnement du Théorème 5 montre immédiatement que l'on a la formule suivante (voir aussi [5]) :

14.2. $A_n(\Pi + \Pi', G) = A_n(\Pi, G) + A_n(\Pi', G)$ pour tout $n \geqslant 0$.

On définit de façon analogue les groupes $A^n(\Pi, G) = H^{n+q}(\Pi; q, G)$, avec $q > n$. Les Théorèmes 2, 3, 4 permettent de déterminer ces groupes lorsque $G = Z_2$, et lorsque $\Pi = Z$, Z_2, ou Z_m avec $m = 2^h$:

Théorème 6. *L'espace vectoriel* $A^n(Z_2, Z_2)$ (*resp.* $A^n(Z_m, Z_2)$, *avec* $m = 2^h$) *admet pour base l'ensemble des éléments* $Sq^I(u)$ (*resp.* $Sq^I_h(u)$), *où* I *parcourt l'ensemble des suites admissibles de degré* n.

(Nous avons noté u l'unique générateur de $A^0(Z_m, Z_2)$).

Par exemple, $A^{10}(Z_2, Z_2)$ admet pour base les six éléments :

$$Sq^{10}u,\ Sq^9Sq^1u,\ Sq^8Sq^2u,\ Sq^7Sq^3u,\ Sq^7Sq^2Sq^1u,\ Sq^6Sq^3Sq^1u\ .$$

Théorème 7. *L'espace vectoriel* $A^n(Z, Z_2)$ *admet pour base l'ensemble des éléments* $Sq^I u$, *où* I *parcourt l'ensemble des suites admissibles dont le dernier terme est* > 1 *et dont le degré est* n.

Par exemple, $A^{10}(Z, Z_2)$ admet pour base les trois éléments : $Sq^{10}u$, Sq^8Sq^2u, Sq^7Sq^3u.

---

[5]) La notation adoptée ici diffère d'une unité de celle de [5].

15. Les groupes stables; cas de l'homologie

Pour passer des groupes de cohomologie modulo 2 aux groupes d'homologie nous aurons besoin du Lemme suivant :

Lemme 2. *Soient X un espace, n un entier > 0. Supposons que $H_n(X, Z)$ ait un nombre fini de générateurs, et que la suite :*

$$H^{n-1}(X, Z_2) \xrightarrow{Sq^1} H^n(X, Z_2) \xrightarrow{Sq^1} H^{n+1}(X, Z_2)$$

soit exacte. Posons $N = dim.\ [H^n(X, Z_2)/Sq^1(H^{n-1}(X, Z_2))]$.

Le groupe $H_n(X, Z)$ est alors somme directe d'un groupe fini d'ordre impair et de N groupes isomorphes à Z_2.

Pour simplifier les notations, nous poserons $L_i = H_i(X, Z)$. D'après la formule des coefficients universels[6]), on a, pour tout groupe abélien G, une suite exacte :

$$0 \to \text{Ext}\,(L_{n-1}, G) \to H^n(X, G) \to \text{Hom}\,(L_n, G) \to 0\ .$$

En appliquant ceci à $G = Z_4$ et à $G = Z_2$, on obtient le diagramme :

$$\begin{array}{ccccccccc} 0 & \to & \text{Ext}\,(L_{n-1}, Z_4) & \to & H^n(X, Z_4) & \to & \text{Hom}\,(L_n, Z_4) & \to & 0 \\ \downarrow & & \varphi\downarrow & & \psi\downarrow & & \chi\downarrow & & \downarrow \\ 0 & \to & \text{Ext}\,(L_{n-1}, Z_2) & \to & H^n(X, Z_2) & \to & \text{Hom}\,(L_n, Z_2) & \to & 0\ . \end{array}$$

D'après la suite exacte 2.7, le noyau Q^n de

$$Sq^1 : H^n(X, Z_2) \to H^{n+1}(X, Z_2)$$

est égal à l'image de ψ. Comme l'application φ est sur (d'après une propriété générale du foncteur Ext), il s'ensuit que Q^n contient $\text{Ext}\,(L_{n-1}, Z_2)$.

Soit d'autre part R^n l'image de Sq^1: $H^{n-1}(X, Z_2) \to H^n(X, Z_2)$. On voit facilement (par calcul direct, par exemple) que toute classe de cohomologie $f \in R^n$ donne 0 dans $\text{Hom}(L_n, Z_2)$. Donc R^n est contenu dans $\text{Ext}\,(L_{n-1}, Z_2)$.

Vu l'hypothèse faite dans le Lemme, on a donc :

$$Q^n = R^n = \text{Ext}\,(L_{n-1}, Z_2)\ .$$

Ainsi l'image de ψ est égale à $\text{Ext}(L_{n-1}, Z_2)$. Il s'ensuit que l'homomorphisme χ est nul ; compte tenu de la structure des groupes abéliens à un nombre fini de générateurs, ceci montre que L_n est somme directe d'un groupe fini d'ordre impair et d'un certain nombre de groupes Z_2.

---

[6]) Voir par exemple *S. Eilenberg and N. E. Steenrod*, Foundations of Algebraic Topology, I., Princeton 1952, p. 161.

Il est clair que le nombre de ces derniers est égal à la dimension de Hom (L_n, Z_2) c'est-à-dire à N.

Théorème 8. *Le groupe* $A_n(Z_2, Z)$ *est somme directe de groupes* Z_2 *en nombre égal au nombre des suites admissibles* $I = \{i_1, \ldots, i_k\}$, *où* i_1 *est pair et où* $n(I) = i_1 + \cdots + i_k$ *est égal à* n.

Nous allons déterminer l'opération Sq^1 dans $A^*(Z_2, Z_2)$, de façon à pouvoir appliquer le Lemme 2.

Rappelons que l'on a $Sq^1 Sq^n = Sq^{n+1}$ si n est pair, et $Sq^1 Sq^n = 0$ si n est impair. On tire de là :

$$Sq^1(Sq^{i_1} \ldots Sq^{i_k} u) = \begin{cases} 0 & \text{si } i_1 \text{ est impair} \\ Sq^{i_1+1} \ldots Sq^{i_k} u & \text{si } i_1 \text{ est pair} . \end{cases}$$

Soit alors B^n (resp. C^n) le sous-espace vectoriel de $A^n(Z_2, Z_2)$ engendré par les $Sq^I(u)$ où i_1 est pair (resp. impair). $A^n(Z_2, Z_2)$ est somme directe de B^n et de C^n ; d'après la formule écrite plus haut, Sq^1 est nul sur C^n et applique isomorphiquement B^n sur C^{n+1}. La suite :

$$A^{n-1}(Z_2, Z_2) \xrightarrow{Sq^1} A^n(Z_2, Z_2) \xrightarrow{Sq^1} A^{n+1}(Z_2, Z_2)$$

est donc exacte, et B^n est isomorphe à $A^n(Z_2, Z_2)/Sq^1 A^{n-1}(Z_2, Z_2)$.

Le théorème résulte alors du Lemme 2, et du fait (démontré dans [8], p. 500), que $A_n(Z_2, Z)$ est un groupe fini d'ordre une puissance de 2.

On démontre de même ;

Théorème 9. *Le groupe* $A_n(Z_m, Z)$, $n > 0$, *est isomorphe à* $A_n(Z_2, Z)$ *lorsque* m *est une puissance de* 2.

Théorème 10. *Le groupe* $A_n(Z, Z)$, $n > 0$, *est un groupe fini dont le 2-composant est somme directe de groupes* Z_2 *en nombre égal au nombre des suites admissibles* $I = \{i_1, \ldots, i_k\}$, *où* i_1 *est pair*, $i_k > 1$, *et où* $n(I) = i_1 + \cdots + i_k$ *est égal à* n.

Remarque. En comparant les Théorèmes 7 et 8, on peut montrer que $A_n(Z_2, Z)$ est isomorphe à $A_n(Z, Z_2)$. De façon générale, on conjecture que $A_n(\Pi, G)$ est isomorphe à $A_n(G, \Pi)$ quels que soient les groupes abéliens G et Π ; il suffirait d'ailleurs de démontrer le cas particulier $\Pi = Z$ pour avoir le cas général (compte tenu des résultats annoncés par Eilenberg-MacLane dans [5], II, ceci vérifie la conjecture en question pour $n = 0, 1, 2, 3$).

Théorème 11. *Pour tout groupe abélien* Π, *le groupe* $A_n(\Pi, Z)$, $n > 0$, *est un groupe de torsion dont le 2-composant est somme directe de groupes isomorphes à* Z_2.

Soient Π_α les sous-groupes de type fini de Π ; puisque Π est limite inductive des Π_α, le complexe $K(\Pi, q)$ est limite inductive des complexes $K(\Pi_\alpha, q)$, et on en conclut que $A_n(\Pi, Z)$ est limite inductive des $A_n(\Pi_\alpha, Z)$ ce qui réduit la question au cas où Π est de type fini.

En utilisant la formule 14.2, on est alors ramené au cas des groupes cycliques, qui est traité dans les Théorèmes 8, 9, 10.

Remarque. Le fait que $A_n(\Pi, Z)$ soit un groupe de torsion résulte aussi de [8], p. 500—501.

§ 4. Opérations cohomologiques

26. Définition des opérations cohomologiques

Soient q et n deux entiers > 0, A et B deux groupes abéliens. Une *opération cohomologique*, relative à $\{q, n, A, B\}$, est une application :

$$C : H^q(X, A) \to H^n(X, B) ,$$

définie pour tout complexe simplicial X, et vérifiant la condition suivante :

26.1. Pour toute application continue f d'un complexe X dans un complexe Y, on a $C \circ f^* = f^* \circ C$.

Remarque. Nous nous sommes placés dans la catégorie des complexes simpliciaux pour des raisons de commodité. On pourrait aussi bien se placer dans la catégorie de tous les espaces topologiques (la cohomologie étant la cohomologie singulière). Cela ne changerait rien, puisque l'on peut remplacer tout espace topologique par le complexe simplicial «réalisation géométrique» de son complexe singulier, et que cette opération ne modifie pas les groupes de cohomologie.

27. Exemples

27.1. Supposons que $n = q$, et donnons-nous un homomorphisme de A dans B. Cela définit un homomorphisme de $H^q(X, A)$ dans $H^q(X, B)$ qui vérifie 26.1.

27.2. Supposons que $n = q + 1$, et donnons-nous une suite exacte :

$$0 \to B \to L \to A \to 0 .$$

Cette suite définit une opération cobord : $H^q(X, A) \to H^{q+1}(X, B)$ qui vérifie 26.1.

27.3. Supposons que $n = 2q$, et donnons-nous une application bilinéaire de A dans B. Au moyen de cette application, on peut définir le cup-carré d'un élément de $H^q(X, A)$, qui est un élément de $H^{2q}(X, B)$, et cette opération vérifie 26.1.

27.4. Les Sq^i, les Sq^I, les puissances réduites de Steenrod (voir [13]), sont des opérations cohomologiques.

28. Caractérisation des opérations cohomologiques

Théorème 1. *Les opérations cohomologiques relatives à* $\{q, n, A, B\}$ *correspondent biunivoquement aux éléments du groupe* $H^n(A ; q, B)$.

Soit T un complexe simplicial qui soit un espace $K(A, q)$. Comme nous l'avons vu au n° 3, $H^q(T, A)$ possède une classe fondamentale u qui correspond dans $\mathrm{Hom}(A, A)$ à l'application identique de A sur A. Si C est une opération cohomologique relative à $\{q, n, A, B\}$, $C(u)$ est un élément bien défini de $H^n(T, B) = H^n(A; q, B)$, élément que nous noterons $\varphi(C)$.

Inversement, soit c un élément de $H^n(T, B)$, et soit $x \in H^q(X, A)$ une classe de cohomologie d'un complexe simplicial arbitraire X. D'après 5.3, il existe une application $g_x : X \to T$ telle que $g_x^*(u) = x$, et cette application g_x est unique, à une homotopie près. L'élément $g_x^*(c) \in H^n(X, B)$ est donc défini sans ambiguïté, et il est immédiat que l'application $x \to g_x^*(c)$ vérifie 26.1. C'est donc une opération cohomologique relative à $\{q, n, A, B\}$, que nous noterons $\psi(c)$.

On a $\varphi \circ \psi = 1$. Soit en effet $c \in H^n(A; q, B)$. Par définition, $\varphi \circ \psi(c)$ est égal à $g_u^*(c)$, où $g_u : T \to T$ est une application telle que $g_u^*(u) = u$. On peut donc prendre pour g_u l'application identique, ce qui donne $\varphi \circ \psi(c) = g_u^*(c) = c$.

Il nous reste à montrer que $\psi \circ \varphi = 1$. Pour cela, soit C une opération cohomologique, et posons $c = \varphi(C) = C(u)$. Pour tout élément $x \in H^q(X, A)$, on a $\psi(c)(x) = g_x^*(c) = g_x^*(C(u)) = C(g_x^*(u)) = C(x)$. Ceci signifie bien que $\psi(c) = \psi \circ \varphi(C)$ est identique à C.

Corollaire. *Soient C_1 et C_2 deux opérations cohomologiques relatives au même système $\{q, n, A, B\}$, et soit u la classe fondamentale de $H^q(A; q, A)$. Si $C_1(u) = C_2(u)$, alors $C_1 = C_2$.*

Remarques. 1) On aurait aussi bien pu définir les opérations cohomologiques pour la cohomologie relative (des complexes simpliciaux, ou bien de tous les espaces topologiques, ce qui revient au même). La démonstration précédente reste valable.

2) On pourrait également définir les opérations cohomologiques $C(x_1, \ldots, x_r)$ de plusieurs variables $x_i \in H^{q_i}(X, A_i)$, à valeurs dans $H^n(X, B)$. Ces opérations correspondent biunivoquement aux éléments de $H^n(K(A_1, q_1) \times \cdots \times K(A_r, q_r), B)$, comme on le voit par le même raisonnement que plus haut. Lorsque les A_i sont de type fini et que B est un corps, il résulte de la formule de Künneth que ces opérations se réduisent à des cup-produits d'opérations cohomologiques à une seule variable.

29. Premières applications

Nous allons appliquer le Théorème 1 à divers cas simples. Nous désignerons par C une opération cohomologique relative à $\{q, n, A, B\}$.

29.1. *Si* $0 < n < q$, *C est identiquement nulle.* En effet, $H^n(A; q, B)$ est alors réduit à 0.

29.2. *Si* $n = q$, *C est associé à un homomorphisme de A dans B* (au sens de 27.1). En effet, $H^q(A; q, B) = \mathrm{Hom}(A, B)$.

29.3. *Si* $q = 1$, $A = Z$, $n > 1$, *C est identiquement nulle.* En effet $H^n(Z; 1, B) = 0$ si $n > 1$, puisqu'un cercle est un espace $K(Z, 1)$.

29.4. *Si* $q = 2$, $A = Z$, *n impair, C est identiquement nulle. Si n est pair, et si* $B = Z$ *ou* Z_m, *on a* $C(x) = k \cdot x^{n/2}$, $k \in B$. En effet, on peut prendre pour espace $K(Z, 2)$ un espace projectif complexe à une infinité de dimensions.

29.5. *Si q est impair,* $A = Z$, $B = Q$ *(corps des rationnels),* $n > q$, *C est identiquement nulle.* En effet, on a $H^n(Z; q, Q) = 0$ si $n > q$, d'après [8], p. 501.

29.6. *Si q est pair,* $A = Z$, $B = Q$, *et si n n'est pas divisible par q, C est identiquement nulle; si n est divisible par q, on a* $C(x) = k \cdot x^{n/q}$, $k \in Q$. En effet, d'après [8], loc. cit., $H^*(Z; q, Q)$ est l'algèbre de polynômes sur Q qui admet u pour unique générateur.

On peut donner bien d'autres applications du Théorème 1. Par exemple lorsque B est un corps, établir une formule de produit :

$$C(x \cdot y) = \Sigma C_i(x) \cdot C_j(y) \ ;$$

lorsque $n < 2q$, montrer que C est un homomorphisme. Etc.

30. Caractérisation des *i*-carrés

Soit i un entier $\geqslant 0$, et supposons donné, pour tout couple (X, Y) de complexes simpliciaux, et tout entier $n \geqslant 0$, des applications

$$A^i : H^q(X, Y; Z_2) \to H^{q+i}(X, Y; Z_2)$$

vérifiant les propriétés 2.1, 2.2 et 2.4, c'est-à-dire telles que $A^i \circ f^* = f^* \circ A^i$, $A^i \circ \delta = \delta \circ A^i$, $A^i(x) = x^2$ si dim. $x = i$, $A^i(x) = 0$ si dim. $x < i$. Nous allons montrer que les A^i coïncide avec les Sq^i [8]).

D'après le Théorème 1 (qui est valable dans le cas de la cohomologie relative, comme nous l'avons remarqué), il suffit de prouver que $A^i(u_q) = Sq^i(u_q)$, u_q désignant le générateur de $H^q(Z_2; q, Z_2)$. Ceci est clair si

---

[8]) R. Thom a obtenu antérieurement une caractérisation analogue.

$q \leqslant i$, à cause de 2.4; pour $q > i$, raisonnons par récurrence sur q. D'après le raisonnement de [8], p. 457 (qui n'utilise que les propriétés 2.1 et 2.2), A^i commute à la transgression τ. On a donc

$$A^i(u_q) = A^i(\tau\, u_{q-1}) = \tau(A^i\, u_{q-1}) = \tau(Sq^i u_{q-1}) = Sq^i u_q \ ,$$

c. q. f. d.

Note. Comme nous l'avons indiqué au n° 26, on peut étendre les A^i à tous les couples (X, Y) d'espaces topologiques, à condition d'utiliser la cohomologie singulière, et les propriétés 2.1, 2.2, 2.4 sont encore vérifiées. C'est ce qui nous a permis d'utiliser les A^i dans la cohomologie de l'espace fibré 5.1, qui relie $K(Z_2, q-1)$ à $K(Z_2, q)$, espace fibré qui n'est pas un complexe simplicial.

On pourrait d'ailleurs remplacer, dans la démonstration précédente, le complexe $K(Z_2, q)$ par le joint de $K(Z_2, q-1)$ avec deux points, et l'on pourrait ainsi demeurer entièrement à l'intérieur de la catégorie des complexes simpliciaux.

31. Opérations cohomologiques en caractéristique 2

Posons $A = B = Z_2$. En combinant le Théorème 1 avec le Théorème 2 du § 2, on obtient :

Théorème 2. *Toute opération cohomologique* $C: H^q(X,Z_2) \to H^n(X,Z_2)$ *est de la forme :*

$$C(x) = P\big(Sq^{I_1}(x), \ldots, Sq^{I_k}(x)\big) \ ,$$

où P *désigne un polynôme (par rapport au cup-produit), et où* $Sq^{I_1}, \ldots, Sq^{I_k}$ *désignent les i-carrés itérés correspondant aux suites admissibles d'excès* $< q$. *En outre, deux polynômes distincts* P *et* P' *définissent des opérations* C *et* C' *distinctes.*

Lorsque $A = Z_m$ ($m = 2^h$), on a un résultat analogue en remplaçant les Sq^I par les Sq^I_h; lorsque $A = Z$, on ne doit considérer que des suites I dont le dernier terme est > 1.

Corollaire. *Si* $n \leqslant 2q$, *les i-carrés itérés* Sq^I, *où* I *parcourt l'ensemble des suites admissibles de degré* $n - q$, *forment une base de l'espace vectoriel des opérations cohomologiques relatives à* $\{q, n, Z_2, Z_2\}$.

32. Relations entre *i*-carrés itérés

Le Corollaire précédent montre que tout i-carré itéré est combinaison linéaire de Sq^I, où I est admissible. Il est naturel de chercher une méthode permettant d'écrire *explicitement* une telle décomposition. Cette question a été résolue par J. Adem [1], qui a démontré la formule suivante (conjecturée par Wu-Wen-Tsün) :

$$\text{Si } a<2b, \quad Sq^a Sq^b = \Sigma_{0\leqslant c\leqslant a/2}\binom{b-c-1}{a-2c} Sq^{a+b-c} Sq^c , \qquad (32.1)$$

où $\binom{k}{j}$ désigne le coefficient binômial $k!/j!(k-j)!$, avec la convention usuelle : $\binom{k}{j} = 0$ si $j>k$.

On voit facilement que cette formule permet de ramener, par des réductions successives, tout i-carré itéré à une somme de Sq^I où I est admissible. Elle répond donc bien à la question posée.

Citons quelques cas particuliers de 32.1 dont nous ferons usage au § 5 :

32.2. $Sq^1 Sq^n = 0$ *si n est impair,* $Sq^1 Sq^n = Sq^{n+1}$ *si n est pair.*

32.3. $Sq^2 Sq^2 = Sq^3 Sq^1$; $Sq^2 Sq^3 = Sq^5 + Sq^4 Sq^1$.

33. Méthode permettant d'obtenir les relations entre *i*-carrés itérés

La démonstration donnée par J. Adem de la formule 32.1 est basée sur une étude directe des i-carrés itérés. Nous allons esquisser une méthode plus indirecte, mais qui conduit plus aisément au résultat[9]).

Soit X l'espace projectif réel à une infinité de dimensions, $Y = X^q$ le produit direct de q espaces homéomorphes à X. L'algèbre de cohomologie $H^*(Y, Z_2)$ est donc l'algèbre de polynômes à q générateurs $x_1, \ldots, x_q$, de degrés 1. Nous noterons W_q le produit $x_1 \ldots x_q$ de ces générateurs : on a $W_q \in H^q(Y, Z_2)$.

Lemme 1. *Soit C une somme de i-carrés itérés, tous de degrés* $\leqslant q$. *Si* $C(W_q) = 0$, *alors C est identiquement nulle.*

Compte tenu du Corollaire au Théorème 2, il suffit de vérifier que les $Sq^I(W_q)$ sont linéairement indépendants lorsque I parcourt l'ensemble des suites admissibles de degré $\leqslant q$. Or, il est très facile de déterminer explicitement les opérations Sq^i dans $H^*(Y, Z_2)$, en utilisant les propriétés 2.3, 2.4, 2.5 ; le résultat cherché s'ensuit par un calcul que nous ne ferons pas ici (voir un article en préparation de R. Thom).

Théorème 3. *Soit C une somme de i-carrés itérés. Supposons que, pour tout espace T, la relation* $C(y) = 0$, $y \in H^*(T, Z_2)$, *entraîne* $C(x \cdot y) = 0$ *pour tout* $x \in H^1(T, Z_2)$. *Alors C est identiquement nulle.*

Prenons pour T l'espace Y défini plus haut (q étant égal au degré maximum des i-carrés itérés qui figurent dans C). On a évidemment

[9]) Cette méthode est d'ailleurs très proche de celle qui avait amené Wu-Wen-Tsün à conjecturer la formule 32.1.

$C(1) = 0$, d'où $C(x_1 \ldots x_i) = 0$ par récurrence sur i, et en particulier $C(W_q) = 0$, d'où $C = 0$ d'après le Lemme 1.

A titre d'exemple, vérifions l'hypothèse du Théorème 3 pour $C = Sq^2Sq^2 + Sq^3Sq^1$. En utilisant 2.3, 2.4, 2.5, on obtient :

$$Sq^2Sq^2(x \cdot y) = x^4 \cdot Sq^1y + x^2 \cdot (Sq^2Sq^1y + Sq^1Sq^2y) + x \cdot Sq^2Sq^2y \ ,$$
$$Sq^3Sq^1(x \cdot y) = x^4 \cdot Sq^1y + x^2 \cdot (Sq^3y + Sq^2Sq^1y) + x \cdot Sq^3Sq^1y \ .$$

Comme $Sq^3 = Sq^1Sq^2$, on tire de là :

$$C(x \cdot y) = x \cdot C(y) \ ,$$

ce qui montre bien que $C(y) = 0$ entraîne $C(x \cdot y) = 0$. D'après le Théorème 3, on a donc $Sq^2Sq^2 + Sq^3Sq^1 = 0$, d'où $Sq^2Sq^2 = Sq^3Sq^1$, et nous avons démontré la première des relations 32.3.

On démontrerait de la même façon la formule 32.1 dans le cas général, en raisonnant par récurrence sur $a + b$. Nous laissons le détail du calcul au lecteur.

BIBLIOGRAPHIE

[1] *J. Adem,* The iteration of the Steenrod squares in algebraic topology, Proc. Nat. Acad. Sci. U. S. A. **38**, 1952, p. 720—726.

[2] *A. Borel,* Sur la cohomologie des espaces fibrés principaux et des espaces homogènes de groupes de Lie compacts, Ann. Math. **57**, 1953, p. 115—207.

[3] *H. Cartan et J.-P. Serre,* Espaces fibrés et groupes d'homotopie. I., *C. R. Acad. Sci. Paris* **234**, 1952, p. 288—290; II., ibid., p. 393—395.

[4] *S. Eilenberg and S. MacLane,* Relations between homology and homotopy groups of spaces, Ann. Math. **46**, 1945, p. 480—509; II., ibid. **51**, 1950, p. 514—533.

[5] *S. Eilenberg and S. MacLane,* Cohomology theory of abelian groups and homotopy theory. I., Proc. Nat. Acad. Sci. U. S. A. **36**, 1950, p. 443—447; II., ibid. p. 657—663; III., ibid. **37**, 1951, p. 307—310; IV., ibid. **38**, 1952, p. 1340—1342.

[6] *P. Hilton,* The Hopf invariant and homotopy groups of spheres, Proc. Cambridge Philos. Soc. **48**, 1952, p. 547—554.

[7] *J. C. Moore,* Some applications of homology theory to homotopy problems, Ann. Math., 1953.

[8] *J.-P. Serre,* Homologie singulière des espaces fibrés. Applications, Ann. Math. **54**, 1951, p. 425—505.

[9] *J.-P. Serre,* Sur les groupes d'Eilenberg-MacLane, C. R. Acad. Sci. Paris **234**, 1952, p. 1243—1245.

[10] *J.-P. Serre,* Sur la suspension de Freudenthal, C. R. Acad. Sci. Paris **234**, 1952, p. 1340—1342.

[11] *J.-P. Serre,* Groupes d'homotopie et classes de groupes abéliens, Ann. Math. 1953.

[12] *N. E. Steenrod,* Products of cocycles and extensions of mappings, Ann. Math. **48**, 1947, p. 290—320.

[13] *N. E. Steenrod*, Reduced powers of cohomology classes, Ann. Math. **56**, 1952, p. 47—67.

[14] *G. W. Whitehead*, Fibre spaces and the Eilenberg homology groups, Proc. Nat. Acad. Sci. U. S. A. **38**, 1952, p. 426—430.

7

The next extract is taken from my own lecture notes. All the material in it is in the literature, but it is scattered around several original papers by Blakers, Massey and Moore. The main prerequisites for reading it are a knowledge of elementary homotopy theory and of homology theory up to spectral sequences, plus Serre's theory of classes of abelian groups (see §§1, 4, 5, 8, 10 of the introduction).

7
ON THE TRIAD CONNECTIVITY THEOREM

J. F. Adams

(from unpublished lecture notes)

I have mentioned the theorem of J. H. C. Whitehead: if the spaces X and Y are 1-connected, and $f_*:H_*(X) \to H_*(Y)$ is iso, then $f_*:\pi_*(X) \to \pi_*(Y)$ is iso. Why shouldn't we have a relative Whitehead theorem? It should say that if $f_*:H_*(X, X') \to H_*(Y, Y')$ is iso, then $f_*:\pi_*(X, X') \to \pi_*(Y, Y')$ is iso. Unfortunately, this theorem is not true. For a counterexample, let E^n_+, E^n_- be the two hemispheres of S^n, meeting in S^{n-1}. Take $X, X' = E^n_+, S^{n-1}$ and $Y, Y' = S^n, E^n_-$; and let f be the injection map. Then $f_*:H_*(X, X') \to H_*(Y, Y')$ is iso. But

$$\pi_r(X, X') \underset{\cong}{\overset{d}{\rightarrow}} \pi_{r-1}(S^{n-1})$$

$$\pi_r(Y, Y') \underset{\cong}{\overset{j}{\leftarrow}} \pi_r(S^n) .$$

For each $n \geq 2$, these are not isomorphic for all r. For, one of $n - 1$, n is even and one is odd; hence one of $\pi_*(S^{n-1})$, $\pi_*(S^n)$ contains an infinite group above the Hurewicz dimension and the other doesn't.

We can, however, have a limited result in the direction of a relative Whitehead theorem. To proceed, we have to introduce new groups, measuring the obstruction to the truth of the theorem. First, if we have a map $f:X, X' \to Y, Y'$, then (by using the mapping-cylinder) we may suppose without loss of generality that $X, X' \subset Y, Y'$ and $X \cap Y' = X'$. We will now rename the spaces, and take a space X with subspaces A, B and $A \cap B = C$. We are interested in the injection $A, C \to X, B$ (or, similarly, $B, C \to X, A$).

We assume that the base-point x_0 in X is taken in C. Let $L(X, B)$ be the space of functions $\omega:[0, 1] \to X$ such that $\omega(0) = x_0$, $\omega(1) \in B$; that is, $L(X, B)$ is the space of paths in X starting at the base-point and ending in B. Define $L(A, C)$ similarly. Then we have $L(A, C) \subset L(X, B)$, and the following diagram is commutative.

$$\begin{array}{ccc} \pi_n(A, C) & \cong & \pi_{n-1}(L(A, C)) \\ \downarrow i_* & & \downarrow i_* \\ \pi_n(X, B) & \cong & \pi_{n-1}(L(X, B)) \end{array}$$

Factor $L(A, C) \to L(X, B)$ through an equivalence and a fibering; the fibre F is given by $F = L(L(X, B), L(A, C))$. To within a homeomorphism F is the space of functions $f:I^2 \to X$ such that

$$f(t, u) \begin{cases} = x_0 & \text{if } t = 0 \\ \in A & \text{if } t = 1 \\ = x_0 & \text{if } u = 0 \\ \in B & \text{if } u = 1 \end{cases}$$

We may show where the various parts of the square map as follows.

(7)

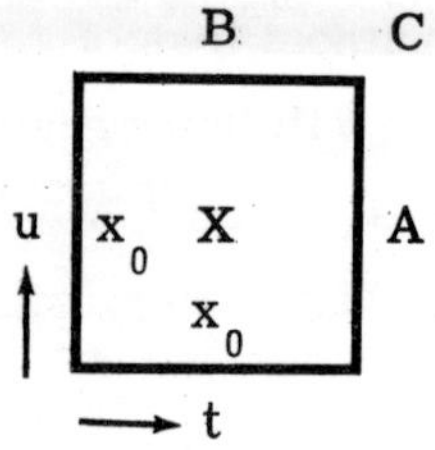

We have the following exact sequence.

$$\ldots \to \pi_{n-1}(F) \to \pi_{n-1}(L(A, C)) \to \pi_{n-1}(L(X, B)) \to \pi_{n-2}(F) \to \ldots$$

We define the triad homotopy groups by $\pi_n(X; A, B) = \pi_{n-2}(F)$. (We use the subscript n because $\pi_n(X; A, B)$ evidently admits an interpretation in terms of maps of the n-cube I^n into X.) We thus obtain the following exact sequence.

$$\ldots \to \pi_{n+1}(X; A, B) \xrightarrow{d} \pi_n(A, C) \xrightarrow{i_*} \pi_n(X, B) \to \pi_n(X; A, B) \to \ldots$$

When the triad groups $\pi_{n+1}(X; A, B)$ and $\pi_n(X; A, B)$ are zero, the map i_* is iso.

There is a natural isomorphism between $\pi_n(X; A, B)$ and $\pi_n(X; B, A)$ given, in terms of maps of I^n, by

$$f(x_1, x_2, x_3, \ldots, x_n) \longleftrightarrow -f(x_2, x_1, x_3, \ldots, x_n).$$

Thus we have another exact sequence, as follows.

$$\ldots \to \pi_{n+1}(X; A, B) \xrightarrow{d} \pi_n(B, C) \xrightarrow{i_*} \pi_n(X, A) \to \pi_n(X; A, B) \to \ldots$$

The following diagram is anticommutative.

$$(7)\qquad \begin{array}{ccc} \pi_{n+1}(X;\ A,\ B) & \xrightarrow{\ d\ } & \pi_n(A,\ C) \\ {\scriptstyle d}\downarrow & & \downarrow{\scriptstyle d} \\ \pi_n(B,\ C) & \xrightarrow{\ d\ } & \pi_{n-1}(C) \end{array}$$

Otherwise the naturality is as expected.

We say that a triad is *excisive* if

$$i_*: H_*(A,\ C) \to H_*(X,\ B)$$

$$i_*: H_*(B,\ C) \to H_*(X,\ A)$$

are iso. Let $\mathscr{C}$ be a class of abelian groups, satisfying the axioms needed for Serre's relative Hurewicz isomorphism theorem mod $\mathscr{C}$ (Spanier p. 511).

Theorem (*Blakers-Massey Triad Theorem* Mod $\mathscr{C}$). *Suppose that* $(X;\ A,\ B)$ *is an excisive triad,* X *is* 1-*connected,* $(X,\ A)$, $(X,\ B)$ *and* $(X,\ C)$ *are all* 2-*connected, and* $(X,\ A)$, $(X,\ B)$ *are* $(q - 1)$, $(p - 1)$-*connected* mod $\mathscr{C}$. *Then* $(X;\ A,\ B)$ *is* $(p + q - 2)$-*connected* mod $\mathscr{C}$ *and the generalised Whitehead product* $\pi_p(A,\ C) \otimes \pi_q(B,\ C) \to \pi_{p+q-1}(X;\ A,\ B)$ *is an isomorphism* mod $\mathscr{C}$.

Before beginning the proof, I must explain about Whitehead products. Recall (e.g. from Hilton (1) p. 42) that we have

$$\pi_r(S^p \vee S^q) \cong \pi_r(S^p) \oplus \pi_r(S^q) \oplus \pi_{r+1}(S^p \times S^q,\ S^p \vee S^q)\ .$$

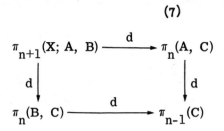

Here the first two summands are embedded by the injections i_p, i_q of S^p, S^q in $S^p \vee S^q$, and the third summand is embedded by d. By the relative Hurewicz theorem, the first non-zero group $\pi_{r+1}(S^p \times S^q, S^p \vee S^q)$ is $\pi_{p+q}(S^p \times S^q, S^p \vee S^q) \cong Z$. Take the two generators in $H_p(S^p)$, $H_q(S^q)$; their product gives a generator in $H_{p+q}(S^p \times S^q)$; let the corresponding generator of $\pi_{p+q}(S^p \times S^q, S^p \vee S^q)$ be g. Then in $\pi_{p+q-1}(S^p \vee S^q)$ we have the element dg.

Proposition. To $\alpha \in \pi_p(X)$ and $\beta \in \pi_q(X)$ we can assign a unique element $[\alpha, \beta] \in \pi_{p+q-1}(X)$ (called the Whitehead product of α and β) to satisfy the following conditions.

(i) The Whitehead product is natural; that is, $f_*[\alpha, \beta] = [f_*\alpha, f_*\beta]$.

(ii) In $S^p \vee S^q$ we have

$$[i_p, i_q] = dg .$$

Proof. Suppose that $\alpha \in \pi_p(X)$ and $\beta \in \pi_q(X)$ are represented by maps $f: S^p \to X$, $g: S^q \to X$. Then we can construct a map $h: S^p \vee S^q \to X$ such that $hi_p = f$, $hi_q = g$. Then

$$[\alpha, \beta] = [h_* i_p, h_* i_q]$$

$$= h_*[i_p, i_q]$$

$$= h_* dg .$$

This proves the uniqueness of $[\alpha, \beta]$.

Conversely, we can define $[\alpha, \beta]$ by $[\alpha, \beta] = h_* dg$, where h is as above; this satisfies (i) and (ii) and proves the existence of the Whitehead product. Another way to express this proof is to say that we first define $[i_p, i_q] = dg$ and then define $[\alpha, \beta]$ for general α and β by naturality.

In this proof, the space $S^p \vee S^q$ is called a universal example (or representing object).

The Whitehead product is bilinear and anticommutative; this can be proved by diagram-chasing with the universal example.

One can give similar treatments of the relative and generalised Whitehead products. These are pairings

(i) $\pi_p(X, A) \otimes \pi_q(A) \rightarrow \pi_{p+q-1}(X, A)$

(ii) $\pi_p(B) \otimes \pi_q(X, B) \rightarrow \pi_{p+q-1}(X, B)$

(iii) $\pi_p(A, C) \otimes \pi_q(B, C) \rightarrow \pi_{p+q-1}(X; A, B)$.

The universal examples are as follows:

(i) $X' = E^p \vee S^q, \quad A' = S^{p-1} \vee S^q$

(ii) $X' = S^p \vee E^q, \quad B' = S^p \vee S^{q-1}$

(iii) $X' = E^p \vee E^q, \quad A' = E^p \vee S^{q-1}$,

$B' = S^{p-1} \vee E^q, \quad C' = S^{p-1} \vee S^{q-1}$.

To define specific elements in their homotopy groups, we employ the following diagrams.

(i) $0 \rightarrow \pi_r(X', A') \rightarrow \pi_{r-1}(A') \rightleftarrows \pi_{r-1}(X')$

(ii) $0 \rightarrow \pi_r(X', B') \rightarrow \pi_{r-1}(B') \rightleftarrows \pi_{r-1}(X')$

(7)

(iii)

$$\begin{array}{ccccccc}
 & & 0 & & 0 & & \\
 & & \downarrow & & \downarrow & & \\
0 & \to & \pi_r(X'; A', B') & \to & \pi_{r-1}(A', C') & \rightleftarrows & \pi_{r-1}(X', B') \\
 & & \downarrow & & \downarrow & & \\
0 & \to & \pi_{r-1}(B', C') & \to & \pi_{r-2}(C') & \rightleftarrows & \pi_{r-2}(B') \\
 & & \downarrow\uparrow & & \downarrow\uparrow & & \\
 & & \pi_{r-1}(X', A') & & \pi_{r-2}(A') & &
\end{array}$$

Here, for example, the map $\pi_{r-1}(X') \to \pi_{r-1}(A')$ in case (i) is induced by the map $X' \to A'$ which maps E^p to the base-point and keeps S^q fixed.

We use these diagrams to deduce the existence in $\pi_r(X', A')$, $\pi_r(X', B')$ and $\pi_r(X'; A', B')$ of elements $[i_p, i_q]$ which map as follows.

(i) $[i_p, i_q] \to [i_{p-1}, i_q] \to 0$

(ii) $[i_p, i_q] \to (-1)^p[i_p, i_{q-1}] \to 0$

(iii)
$$\begin{array}{ccccc}
[i_p, i_q] & \to & (-1)^p[i_p, i_{q-1}] & \to & 0 \\
\downarrow & & & & \\
[i_{p-1}, i_q] & & & & \\
\downarrow & & & & \\
0 & & & &
\end{array}$$

From these elements we define $[\alpha, \beta]$ for general α and β by naturality, as before.

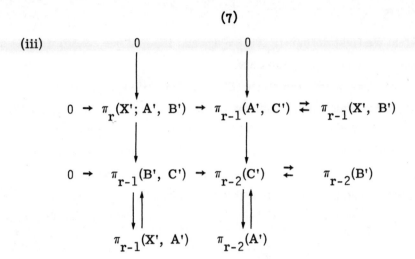

(7)

We further deduce from diagram (iii) that the first non-zero group $\pi_r(X'; A', B')$ is Z for $r = p + q - 1$, and that the element $[i_p, i_q]$ is a generator.

The relative and generalised Whitehead products are bilinear and anticommutative; this is deduced by diagram-chasing from the bilinearity and anticommutativity of the Whitehead product.

Proof of the Theorem. **Step 1.** By a functor, we may replace $(X; A, B)$ by $(X_1; A_1, B_1)$ so that X_1 is contractible, (X_1, A_1, B_1) is excisive and we do not change the relative or triad homotopy groups.

In fact, we may take $X_1 = L(X, X)$, $A_1 = L(X, A)$, $B_1 = L(X, B)$, $C_1 = L(X, C)$. We have $C_1 = A_1 \cap B_1$. We have the following diagram.

$$\begin{array}{ccc} A_1, C_1 & \longrightarrow & X_1, B_1 \\ \downarrow & & \downarrow \\ A, C & \longrightarrow & X, B \end{array}$$

All the vertical arrows are fiberings with the same fibre $F = \Omega(X)$ and trivial operations of $\pi_1(X)$ on $H_*(F)$. We are given that $H_*(A, C) \to H_*(X, B)$ is iso; by an obvious spectral-sequence argument, $H_*(A_1, C_1) \to H_*(X_1, B_1)$ is iso. Similarly for $H_*(B_1, C_1) \to H_*(X_1, A_1)$.

Step 2. We will define a homeomorphism from the function space $L(X; A, B) = L(L(X, B), L(A, C))$ to another. This is done by dividing the square I^2 into two triangles along the line $t = u$.

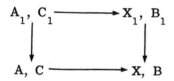

(7)

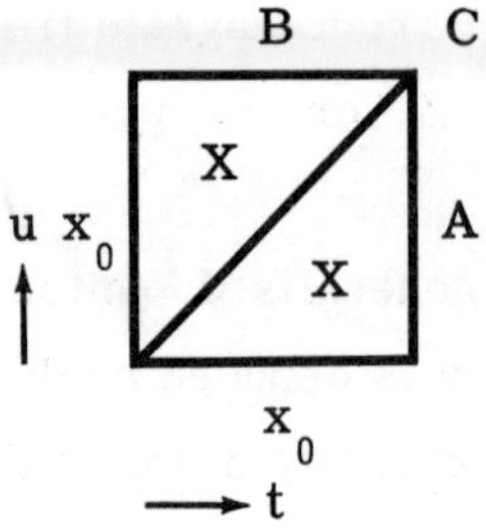

The two triangles are homeomorphic to two squares, and we have to consider pairs of maps which map the edges of the two squares as follows.

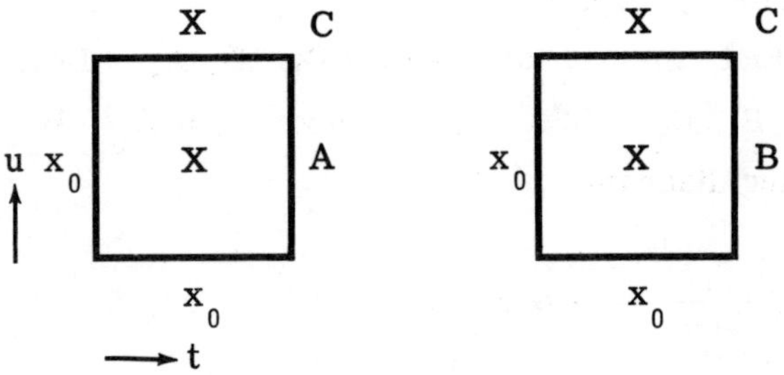

However, the maps must also agree along the top edges $u = 1$ of the two squares. Thus we have

$$L(X; A, B) = L(X \times X, A \times B, D)$$

where D is the diagonal in $X \times X$. Set

$$X_2 = X_1 \times X_1, \quad A_2 = A_1 \times B_1, \quad B_2 = D_1 ;$$

then A_2 meets B_2 in a homeomorph of C_1. We have

$$\pi_r(X_1; A_1, B_1) \cong \pi_r(X_2; A_2, B_2) .$$

(7)

Also since X_1 is contractible we have $\pi_*(X_1) = 0$, $\pi_*(X_2, B_2) = 0$ and

$$\pi_{r+1}(X_2; A_2, B_2) \underset{\cong}{\overset{d}{\rightarrow}} \pi_r(A_2, C_2)$$

$$\pi_{r+1}(X_2, A_2) \underset{\cong}{\overset{d}{\rightarrow}} \pi_r(A_2)$$

$$\pi_{r+1}(X_2, B_2) \underset{\cong}{\overset{d}{\rightarrow}} \pi_r(B_2) .$$

Step 3. Since X_1 is contractible, the Mayer-Vietoris sequence gives us

$$H_r(C_1) \overset{\cong}{\rightarrow} H_r(A_1) \oplus H_r(B_1) ,$$

where the map is induced by the two injections. Thus we obtain the following diagram.

$$\begin{array}{ccccc} H_r(C_2) & \overset{i}{\rightarrow} & H_r(A_1 \times B_1) & \overset{j}{\rightarrow} & H_r(A_1 \times B_1, C_2) \\ & \underset{\cong}{\searrow} & \downarrow \lambda & & \\ & & H_r(A_1) \oplus H_r(B_1) & & \end{array}$$

(Here λ is induced by the two projections.) Thus j gives an isomorphism from $\operatorname{Ker} \lambda$ to $H_r(A_2, C_2)$. Therefore $H_r(A_2, C_2) = 0$ mod $\mathscr{C}$ for $r < p + q - 2$, and

$$j\mu : H_{q-1}(A_1) \otimes H_{p-1}(B_1) \rightarrow H_{p+q-2}(A_2, C_2)$$

is an isomorphism mod $\mathscr{C}$. Since $\pi_1(C_2) = 0$, $\pi_2(A_2, C_2)$ is abelian; since A_1 and B_1 are 1-connected by the data, $H_2(A_2, C_2) = 0$, and (A_2, C_2) is 2-connected. By the relative Hurewicz isomorphism mod $\mathscr{C}$, (A_2, C_2) is $(p + q - 2)$-connected mod $\mathscr{C}$. That is, $(X; A, B)$ is $(p + q - 2)$-connected mod $\mathscr{C}$. We may assume $p \geq 3$, $q \geq 3$, and therefore

$$\pi_p(A, C) \to \pi_p(X, B)$$

$$\pi_q(B, C) \to \pi_q(X, A)$$

are isomorphisms mod $\mathscr{C}$. We now have a diagram of natural maps which are isomorphisms mod $\mathscr{C}$.

$$\begin{array}{ccc}
\pi_p(A, C) \otimes \pi_q(B, C) & & \pi_{p+q-1}(X; A, B) \\
\downarrow & & \downarrow \\
\pi_p(X, B) \otimes \pi_q(X, A) & & \pi_{p+q-1}(X_1; A_1, B_1) \\
\downarrow & & \downarrow \\
\pi_{p-1}(B_1) \otimes \pi_{q-1}(A_1) & & \pi_{p+q-2}(A_2, C_2) \\
\downarrow & & \downarrow \\
H_{p-1}(B_1) \otimes H_{q-1}(A_1) & \longrightarrow & H_{p+q-2}(A_2, C_2)
\end{array}$$

When we consider the universal example and take $\mathscr{C} = 0$, $i_p \otimes i_q$ must correspond either to $+[i_p, i_q]$ or to $-[i_p, i_q]$. Since all the maps are natural, when we insert the generalised Whitehead product

$$P: \pi_p(A, C) \otimes \pi_q(B, C) \to \pi_{p+q-1}(X; A, B)$$

the diagram is commutative with a fixed sign, either + or -. Hence P is an isomorphism mod $\mathscr{C}$. This proves the theorem.

8&9

The next two extracts are from works by G. W. Whitehead and I. M. James, which were of considerable importance in the historical development of suspension-theory. For reasons of space, the extracts have been confined to the statements of the main theorems. The prerequisite is a certain knowledge of homotopy theory (see §10 and paper 7).

8

ON THE FREUDENTHAL THEOREMS

By George W. Whitehead

(Received February 18, 1952)

1. Introduction

The suspension homomorphism $E:\pi_q(S^n) \rightarrow \pi_{q+1}(S^{n+1})$ is an important tool in the calculation of homotopy groups of spheres. This homorphism was introduced by Freudenthal [1], who proved that E is onto if $q \leqq 2n - 1$ and an isomorphism if $q < 2n - 1$. Freudenthal also determined the image of E for $q = 2n$ and obtained some partial results on the kernel of E for $q = 2n - 1$. The latter results were completed by the author in [2]. Freudenthal's proofs are very complicated, relying heavily on geometrical arguments.

Recently Blakers and Massey [3] have introduced the notion of homotopy groups of a triad and have shown that the "easy" Freudenthal theorems are a consequence of a general theorem on triad homotopy groups. They have also announced [4] further results on homotopy groups of triads which imply the Freudenthal theorems in the critical dimensions. Their proofs are very much simpler than those of Freudenthal.

The group $\pi_{q+1}(S^{n+1})$ is isomorphic with $\pi_q(\Omega^{n+1})$, where Ω^{n+1} is the space of loops in S^{n+1}. The n-sphere S^n can be imbedded in Ω^{n+1} in such a way that the homomorphism of $\pi_q(S^n)$ into $\pi_{q+1}(S^{n+1})$ induced by the inclusion-map is the suspension. Thus one can obtain results on the suspension by calculating the relative homotopy groups $\pi_q(\Omega^{n+1}, S^n)$. In this way we first obtain a very simple proof of the "easy" Freudenthal theorems. This proof does not make use of the Blakers-Massey theorems, which, however, appear in a different way in the calculation of the groups $\pi_q(\Omega^{n+1}, S^n)$ for $q \geqq 2n$. Our main result, which was conjectured in [2], is that there is an exact sequence

$$\pi_{3n-1}(S^n) \xrightarrow{E} \cdots \rightarrow \pi_q(S^n) \xrightarrow{E} \pi_{q+1}(S^{n+1}) \xrightarrow{H'} \pi_{q-1}(S^{2n-1}) \xrightarrow{P} \pi_{q-1}(S^n) \rightarrow \cdots,$$

where H' is essentially the generalized Hopf homomorphism of [2] and P is induced by composition with $[\iota_n, \iota_n]$, where ι_n generates $\pi_n(S^n)$. We further show that, if $q \leqq 3n - 2$, then every element of $\pi_{q+1}(S^{n+1})$ can be obtained by Hopf's construction.

The results of this paper overlap with some recently announced by Pitcher [5]. Pitcher also makes use of the space Ω^{n+1} (or rather of the space of Frechet equivalence classes of loops in S^{n+1}). However, Pitcher makes use of the Morse critical point theory, while our proofs are more elementary.

Bibliography

1. H. Freudenthal, *Über die Klassen der Sphärenabbildungen*, Compositio Math. 5 (1937), 299–314.
2. G. W. Whitehead, *A generalization of the Hopf invariant*, Ann. of Math. 51 (1950), 192–237.
3. A. L. Blakers and W. S. Massey, *The homotopy groups of a triad I*, Ann. of Math. 53 (1951), 161–205.
4. A. L. Blakers and W. S. Massey, Bull. Am. Math. Soc. 56 (1950), abstract 468t, p. 540.
5. Everett Pitcher, Bull. Amer. Math. Soc. 56 (1950), abstract 38, p. 44.
6. W. Hurewicz and N. E. Steenrod, *Homotopy relations in fibre spaces*, Proc. Nat. Acad. Sci. U. S. A. 27 (1941), 60–64.
7. W. Hurewicz, Beiträge zur Topologie der Deformationen, Neder. Akad. Wetensch. 38 (1935), 112–119, 521–528; 39 (1936), 117–126, 215–224.
8. J. H. C. Whitehead, *On adding relations to homotopy groups*, Ann. of Math. 42 (1941), 409–428.
9. S. T. Hu, *An exposition of the relative homotopy theory*, Duke Math. J. 14 (1947), 991–1033.
10. J.-P. Serre, *Homologie singulière des espaces fibrés*, Ann. of Math. 54 (1951), 425–505.
11. H. C. Wang, *The homology groups of the fibre bundles over a sphere*, Duke Math. J. 16 (1949), 33–38.
12. S. Eilenberg and N. E. Steenrod, Foundations of algebraic topology, Princeton University Press, 1952.
13. L. Pontrjagin, *Homologies in compact Lie groups*, Rec. Math. (Math. Sbornik) N. S. 6 (1939), 389–422.
14. A. L. Blakers and W. S. Massey, *The homotopy groups of a triad II*, Ann. of Math. 55 (1952), 192–201.
15. A. L. Blakers and W. S. Massey, Bull. Amer. Math. Soc. 57 (1951), abstract 165t, p. 142.

9

THE SUSPENSION TRIAD OF A SPHERE

By I. M. James

(Received September 9, 1954)
(Revised June 17, 1955)

1. Introduction

This paper is the third in a series of four. The constructions which were developed in *Reduced product spaces* [5],[1] and *On the suspension triad* [6], are applied to the suspension triad of a sphere, with the help of the method of spectral sequences. There are many applications of the theorems which we shall prove to the tneory of homotopy groups of spheres, and these will be the subject of a concluding article, *On the suspension sequence*.

The definitions and notations of [6] are carried over to the present work. In particular, S^n denotes the euclidean n-sphere

$$x_0^2 + x_1^2 + \cdots + x_n^2 = 1 \qquad (n = 1, 2, \cdots).$$

We study the suspension triad of S^n, i.e. the triad

$$(S^{n+1}; E_+, E_-),$$

where E_+, E_- are the two hemispheres into which S^n divides S^{n+1}, so that

$$S^{n+1} = E_+ \cup E_-, \qquad S^n = E_+ \cap E_-.$$

The point

$$e = (-1, 0, 0, \cdots)$$

acts as basepoint in these various spaces. We have an exact sequence

$$(1.1) \qquad \cdots \to \pi_r(S^n) \xrightarrow{E} \pi_{r+1}(S^{n+1}) \to \pi_{r+1}(S^{n+1}; E_+, E_-) \to \pi_{r-1}(S^n) \to \cdots,$$

the suspension sequence of S^n, which refers the behaviour of Freudenthal's suspension operator, E, to the properties of the triad homotopy group

$$\pi_{r+1}(S^{n+1}; E_+, E_-).$$

This group has been studied by various authors for values of r less than $4n$, especially by Serre in [9], Toda in [12], and G. Whitehead in [14]. In particular, the group is known to be zero if $r < 2n$, and to be cyclic infinite if $r = 2n$ (cf. Theorem 1 of [2], and the suspension theorems of Freudenthal).[2] Moreover the group is certainly finite if $r > 2n$, by the exactness of (1.1), since the homotopy groups of spheres are finite with the exceptions of $\pi_m(S^m)$ and $\pi_{4m-1}(S^{2m})(m \geqq 1)$ (Proposition 5 on p. 498 of [8]).

---

[1] Numbers in square brackets refer to the bibliography at the end of this article.
[2] There is a convenient summary of these theorems in §3 of [13].

Consider the natural homomorphism

$$h: \pi_r(S^{n+1}; E_+, E_-) \to \pi_r(S^{2n+1}),$$

which is defined in §15 of [6]. We shall prove the following three theorems in Part I of this paper.

THEOREM (1.2). *Let n be odd. Then*

$$h: \pi_r(S^{n+1}; E_+, E_-) \approx \pi_r(S^{2n+1}).$$

THEOREM (1.3). *Let n be even. Then the order of the kernel of h is odd, and has no prime factor greater than r/n. Also the index of the subgroup*

$$h\pi_r(S^{n+1}; E_+, E_-) \subset \pi_r(S^{2n+1})$$

is odd, and has no prime factor greater than r/2n.

THEOREM (1.4). *Let n be even, and let p be an odd prime number. If* $r \leqq 2pn - 2$, *then the p-primary component of the kernel of*

$$h: \pi_r(S^{n+1}; E_+, E_-) \to \pi_r(S^{2n+1})$$

is a direct summand of $\pi_r(S^{n+1}; E_+, E_-)$. *If* $r \leqq 2pn - 4$, *then this direct summand is isomorphic to the direct sum*

$$\pi_r(S^{pn}) \otimes Z_p + \pi_r(S^{pn+1}) * Z_p .$$

In the last assertion of (1.4), the symbols $\otimes$ and $*$ denote the tensor and torsion products, respectively, and Zp denotes the cyclic group of order p.

BIBLIOGRAPHY

1. M. G. BARRATT, *Homotopy ringoids and homotopy groups*, Quart. J. Math. Oxford Ser., 5 (1954), pp. 271–290.
2. A. L. BLAKERS and W. S. MASSEY, *The homotopy groups of a triad*, III, Ann. of Math., 58 (1953), pp. 409–417.
3. P. J. HILTON, *The Hopf invariant and homotopy groups of spheres*, Proc. Cambridge Philos. Soc., 48 (1952), pp. 547–554.
4. I. M. JAMES, *On the homotopy groups of certain pairs and triads*, Quart. J. Math. Oxford Ser., 5 (1954), pp. 260–270.
5. ———, *Reduced product spaces*, Ann. of Math., 62 (1955), pp. 170–197.
6. ———, *On the suspension triad*, Ann. of Math., 63 (1956), pp. 191–247.
7. ——— and J. H. C. WHITEHEAD, *Note on fibre spaces*, Proc. London Math. Soc. (3), 4 (1954), pp. 129–137.
8. J.-P. SERRE, *Homologie singulière des espaces fibrés*, Ann. of Math., 54 (1951), pp. 425–505.
9. ———, *Sur la suspension de Freudenthal*, C. R. Acad. Sci. Paris, 234 (1952), pp. 1340–1342.
10. ———, *Groupes d'homotopie et classes de groupes abéliens*, Ann. of Math., 58 (1953), pp. 258–294.
11. ———, *Cohomologie modulo 2 des complexes d'Eilenberg-MacLane*, Comment. Math. Helv., 27 (1953), pp. 198–231.
12. HIROSI TODA, *Topology of standard path spaces and homotopy theory*, I, Proc. Japan Acad., 29 (1953), pp. 299–304.
13. G. W. WHITEHEAD, *A generalization of the Hopf invariant*, Ann. of Math., 51 (1950), pp. 192–237.
14. ———, *On the Freudenthal theorems*, Ann. of Math., 57 (1953) pp. 209–228.

10

The next paper is an unpublished one by Milnor, to which many other authors have referred. The work on semi-simplicial loop-spaces has been carried further by Kan; however, Milnor's work remains the standard reference for the generalisation of Hilton's theorem. The prerequisite is a knowledge of semi-simplicial complexes; see the remarks in §3 and the book by May. In particular, for the 'geometrical realization' appearing in Lemma 1, see May chap. III or Milnor, 'The geometrical realization of a semi-simplicial complex', Annals of Math. 65 (1957), 357-362.

10
ON THE CONSTRUCTION FK

John Milnor

(lecture notes from Princeton University, 1956)

1. Introduction

The reduced product construction of Ioan James [5] assigns to each CW-complex a new CW-complex having the same homotopy type as the loops in the suspension of the original. This paper will describe an analogous construction proceeding from the category of semi-simplicial complexes to the category of group complexes. The properties of this construction **FK** are studied in §2.

A theorem of Peter Hilton [4] asserts that the space of loops in a union $S_1 \vee \ldots \vee S_r$ of spheres splits into an infinite direct product of loops spaces of spheres. In §3 the construction of **FK** is applied to prove a generalization (Theorem 4) of Hilton's theorem in which the spheres may be replaced by the suspensions of arbitrary connected (semi-simplicial) complexes.

The author is indebted to many helpful discussions with John Moore.

2. The construction

It will be understood that with every semi-simplicial complex there is to be associated a specified base point.

Let K be a semi-simplicial complex with base point b_0. Denote $S_0^n b_0$ by b_n. Let FK_n denote the free group generated by the elements of K_n with the single relation $b_n = 1$. Let the face and degeneracy operations ∂_i, s_i in $FK = \cup FK_n$ be the unique homomorphisms which carry the generators k_n into $\partial_i k_n$, $s_i k_n$ respectively. Thus each complex K determines a group complex FK.

It will be shown that FK is a loop space for EK, the suspension of K. (Definitions will be given presently.)

Alternatively let $F^+K_n \subset FK_n$ be the free monoid (= associative semi-group with unit) generated by K_n, with the same relation $b_n = 1$. Then the monoid complex F^+K is also a loop space for EK. This construction is the direct generalization of James' construction. (See Lemma 4.)

The <u>suspension</u> EK of the semi-simplicial complex K is defined as follows. For each simplex k_n, other than b_n, of K there is to be a sequence (Ek_n), $(s_0 Ek_n)$, $(s_0^2 Ek_n)$, ... of simplexes of EK having dimensions $n + 1$, $n + 2$, ... In addition there is to be a base point (b_0) and its degeneracies (b_n). The symbols $(s_0^i Eb_n)$ will be identified with (b_{n+i+1}). The face and degeneracy operations in EK are given by

$$\partial_j(Ek_n) = (E\partial_{j-1}k_n) \qquad (j > 0,\ n > 0)$$

$$s_j(Ek_n) = (Es_{j-1}k_n) \qquad (j > 0)$$

$$\partial_0(Ek_n) = (b_n)\,, \qquad \partial_1(Ek_0) = (b_0)$$

$$s_0(Ek_n) = (s_0 Ek_n)\,.$$

The face and degeneracy operations on the remaining simplexes

$(s_0^i Ek_n) = s_0^i(Ek_n)$ are now determined by the identities

$$\partial_j s_0^i = \begin{cases} s_0^i \partial_{j-1} & (j > i) \\ s_0^{i-1} & (j \le i \ne 0) \end{cases}$$

$$s_j s_0^i = \begin{cases} s_0^i s_{j-1} & (j > i) \\ s_0^{i+1} & (j \le i) . \end{cases}$$

It is not hard to show that this defines a semi-simplicial complex. The following lemma will justify calling it the suspension of K. Recall that the suspension of a topological space A with base point a_0 is the identification space of $A \times I$ obtaining by collapsing $(A \times \dot{I}) \cup (a_0 \times I)$ to a point.

Lemma 1. The geometric realization $|EK|$ is canonically homeomorphic to the suspension of $|K|$.

(For the definition of realization see [6]. In fact the required homeomorphism is obtained by mapping the point $(|k_n, \delta_n|, 1-t)$ of the suspension of $|K|$, where δ_n has barycentric coordinates $(t_0, \ldots, t_n)$ into the point $|(Ek_n), \delta_{n+1}| \in |EK|$, where δ_{n+1} has barycentric coordinates $(1-t, tt_0, \ldots, tt_n)$.)

Next the space of loops on a semi-simplicial complex K will be discussed. If K satisfies the Kan extension condition then ΩK can be defined as in [7]. This definition has two disadvantages:

(1) Many interesting complexes do not satisfy the extension condition. In particular EK does not.

(2) There is no natural way (and in some cases† no possible way) of defining a group structure in ΩK.

The following will be more convenient. A group complex G, or more generally a monoid complex, will be called a loop space for K if there exists a (semi-simplicial) principal bundle with base space K, fibre G, and with contractible total space T.

(By a principal bundle is meant a projection p of T onto K together with a left translation $G \times T \to T$ satisfying

$$(g_n \cdot g'_n) \cdot t_n = g_n \cdot (g'_n \cdot t_n)$$

where $g_n \cdot t_n = t_n$ if and only if $g_n = 1_n$; and where $g_n \cdot t_n = t'_n$ for some g_n if and only if $p(t_n) = p(t'_n)$. A complex is called contractible if its geometric realization is contractible. This is equivalent to requiring that the integral homology groups and the fundamental group be trivial.)

The existence of such a loop space for any connected complex K has been shown in recent work of Kan, which generalizes the present paper. The following Lemma is given to help justify the definition.

Lemma 2. If K satisfies the extension condition, and the group complex G is a loop space for K, then there is a homotopy equivalence $\Omega K \to G$.

---

† Let K be the minimal complex of the n-sphere $n \geq 2$. Then it can be shown that there is no group complex structure in ΩK having the correct Pontrjagin ring.

The proof is based on the following easily proven fact (compare [7] p. 2-10): Every principal bundle can be given the structure of a twisted cartesian product. That is one can find a one-one function

$$\eta : G \times K \to T$$

satisfying $\partial_i \eta = \eta \partial_i$ for $i > 0$ and $s_i \eta = \eta s_i$ for all i, where $\partial_0 \eta$ is given by an expression of the form

$$\partial_0 \eta(g_n k_n) = \eta((\partial_0 g_n) \cdot (\tau k_n), \partial_0 k_n) .$$

(For this assertion the fibre must be a monoid complex satisfying the extension condition.) Thus the bundle is completely described by G and K together with the 'twisting function' $\tau : K_n \to G_{n-1}$; where τ satisfies the identities

$$s_i \tau = \tau s_{i+1} \quad (i \geq 0), \qquad \partial_i \tau = \tau \partial_{i+1} \quad i \geq 1 ,$$

$$\tau s_0 k_n = 1_n , \qquad (\partial_0 \tau k_n) \cdot (\tau \partial_0 k_n) = \tau \partial_1 k_n .$$

Now a map $\overline{\tau} : \Omega K_{n-1} \to G_{n-1}$ is defined by $\overline{\tau}(k_n) = \tau(k_n)$. From the definition of ΩK and the above identities it follows that $\overline{\tau}$ is a map. From the homotopy sequence of the bundle it is easily verified that $\overline{\tau}$ induces isomorphisms of the homotopy groups, which proves Lemma 2.

To define a principal bundle with fibre FK and base space EK it is sufficient to define twisting functions $\tau : EK_{n+1} \to FK_n$. These will be given by

$$\tau(Ek_n) = k_n , \qquad \tau(s_0^i Ek_{n-i}) = 1_n \qquad (i > 0) .$$

Theorem 1. FK is a loop space for EK. In fact the twisted cartesian product $\{FK, EK, \tau\}$ has a contractible total space.

It is easy to verify that τ satisfies the conditions for a twisting function. Hence we have defined a twisted cartesian product, and therefore a principal bundle. Let T denote its total space. Note that T may be identified with $FK \times EK$ except that ∂_0 is given by

$$\partial_0(g_n, (Ek_{n-1})) = (\partial_0 g_n \cdot k_{n-1}, (b_{n-1}))$$

$$\partial_0(g_n, (s_0^i Ek_{n-i-1})) = (\partial_0 g_n, (s_0^{i-1}(Ek_{n-i-1})) \quad (i \geq 1) .$$

It will first be shown that the homology groups of T are trivial. This will be done by giving a contracting homotopy S for the chain complex C(T).

Lemma 3. Let G be the free group on generators x_α. Then the integral group ring ZG has as basis (over Z) the elements $gx_\alpha - g$, where g ranges over all elements of G; together with the element 1.

The proof is not difficult. Now define S by the rules

$$S(1_n, (b_n)) = \begin{cases} 0 & (n \text{ even}) \\ (1_{n+1}, (b_{n+1})) & (n \text{ odd}) \end{cases}$$

$$S[(g_n \cdot k_n, (b_n)) - (g_n, (b_n))]$$

$$= \sum_{i=0}^{n} (-1)^i [(s_i g_n, (s_0^i E \partial_0^i k_n)) - (s_i g_n, (b_{n+1}))]$$

$$S[(g_n, (s_0^{r-1} E k_{n-r})) - (g_n, (b_n))]$$

$$= \sum_{j=r}^{n} (-1)^j [(s_j g_n, (s_0^j E \partial_0^{j-r} k_{n-r})) - (s_j g_n, (b_{n+1}))]$$

where g_n ranges over all elements of the group FK_n.

It follows easily from Lemma 3 that the elements for which S has been defined form a basis for $C(T)$, providing that k_n, k_{n-r} are restricted to elements other than b_n, b_{n-r}. However the above rules reduce to the identity $0 = 0$ if we substitute $k_n = b_n$ or $k_{n-r} = b_{n-r}$. This shows that S is well defined.

The necessary identity $Sd + dS = 1 - \varepsilon$, where

$dx_n = \sum_{i=0}^{n} (-1)^i \partial_i x_n$ and where $\varepsilon : C(T) \to C(T)$ is the augmentation $(\varepsilon \sum \alpha_i (g_0, b_0) = \sum \alpha_i (1_0, b_0))$ can now be verified by direct computation. Since this computation is rather long it will not be given here.

Proof that $|T|$ is simply connected. A maximal tree in the CW-complex $|T|$ will be chosen. Then $\pi_1(|T|)$ can be considered as the group with one generator corresponding to each 1-simplex not in the tree, and one relation corresponding to each 2-simplex.

As maximal tree take all 1-simplexes of the form $(s_0 g_0, (Ek_0))$. Then as generators of $\pi_1(|T|)$ we have all elements $(g_1, (Ek_0))$ such that g_1 is non-degenerate. The relation $\partial_1 x = (\partial_2 x) \cdot (\partial_0 x)$ where $x = (s_1 g_1, (s_0 E k_0))$ asserts that

$$\begin{aligned}(g_1, (Ek_0)) &= (g_1, (b_1)) \cdot (s_0 \partial_0 g_1, (Ek_0)) \\ &= (g_1, (b_1)) .\end{aligned}$$

From the 2-simplex $(s_0 g_1, (Ek_1))$ we obtain

$$(g_1, (E\partial_0 k_1)) = (s_0 \partial_1 g_1, (E\partial_1 k_1)). (g_1 k_1, (b_1))$$

$$= (g_1 k_1, (b_1)) .$$

Combining these two relations we have

$$(g_1, (b_1)) = (g_1 k_1, (b_1)) ,$$

from which it follows easily that

$$(g_1, (b_1)) = 1$$

for all g_1. In view of the first relation, this shows that $|T|$ is simply connected, and completes the proof of theorem 1.

The following theorem shows that FK is essentially unique.

Theorem 2. Any principal bundle over EK with any group complex G as fibre is induced from the above bundle by a homomorphism FK → G.

Proof. We may assume that this bundle is a twisted cartesian product with twisting function $\tau:(EK)_{n+1} \to G_n$. Define the homomorphism $\bar{\tau}:FK \to G$ by $\bar{\tau}(k_n) = \tau(Ek_n)$. Since $\bar{\tau}(b_n) = \tau(Eb_n) = \tau(s_0(b_n)) = 1_n$ this defines a homomorphism. It is easy to verify that $\bar{\tau}$ commutes with the face and degeneracy operations, and induces a map between the two twisted cartesian products.

Corollary. *If* G *is also a loop space for* EK *then there is a homomorphism* $FK \to G$ *inducing an isomorphism between the Pontrjagin rings.*

This follows easily using [7], IV Theorem B.

Analogues of theorems 1 and 2 for the construction $F^+(K)$ can be proved using exactly the same formulas. The following shows the relationship between $F^+(K)$ and the construction of James.

Lemma 4. *If* K *is countable then the realization* $|F^+K|$ *is homeomorphic to the reduced product of* $|K|$.

In fact the product $(k_n, k'_n, k''_n, \ldots) \to k_n . k' . k'' . \ldots$ maps $K x \ldots x K$ into F^+K. Taking realizations we obtain a map $|K| x \ldots x |K| \to |F^+K|$. From these maps it is easy to define a map of the reduced product of $|K|$ into $|F^+K|$, and to show that it is a homeomorphism.

3. A theorem of Hilton

If A, B are semi-simplicial complexes with base points a_0, b_0 let $A \vee B$ denote the subcomplex $A \times [b_0] \cup [a_0] \times B$ of $A \times B$. Let $A \ast B$ denote the complex obtained from $A \times B$ by collapsing $A \vee B$ to a point. The notation $A^{(k)}$ will be used for the k-fold 'collapsed product' $A \ast \ldots \ast A$.

The free product $G \ast H$ of two group complexes is defined by $(G \ast H)_n = G_n \ast H_n$. There is clearly a canonical isomorphism between the group complexes $F(A \vee B)$ and $FA \ast FB$.

Lemma 5. The complex $F(A \vee B)$ is isomorphic (ignoring group structure) to $FA \times F(B \vee (B \ast FA))$.

In fact we will show that $F(A \vee B)$ is a split extension:

$$I \to F(B \vee (B \ast FA)) \to F(A \vee B) \to FA \to I .$$

The collapsing map $A \vee B \xrightarrow{c} A$ induces a homomorphism c' of $F(A \vee B)$ onto FA. Denote the kernel of c' by F'. The inclusion $A \xrightarrow{i} A \vee B$ induces a homomorphism $i':FA \to F(A \vee B)$. Since $c'i'$ is the identity it follows that $F(A \vee B)$ is a split extension of F' by FA.

We will determine this kernel F'_n for some fixed dimension n. Let a, b, ϕ range over the n-simplexes other than the base point of A, B, and FA respectively. Then $F(A \vee B)_n$ is the free group $\{a, b\}$ and F'_n is the normal subgroup generated by the b. By the Reidemeister-Schreier theorem (see [8]) F'_n is freely generated by the elements $w(a)bw(a)^{-1}$ where $w(a)$ ranges over all elements of the free group $\{a\} = FA_n$. Thus

$$F'_n = \{b, \phi b \phi^{-1}\} .$$

Now setting $[b, \phi] = b\phi b^{-1}\phi^{-1}$ and making a simple Tietze transformation (see for example [1]) we obtain

$$F'_n = \{b, [b, \phi]\} .$$

Identify $[b, \phi]$ with the simplex $b \ast \phi$ of $B \ast F(A)$. Then we can identify F'_n with $F(B \vee (B \ast FA))$. Since this identification commutes with face and degeneracy operations, this proves Lemma 5.

Lemma 6. *The group complex* $F(B \divideontimes FA)$ *is isomorphic to*

$$F((B \divideontimes A) \vee (B \divideontimes A \divideontimes FA)) .$$

The inclusion $A \to FA$ induces a homomorphism

$$F(B \divideontimes A) \to F(B \divideontimes FA) .$$

A homomorphism

$$F(B \divideontimes A \divideontimes FA) \to F(B \divideontimes FA)$$

is defined by

$$b \divideontimes a \divideontimes \phi \to (b \divideontimes a)(b \divideontimes \phi a)^{-1} (b \divideontimes \phi) .$$

(This is motivated by the group identity $[[b, a], \phi] = [b, a] [b, \phi a]^{-1}[b, \phi]$.)

Combining these we obtain a homomorphism

$$F(B \divideontimes A) \ast \tau F(B \divideontimes A \divideontimes FA) \to F(B \divideontimes FA)$$

which is asserted to be an isomorphism.

Using the same notation as in Lemma 5 and identifying $b \divideontimes a \divideontimes \phi$ with $[[b, a], \phi]$ it is evidently sufficient to prove the following.

Lemma 7. *In the free group* $\{a, b\}$ *the subgroup freely generated by the elements* $[b, \phi]$ *is also freely generated by the elements* $[b, a]$ and $[[b, a], \phi]$.

The proof consists of a series of Tietze transformations. Details will not be given.

As a consequence of Lemma 6 we have:

Lemma 8. For each m the group complex $F(B \divideontimes FA)$ is isomorphic to

$$F(B \divideontimes A) \ast F(B \divideontimes A \divideontimes A) \ast \ldots \ast F(B \divideontimes A^{(m)}) \ast$$

$$F(B \divideontimes A^{(m)} \divideontimes FA) .$$

Proof by induction on m. For $m = 1$ this is just Lemma 6. Given this assertion for the integer $m - 1$ it is only necessary to show that $F(B \divideontimes A^{(m-1)} \divideontimes FA)$ is isomorphic to $F(B \divideontimes A^{(m)}) \ast F(B \divideontimes A^{(m)} \divideontimes FA)$. But this follows immediately from Lemma 6 by substituting $B \divideontimes A^{(m-1)}$ in place of B.

Theorem 3. If A and B are semi-simplicial complexes with A connected, then there is an inclusion homomorphism

$$F(\vee_{i=1}^{\infty} B \divideontimes A^{(i)}) \to F(B \divideontimes F(A))$$

which is a homotopy equivalence.

Proof. Every element of $F(\vee_{i=1}^{\infty} B \divideontimes A^{(i)})$ is already contained in

$$F(\vee_{i=1}^{\infty} B \divideontimes A^{(i)}) = F(B \divideontimes A) \ast \ldots \ast F(B \divideontimes A^{(m)})$$

for some m. Hence by Lemma 8 it may be identified with an element of $F(B \divideontimes FA)$. Since A is connected, the 'remainder term' $B \divideontimes A^{(m)} \divideontimes FA$ has trivial homology groups in dimensions less than m. From this it follows easily that the above inclusion induces isomorphisms of the homotopy groups in all dimensions.

Remark. The complex B may be eliminated from Theorem 3 by taking B as the sphere S^0, and noting the identity $S^0 \divideontimes K = K$.

Combining theorem 3 with Lemma 5 we obtain the following

Corollary. If A is connected then there is a homotopy equivalence

$$F(A) \times F(\vee_{i=0}^{\infty} B \divideontimes A^{(i)}) \subset F(A \vee B)\ .$$

This corollary will be the basis for the following.

Theorem 4. Let $A_1, \ldots, A_r$ be connected complexes. Then $F(A_1 \vee \ldots \vee A_r)$ has the same homotopy type as a weak infinite cartesian product $\Pi_{i=1}^{\infty} F(A_i)$ where each A_i, $i > r$, has the form

$$A_1^{(n_1)} \divideontimes \ldots \divideontimes A_r^{(n_r)}\ .$$

The number of factors of a given form is equal to the Witt number

$$\phi(n_1, \ldots, n_r) = \frac{1}{n} \sum_{d|\delta} \frac{\mu(d)(n/d)!}{(n_1/d)! \ldots (n_r/d)!}$$

where $n = n_1 + \dots + n_r$, $\delta = \mathrm{GCD}(n_1, \dots, n_r)$.

Proof. For $n = 1, 2, 3, \dots$ define complexes A_i, to be called 'basic products of weight n' as follows, by induction on n. The given complexes $A_1, \dots, A_r$ are the basic products of weight 1. Suppose that

$$A_1, \dots, A_r, \dots, A_\alpha$$

are the basic products of weight less than n. To each $i = 1, \dots, r, \dots, \alpha$ assume we have defined a number $e(i) < i$, where $e(1) = \dots = e(r) = 0$. Then as basic products of weight n take all expressions $A_i \divideontimes A_j$ where weight A_i + weight $A_j = n$ and $e(i) \le j < i$. Call these new complexes $A_{\alpha+1}, \dots, A_\beta$ in any order. If $A_h = A_i \divideontimes A_j$ define $e(h) = j$. (For this discussion we must consider complexes such as $(A \divideontimes B) \divideontimes C$ and $A \divideontimes (B \divideontimes C)$ to be distinct!) This completes the construction of the A_i.

For each $m \ge 1$ define

$$R_m = F(\vee_{\substack{h \ge m \\ e(h) < m}} A_h) .$$

Thus $R_1 = F(A_1 \vee \dots \vee A_r)$.

Lemma 9. *There is a homotopy equivalence*

$$F(A_m) \times R_{m+1} \subset R_m .$$

Note that $R_m = F(A_m \vee B)$, where $B = \vee_{\substack{h > m \\ e(h) < m}} A_h$.

By the corollary to theorem 3 there is a homotopy equivalence

$$(F(A_m) \times F(\vee_{i=0}^{\infty} B \divideontimes A_m^{(i)}) \subset F(A_m \vee B) = R_m .$$

Substituting in the definition of B and using the distributive law

$$(A \vee B) \divideontimes C = (A \divideontimes C) \vee (B \divideontimes C) ,$$

the second factor of the first expression becomes

$$F(\vee_{i=0}^{\infty} \vee_{\substack{h > m \\ e(h) < m}} A_h \divideontimes A_m^{(i)}) .$$

But (filling in parentheses correctly) this is just

$$F(\vee_{\substack{h > m \\ e(h) \leq m}} A_h) = R_{m+1} ,$$

which proves Lemma 9.

Now it follows by induction that there is a homotopy equivalence

$$F(A_1) \times F(A_2) \times \ldots \times F(A_m) \times R_{m+1} \subset R_1 =$$

$$F(A_1 \vee \ldots \vee A_r) .$$

This defines an inclusion of the weak infinite cartesian product

$\Pi_{i=1}^{\infty} F(A_i)$ into R_1. Since $A_1, \ldots, A_r$ are connected, it follows easily that the 'remainder terms' R_m are k-connected where $k \to \infty$ as $m \to \infty$. From this it follows that the above inclusion map induces isomorphisms of the homotopy groups in all dimensions. This proves the first part of theorem 4.

Let $\phi(n_1, \ldots, n_r)$ denote the number of A_h having the form $A_1^{(n_1)} * \ldots * A_r^{(n_r)}$. To compute these numbers consider the free Lie ring L on generators $\alpha_1, \ldots, \alpha_r$. Corresponding to each 'basic product' $A_h = A_i * A_j$ define an element $\alpha_h = [\alpha_i, \alpha_j]$ of L, for $h = r+1, r+2, \ldots$. Then the elements α_h obtained in this way are exactly the standard monomials of M. Hall [2] and P. Hall [3]. M. Hall has proved that these elements form an additive basis for L.

The number of linearly independent elements of L which involve each of the generators $\alpha_1, \ldots, \alpha_r$ a given number $n_1, \ldots, n_r$ of times has been computed by Witt [9]. Since his formula is the same as that in theorem 4, this completes the proof.

In conclusion we mention one more interesting consequence of theorem 3.

Theorem 5. If A is connected then the complex EFA has the same homotopy type as $\vee_{i=1}^{\infty} EA^{(i)}$.

The proof is based on the following lemma, which depends on Theorem 1.

Lemma 10. If A is connected, there is a homotopy equivalence

$$EA \subset \overline{W}FA .$$

In fact the inclusion is defined by $(s_0^i Ea_n) \to s_0^i(a_n, 1_{n-1}, \ldots, 1_0)$. It is easily verified that this is a map, and that it induces a map of the twisted cartesian product T into the twisted cartesian product W. Since both total spaces are acyclic, it follows from [7], IV Theorem A that the homology groups of EA map isomorphically into those of $\overline{W}FA$. Since both spaces are simply connected, this completes the proof of Lemma 10.

Now from Theorem 3 we have a homotopy equivalence

$$\overline{W}F(\vee_{i=1}^{\infty} A^{(i)}) \subset \overline{W}FFA .$$

In view of Lemma 10, and the identity

$$E(A \vee B) = EA \vee EB ,$$

this completes the proof.

References

1. R. H. Fox, Discrete groups and their presentations (lecture notes) Princeton (1955).
2. M. Hall, 'A basis for free Lie rings and higher commutators in free groups', Proc. A. M. S. 1 (1950) 575-581.
3. P. Hall, 'A contribution to the theory of groups of prime power order', Proc. London Math. Soc. 36 (1934) 29-95.
4. P. J. Hilton, 'On the homotopy groups of the union of spheres', Jour. London Math. Soc. 30 (1955) 154-172.
5. I. M. James, 'Reduced product spaces', Ann. of Math. 62 (1955) 170-197.

6. J. Milnor, The geometric realization of a semi-simplicial complex (mimeographed) Princeton (1955).

7. J. Moore, Algebraic homotopy theory (lecture notes) (1955-56).

8. K. Reidemesiter, Eifuhrung in die kombinatorische Topologie, Braunschweig (1932).

9. E. Witt, 'Treue Darstellung Liescher Ringe', J. Reine Angew. Math. 177 (1937) 152-160.

11

The next extract, from one of my own papers, merely proposes a notation for the spaces which arise in the method of killing homotopy groups (see §10).

11
ON CHERN CHARACTERS AND THE STRUCTURE OF THE UNITARY GROUP

J. F. Adams

Killing homotopy groups. We will briefly recall some ideas from homotopy theory, in order to fix the notation needed below.

Let X, Y be connected CW-complexes. We call Y a *space of type* $X(1, \ldots, n)$ if:

(1) There is a map f: $X \to Y$ such that f_*: $\pi_r(X) \to \pi_r(Y)$ is an isomorphism for $1 \leqslant r \leqslant n$, and

(2) $\pi_r(Y) = 0$ for $r > n$.

For each X and n, there is at least one such space Y. Moreover, any such Y has the universal property indicated by the following diagram.

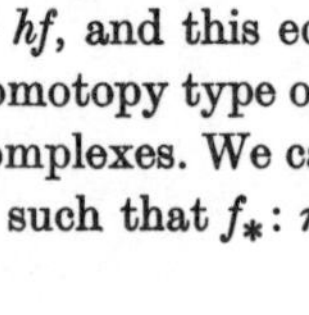

In full, let Y' be another space such that $\pi_r(Y') = 0$ for $r > n$; then any map g: $X \to Y'$ can be factored in the form $g \sim hf$, and this equation determines h: $Y \to Y'$ up to homotopy. It follows that the homotopy type of Y is determined by that of X.

Let W, X be connected CW-complexes. We call W a *space of type* $X(m, \ldots, \infty)$ if:

(1) There is a map f: $W \to X$ such that f_*: $\pi_r(W) \to \pi_r(X)$ is an isomorphism for $r \geqslant m$, and

(2) $\pi_r(W) = 0$ for $1 \leqslant r < m$.

This notion is justified in a manner similar to the preceding one.

Now suppose that W, X, Y, Z are connected CW-complexes, and that Y is of type $X(1, \ldots, n)$, W is of type $X(m, \ldots, \infty)$, where $m \leqslant n$. We call Z a space of type $X(m, \ldots, n)$ if it is

(1) a space of type $W(1, \ldots, n)$, or

(2) a space of type $Y(m, \ldots, \infty)$.

These two conditions are equivalent, since each leads to a diagram of the following form.

$$\begin{array}{ccc} X & \leftarrow & W \\ \downarrow & & \downarrow \\ Y & \leftarrow & Z \end{array}$$

For clarity, we display the types of the four spaces involved in the diagram above:

$$\begin{array}{ccc} X(1, \ldots, \infty) & \leftarrow & X(m, \ldots, \infty) \\ \downarrow & & \downarrow \\ X(1, \ldots, n) & \leftarrow & X(m, \ldots, n) \end{array}$$

These notions have been presented for CW-complexes, but by applying them to singular complexes we obtain corresponding singular notions for arbitrary connected spaces. In particular, given X and given $r \leqslant s < t$, we can construct a Serre fibring $F \to E \to B$ so that F, E and B are spaces of singular type $X(s+1, \ldots, t)$, $X(r, \ldots, t)$ and $X(r, \ldots, s)$.

We will allow ourselves to use the symbol $X(m, \ldots, n)$ to denote some space, not specified, of type $X(m, \ldots, n)$.

12

The next paper, by Cartan and Serre, are the first announcement of the results of the French school on the method of killing homotopy groups (see §10). The end of the second note gives the flavour of practical calculations; the desire to be able to make calculations is important for motivation in this area. The prerequisites are a knowledge of elementary homotopy theory and of homology theory up to spectral sequences (see §§1, 4, 5 of the introduction).

TOPOLOGIE. — *Espaces fibrés et groupes d'homotopie*. I. *Constructions générales*. Note de MM. **Henri Cartan** et **Jean-Pierre Serre**, présentée par M. Jacques Hadamard.

Construction d'espaces fibrés (1) permettant de « tuer » le groupe d'homotopie $\pi_n(X)$ d'un espace X dont les $\pi_i(X)$ sont nuls pour $i < n$. Cette méthode généralise celle qui consiste, pour $n = 1$, lorsque X est connexe, à « tuer » le groupe fondamental $\pi_1(X)$ en passant au revêtement universel de X.

1. Soient X un espace connexe par arcs, $x \in X$, $\mathcal{S}(X)$ le complexe singulier de X. Pour tout entier $q \geq 1$, soit $\mathcal{S}(X; x, q)$ le sous-complexe engendré par les simplexes dont les $(q-1)$-faces sont en x. Les groupes d'homologie (resp. cohomologie) de $\mathcal{S}(X; x, q)$ à coefficients dans G sont les *groupes d'Eilenberg* (2) de l'espace X en x; on les notera $H_i(X; x, q, G)$, resp. $H^i(X; x, q, G)$. Ils forment des systèmes locaux. Rappelons (2) que $\pi_q(X; x) \approx H_q(X; x, q, Z)$ pour $q \geq 2$.

Définition. — Un espace Y, muni d'une application continue f de Y dans X,

(1) L'expression « espace fibré » est prise dans le sens général défini par Serre (*Ann. of Math.*, 54, 1951, p. 425-505). Ce Mémoire sera désigné par [S].

(2) *Ann. of Math.*, 45, 1944, p. 407-447; *voir* § 32.

tue les groupes d'homotopie $\pi_i(X)$ pour $i \leqq n (n \geqq 1)$ si $\pi_i(Y) = 0$ pour $i \leqq n$ et si f définit un isomorphisme de $\pi_i(Y)$ sur $\pi_i(X)$ pour $i > n$.

THÉORÈME 1. — *Si un espace* Y *tue les* $\pi_i(X)$ *pour* $i \leqq n$, *les groupes d'homologie* $H_j(Y)$ *sont isomorphes aux groupes d'Eilenberg* $H_j(X; x, n+1)$; *de même pour la cohomologie.*

Cela résulte du :

LEMME 1. — *Si une application* f *d'un* Y *dans un* X *applique* $y \in Y$ *en* $x \in X$ *et définit, pour tout* $i > n$, *un isomorphisme* $\pi_i(Y; y) \approx \pi_i(X; x)$, *l'homomorphisme* $\mathcal{S}(Y; y, n+1) \to \mathcal{S}(X; x, n+1)$ *défini par* f *est une chaîne-équivalence. (En considérant le « mapping cylinder » de* f, *on se ramène au cas où* Y *est plongé dans* X; *le lemme s'obtient alors par un procédé standard de déformation.)*

Le théorème 1 justifie la notation $(X, n+1)$ pour n'importe quel espace qui tue les $\pi_i(X)$ pour $i \leqq n$.

2. THÉORÈME 2. — *A tout* X *connexe par arcs, on peut associer une suite d'espaces* (X, n) [où $n = 1, 2, \ldots$ et $(X, 1) = X$] *et d'applications continues* f_n : $(X, n+1) \to (X, n)$, *de manière que* $(X, n+1)$ *tue les* $\pi_i(X, n)$ *pour* $i \leqq n$, *et que :*

(I) *l'application* f_n *munisse* $(X, n+1)$ *d'une structure d'espace fibré* (1) *de base* (X, n), *ayant pour fibre un espace* $\mathcal{K}[\pi_n(X), n-1]$(3);

(II) *il existe un espace* X'_n *de même type d'homotopie que* (X, n), *et une fibration de* X'_n, *de fibre* $(X, n+1)$, *ayant pour base un* $\mathcal{K}[\pi_n(X), n]$.

Il suffira de dire comment $(X, n+1)$, f_n et X'_n se construisent à partir de (X, n). On utilise d'abord deux lemmes, déjà employés par certains auteurs (4) :

LEMME 2. — *Étant donné un espace connexe* A *et un entier* $k \geqq 1$, *on peut plonger* A *dans un espace* U *de manière que* $\pi_i(A) \to \pi_i(U)$ *soit un isomorphisme (sur) pour* $i < k$, *et* $\pi_k(U) = 0$. [*Pour tout* $\alpha \in \pi_k(A)$ *on choisit un représentant* $g_\alpha : S_\alpha \to A$, *où* S_α *est une sphère de dimension* k, *frontière d'une boule* E_α *de dimension* $k+1$; *on « attache » à* A *les boules* E_α *au moyen des applications* g_α].

LEMME 3. — *Étant donné un espace* A *tel que* $\pi_i(A) = 0$ *pour* $i < n$, *on peut plonger* A *dans un espace* V *de manière que* $\pi_n(A) \to \pi_n(V)$ *soit un isomorphisme sur, et* $\pi_i(V) = 0$ *pour* $i \neq n$. (*Se déduit du lemme* 2 *par itération, en prenant l'espace-réunion.*)

Constructions. — Étant donné une application continue φ d'un espace A dans un $V = \mathcal{K}(\pi, n)$, soit A′ l'espace des couples (a, ω) où $a \in A$ et ω est un

(3) Rappelons (*cf.* EILENBERG-MACLANE, *Ann. of Math.*, 46, 1945, p. 480-509, § 17; *ibid.*, 51, 1950, p. 514-533) que si un espace V satisfait à $\pi_i(V) = 0$ pour tout $i \neq n$, $\pi_n(V) = \pi$, le complexe $\mathcal{S}(V)$ a même type d'homotopie qu'un complexe $K(\pi, n)$ explicité par ces auteurs, et qui dépend seulement de n et du groupe π (abélien si $n \geqq 2$). D'un tel espace V, nous dirons que c'est un espace $\mathcal{K}(\pi, n)$; ses groupes d'homologie $H_i(\pi; n)$ (resp. de cohomologie) sont les *groupes d'Eilenberg-MacLane* du groupe π, pour l'entier n.

(4) *Voir*, par exemple, J. H. C. WHITEHEAD, *Ann. of Math.*, 50, 1949, p. 261-263.

chemin [5] de V d'extrémité $\varphi(a)$; A′ se rétracte sur A, identifié à l'espace des couples (a, ω) tels que ω soit ponctuel en $\varphi(a)$. L'application g qui, à (a, ω), associe l'origine de ω, définit A′ comme espace fibré de base V. Soit B la fibre au-dessus de $\varphi(a_0)(a_0$, point fixé de A); l'application f qui, à (a, ω), associe a, définit B comme espace fibré de base A, de fibre l'espace W des lacets sur V, qui est un $\mathcal{K}(\pi, n-1)$.

Appliquons ces constructions à l'espace $A = (X, n)$ supposé déjà obtenu, au groupe $\pi = \pi_n(X)$ et à l'injection φ de A dans V (lemme 3). La suite exacte d'homotopie des espaces fibrés montre que B tue les $\pi_i(A)$ pour $i \leq n$; on peut donc prendre $(X, n+1) = B$, $f_n = f$, $X'_n = A'$, et le théorème 2 est démontré.

3. *Utilisation.* — Chacune des fibrations (I) et (II) définit (pour chaque n) une suite spectrale [6]. Dans la mesure où l'on connaît les groupes d'Eilenberg-MacLane d'un groupe π donné, on obtient une méthode de calcul (partiel) des groupes d'Eilenberg de X, et notamment des groupes d'homotopie de X.

La méthode utilisée par Hirsch [7] pour étudier $\pi_3(X)$ quand $\pi_1(X) = 0$ et que $\pi_2(X)$ est libre de base finie, rentre dans notre méthode générale; elle revient à prendre au-dessus de X une fibre $\mathcal{K}(\pi_2, 1)$ qui est ici un produit de cercles.

En vue des applications, la remarque suivante est utile : l'espace $W = \mathcal{K}[\pi_n(X), n-1]$ opère à gauche dans $B = (X, n+1)$, et par suite chaque $\alpha \in H_i[\pi_n(X), n-1]$ définit un endomorphisme λ_α de la suite spectrale d'homologie de la fibration (I); on démontre que λ_α commute avec toutes les différentielles de cette suite spectrale.

(5) Pour tout ce qui concerne les espaces de chemins, *voir* [S], Chap. IV.

(6) Il s'agit de la suite spectrale en homologie (resp. cohomologie) singulière; *voir* [S], Chap. I et II.

(7) *Comptes rendus*, **228**, 1949, p. 1920.

TOPOLOGIE. — *Espaces fibrés et groupes d'homotopie*. II. *Applications*. Note de MM. **Henri Cartan** et **Jean-Pierre Serre**, présentée par M. Jacques Hadamard.

Applications de la méthode générale exposée dans une Note précédente [1]. On retrouve la plupart des relations connues entre homologie et homotopie; les résultats nouveaux concernent notamment les groupes d'homotopie des groupes de Lie et des sphères.

Dans toute la suite X désignera un espace *connexe par arcs*.

Considérons la fibration (II) de la Note [1], pour $n \geq 2$; en lui appliquant la Proposition 5 du Chapitre III de [S], on obtient :

Proposition 1. — *Pour tout espace* X *et tout* $n \geq 2$ [2], *on a une suite exacte :*

$$(1)\left\{\begin{array}{l} H_{2n}(X, n+1) \to H_{2n}(X, n) \to H_{2n}(\pi_n(X); n) \to H_{2n-1}(X, n+1) \to H_{2n-1}(X, n) \to \ldots \\ \ldots \to H_{n+2}(X, n+1) \to H_{n+2}(X, n) \to H_{n+2}(\pi_n(X); n) \to \pi_{n+1}(X) \to H_{n+1}(X, n) \to 0. \end{array}\right.$$

Compte tenu de ce que $H_{n+2}(\pi; n) = \pi/2\pi\ (n \geq 2)$ et $H_{n+3}(\pi; n) = {}_2\pi\ (n \geq 3)$, on retrouve des résultats de G. W. Whitehead [3].

Corollaire 1. — *Les groupes d'homologie relatifs* $H_i[\mathcal{S}(X; x, n), \mathcal{S}(X; x, n+1)]$ (*où* x *est un point de* X) *sont isomorphes aux groupes d'Eilenberg-MacLane* $H_i(\pi_n(X); n)$ *pour* $1 \leq i \leq 2n$.

Ce résultat semble en rapport étroit avec une suite spectrale annoncée récemment par W. Massey et G. W. Whitehead (lorsque X est une sphère) [4].

Corollaire 2. — *Si* $\pi_i(X) = 0$ *pour* $i < n$ *et* $H_j(X) = 0$ *pour* $n < j \leq 2n$ (*en particulier si* X *est une sphère* $\mathbf{S}_n$), *on a des isomorphismes :*

$$H_j(X, n+1) \approx H_{j+1}(\pi_n(X); n) \quad \textit{pour} \quad n \leq j \leq 2n-1 \quad (n \geq 2).$$

On notera que, si $j < 2n-1$, les groupes $H_{j+1}(\pi; n)$ sont « stables » et isomorphes aux groupes $A_{j-n+2}(\pi)$ introduits par Eilenberg-Mac Lane [5], ce qui fournit une interprétation géométrique de ces derniers groupes.

Proposition 2. — *Si* $\pi_i(X) = 0$ *pour* $i < n$ *et* $n < i < m$ (*n et m étant deux entiers tels que* $0 < n < m$), *on a une suite exacte* :

$$H_{m+1}(X) \to H_{m+1}(\pi_n(X); n) \to \pi_m(X) \to H_m(X) \to H_m(\pi_n(X); n) \to 0.$$

Ceci se démontre au moyen de la fibration (II) et complète des résultats

(1) *Comptes rendus*, **234**, 1952, p. 288. Nous renvoyons à cette Note dont nous conservons la terminologie et les notations.

(2) Le cas $n = 1$ est spécial et n'apporte d'ailleurs rien de nouveau.

(3) *Proc. Nat. Acad. Sc. USA*, **34**, 1948, p. 207-211.

(4) *Bull. Amer. Math. Soc.*, **57**, 1951, Abstracts 544 et 545.

(5) *Proc. Nat. Acad. Sc. USA*, **36**, 1950, p. 657-663.

d'Eilenberg-MacLane ([6]) (à l'exception, toutefois, de ceux relatifs à l'invariant $\mathbf{k}$).

Proposition 3. — *Supposons que* $\pi_1(X) = 0$, *que les nombres de Betti de* X *soient finis en toute dimension et que l'algèbre de cohomologie* $H^*(X, Q)$ (Q *désignant le corps des rationnels*) *soit le produit tensoriel d'une algèbre extérieure engendrée par des éléments de degrés impairs et d'une algèbre de polynomes engendrée par des éléments de degrés pairs; si* d_n *désigne le nombre des générateurs de degré* n, *on a*

$$\text{rang } ({}^7) \text{ de } \pi_n(X) = d_n \qquad \text{pour tout } n.$$

On utilise la fibration (I), et le calcul des algèbres de cohomologie d'Eilenberg-MacLane à coefficients dans Q; on montre par récurrence sur n que $H^*(X; n, Q)$ est l'algèbre quotient de $H^*(X, Q)$ par l'idéal engendré par les générateurs de degrés $< n$.

Remarques. — **1.** La démonstration montre aussi que le noyau de l'homomorphisme $\pi_n(X) \to H_n(X)$ est un groupe de torsion.

2. La proposition subsiste même si $\pi_1(X) \neq 0$, pourvu que $\pi_1(X)$ soit abélien et opère trivialement dans $H^*(X; 2, Q)$.

3. La proposition 3 s'applique notamment : *a.* à une sphère de dimension impaire; *b.* à un espace de lacets sur un espace simplement connexe dont les nombres de Betti sont finis; *c.* à un groupe de Lie. En particulier, *les groupes d'homotopie d'un groupe de Lie sont finis en toute dimension où il n'y a pas d'élément « primitif »* (donc en toute dimension *paire*).

Proposition 4. — *Soit* X *tel que* $\pi_1(X) = 0$, *et* q *un entier. Si* $H_i(X)$ *est un groupe de torsion pour* $1 < i < q$, *il en est de même du noyau et du conoyau* ([8]) *de l'homomorphisme* $\varphi_j : H_j(X, q) \to H_j(X)$ *pour tout* j. *Si en outre la composante p-primaire* (*p premier*) *de* $H_i(X)$ *est nulle pour* $1 < i < q$, *il en est de même du noyau et du conoyau de* φ_j. *Ceci vaut notamment pour* $\varphi_q : \pi_q(X) \to H_q(X)$.

Proposition 5. — *Les groupes d'homologie de la sphère* $\mathbf{S}_3$ *dont on a tué le troisième groupe d'homotopie sont les suivants :*

$$H_i(\mathbf{S}_3, 4) = 0 \quad \text{pour } i \text{ impair} \qquad \text{et} \qquad H_{2q}(\mathbf{S}_3, 4) = Z/qZ$$

(Les premiers groupes d'homologie sont donc : Z, 0, 0, 0, Z_2, 0, Z_3, 0, Z_4, ...).

Corollaire. — *La composante p-primaire de* $\pi_{2p}(\mathbf{S}_3)$ *est* Z_p ([9]).

La proposition 5 permet de retrouver aisément les résultats connus sur les $\pi_i(\mathbf{S}_3)$, $i = 4, 5, 6$: pour $i = 4$, c'est évident; appliquant la suite (1)

([6]) *Ann. of Math.*, 51, 1950, p. 514-533.

([7]) Le *rang* d'un groupe G est la dimension du Q-espace vectoriel $Q \otimes G$.

([8]) Le *conoyau* d'un homomorphisme $A \to B$ est le quotient de B par l'image de A.

([9]) Notre méthode montre également que l'homomorphisme $f_p : \pi_{2p}(\mathbf{S}_3) \to Z_p$ introduit par N. E. Steenrod est *sur*.

pour $n = 4$, et utilisant le fait que $H_7(Z_2; 4) = Z_2$, on obtient $\pi_5(\mathbf{S}_3) = Z_2$ et $H_6(\mathbf{S}_3, 5) = Z_6$; en appliquant la suite (1) pour $n = 5$ on obtient une suite exacte : $\pi_5(\mathbf{S}_3) \to \pi_6(\mathbf{S}_3) \to Z_6 \to 0$, qui montre que $\pi_6(\mathbf{S}_3)$ a 6 ou 12 éléments ([4]).

Proposition 6. — *Les groupes* $\pi_7(\mathbf{S}_3)$ *et* $\pi_8(\mathbf{S}_3)$ *sont des groupes 2-primaires;* $\pi_9(\mathbf{S}_3)$ *est somme directe de* Z_3 *et d'un groupe 2-primaire.*

On utilise le fait que $H_i(Z_3; 5) = 0$ pour $i = 7$, 8, et $H_9(Z_3; 5) = Z_3$ ([5]).

Enfin, si l'on admet les résultats sur les groupes d'Eilenberg-MacLane obtenus par H. Cartan au moyen de calculs dont le fondement théorique n'a pas encore reçu de justification complète, on obtient les résultats suivants (que nous donnons donc comme *conjecturaux*) : *pour n impair* ≥ 3, *et p premier, la composante p-primaire de* $\pi_i(\mathbf{S}_n)$ *est* Z_p *si* $i = n + 2p - 3$, *nulle si* $n + 2p - 3 < i < n + 4p - 6$; *celle de* $\pi_{4p-3}(\mathbf{S}_3)$ *est* Z_p, *de même (si* $p \neq 2$*) que celle de* $\pi_{4p-2}(\mathbf{S}_3)$. *Par exemple,* $\pi_{10}(\mathbf{S}_3)$ *est somme directe de* Z_{15} *et d'un groupe 2-primaire.*

13

The next piece is a summary on generalised homology and cohomology written by me especially for the present work.

13

GENERALISED HOMOLOGY AND COHOMOLOGY THEORIES

A summary by J. F. Adams

It is generally accepted that a functor can be called an ordinary homology or cohomology theory if it satisfies the seven axioms of Eilenberg and Steenrod (see Paper no. 2). These axioms serve to describe the behaviour of the functor on finite complexes. If we want to describe the behaviour of the functor on infinite complexes, we add suitable axioms about limits (see Paper no. 15).

A generalised homology or cohomology theory is a functor which satisfies all of the axioms of Eilenberg and Steenrod except for the dimension axiom. In addition, we may impose suitable axioms about limits. In recent years several such functors have been found useful in algebraic topology. The most important are K-theory and various sorts of bordism and cobordism theory (see Papers no. 19, 20, 21, 23, 24). Various forms of stable homotopy and cohomotopy also satisfy the axioms, but are hard to calculate.

Obviously the beginning of the subject involves one in setting up various such examples and proving that they satisfy the axioms. This overlaps with topics to be considered below.

The next part of the subject, which is rather formal, consists in exploiting the consequences of the six axioms of Eilenberg and Steenrod which one does assume. For example, one has the Mayer-Vietoris sequence, and one has the spectral sequence of Atiyah and Hirzebruch (see §12 and Papers no. 14, 19). If one wishes to study also infinite complexes, one exploits the consequences of whatever axiom one has on limits (see §12 and Papers no. 15, 16, 17).

The next part of the subject is the relation between generalised homology and cohomology theories and homotopy theory. We will tackle this now, and then we will go on to the final part of the subject, which is to introduce into generalised homology and cohomology all the apparatus with which we are familiar in ordinary homology and cohomology: various products, cohomology operations, universal coefficient theorems ... .

We may start from the observation that for the ordinary cohomology groups of a CW-complex X we have

$$H^n(X; \Pi) \cong [X, K(\Pi, n)] ,$$

where the symbol $K(\Pi, n)$ means an Eilenberg-MacLane space of type (Π, n) (see §7). This isomorphism is natural for maps of X. However, the given structure of a cohomology theory includes not only groups H^n and induced homomorphisms f^*, but also coboundary homomorphisms δ, or equivalently, suspension isomorphisms

$$\sigma : H^n(X; \Pi) \xrightarrow{\cong} H^{n+1}(SX; \Pi) ,$$

where SX denotes the suspension of X. In terms of Eilenberg-MacLane spaces, this isomorphism is induced by a canonical homotopy equivalence

$$K(\Pi, n) \xrightarrow{\cong} \Omega K(\Pi, n+1) .$$

(Here ΩY means the space of loops in Y, as usual.)

If we begin from this example, then, the obvious generalisation is to consider a sequence of spaces E_n provided with homotopy equivalences $E_n \to \Omega E_{n+1}$. Such a sequence is called an Ω-spectrum. For example, the Eilenberg-MacLane spectrum $K(\Pi)$ is that for which $E_n = K(\Pi, n)$. This notion is sufficiently general to handle the case of K-theory, provided we agree that E_n may be equivalent to one pathwise-component of ΩE_{n+1}. For example, we define the spectrum $\underline{BU}$ by taking E_{2n} to be the space BU and E_{2n+1} to be the space U. The equivalence $U \to \Omega BU$ is the usual one; and the Bott Periodicity Theorem provides an equivalence $\Omega U \to Z \times BU$.

However, the notion of an Ω-spectrum is too restrictive to be convenient for other applications, such as the theory of cobordism (see Paper no. 23). We therefore define a spectrum to be a sequence of CW-complexes E_n (with base-point) and maps $e_n\colon SE_n \to E_{n+1}$. Such a map e_n determines an 'adjoint' map $e'_n\colon E_n \to \Omega E_{n+1}$, but e'_n need not be a homotopy equivalence. For example, we define the spectrum $\underline{MU}$ so that its term E_{2n} is the Thom complex MU_n (see Paper no. 23). We have a map $S^2 MU_n \to MU_{n+1}$, but its adjoint is not a homotopy equivalence.

Again, if X is any CW-complex, we can construct a corresponding spectrum $\underline{X}$ so that

$$\underline{X}_n = \begin{cases} S^n X & (n \geq 0) \\ P & (n < 0) \end{cases} .$$

(Here P means a point.) Similarly, if m is any integer (positive, negative or zero), we can define a spectrum $\underline{S}^m$ by

$$(\underline{S}^m)_n = \begin{cases} S^{n+m} & \text{for } n+m \geq 0 \\ P & \text{for } n+m < 0 . \end{cases}$$

We interpret this spectrum as the 'sphere of stable dimension m'. In both these examples e_n is an equivalence (at least for sufficiently large n), but its adjoint e'_n is not.

Given a spectrum $\underline{E}$, we can construct from it a generalised homology theory and a generalised cohomology theory. On this subject, the definitive paper is G. W. Whitehead, 'Generalised homology theories', Trans. Amer. Math. Soc. 102 (1962), 227-283. This paper is highly recommended, but it is rather long to reprint here. We take first the construction of homology groups. For convenience of notation, we will use the same letter for the homology functor as for the spectrum E which determines it. If X is a CW-complex with base-point, we define its (reduced) E-homology groups by

$$\tilde{E}_n(X) = \operatorname*{DirLim}_{m \to \infty} \pi_{n+m}(E_m \wedge X) .$$

Here the 'smash' product $W \wedge X$ is defined to be $W \times X / W \vee X$, as usual. The notation $\operatorname*{DirLim}_{m \to \infty}$ implies that we are given a direct system of groups and homomorphisms; the homomorphisms are the obvious ones:

$$\pi_{n+m}(E_m \wedge X) \xrightarrow{S} \pi_{1+n+m}(SE_m \wedge X) \xrightarrow{(e_m \wedge 1)_*} \pi_{1+n+m}(E_{m+1} \wedge X) .$$

The relative homology groups are defined by

$$E_n(X, Y) = \tilde{E}_n(X/Y) .$$

Here X/Y is interpreted to mean X with Y identified to a new point, which becomes the base-point. This interpretation covers the case $Y = \emptyset$.

The definitions of the induced homomorphisms f_* and the suspension isomorphism σ or the boundary map ∂ are fairly obvious, and I omit them here.

It can be shown without too much trouble that the functor thus defined satisfies all the required axioms, including an axiom on behaviour under limits. It can also be shown that any functor which satisfies these axioms can be obtained in this way from a suitable spectrum E. In G. W. Whitehead's original paper (loc. cit.) this is proved under the additional assumption that the coefficient groups of the functor are countable; but this restriction can be avoided by using more recent work.

We turn now to the construction of cohomology groups. If X is a finite CW-complex with base-point, we define its (reduced) E-cohomology groups by

$$\tilde{E}^n(X) = \operatorname*{DirLim}_{m \to \infty} [S^{m-n}X, E_m] .$$

It might appear tempting to use the same definition also for infinite CW-complexes, but it can be shown that this would lead to the wrong behaviour with respect to limits. Instead, it seems best to construct a suitable category whose objects are spectra. We can then define the E-cohomology of a spectrum X by

$$E^n(X) = [S^{-n}X, E] .$$

(The suspension or desuspension of a spectrum X can be constructed, for example, by re-indexing the terms X_n, so that

$(SX)_n = X_{n+1}$.) The cohomology of a spectrum is analogous to the reduced cohomology of a complex. If we want the (reduced) E-cohomology of a CW-complex X, we take the E-cohomology of the corresponding spectrum $\underline{X}$, where $\underline{X}_n = S^n X$ for $n \geq 0$, as above.

The construction of the category of spectra takes a little care to get right; we will follow unpublished work of Boardman. We restrict attention to spectra such that e_n is a cellular homeomorphism between SE_n and a subcomplex of E_{n+1}. (There is no essential loss of generality in this.) A <u>subspectrum</u> $E' \subset E$ is defined in the obvious way: it is a set of subcomplexes $E'_n \subset E_n$ such that e_n maps SE'_n into E'_{n+1}. A subspectrum E' is <u>cofinal</u> if for any cell e_α in any E_n there exists an m such that $S^m e_\alpha \subset E'_{m+n}$. In other words, each cell of E gets into E' in the end, but we don't say when. A function $g : E \to F$ between spectra is a sequence of cellular maps $g_n : E_n \to F_n$ such that the following diagram is (strictly) commutative for each n.

$$\begin{array}{ccc} SE_n & \xrightarrow{e_n} & E_{n+1} \\ \Big\downarrow{\scriptstyle Sg_n} & & \Big\downarrow{\scriptstyle g_{n+1}} \\ SF_n & \xrightarrow{f_n} & F_{n+1} \end{array}$$

However, the notion of a function is rather restrictive; so given E and F, we consider functions $g : E' \to F$, where E' varies over the cofinal subspectra of E. In other words, even if a cell e_α exists in E_n, we consider functions which are allowed to come into existence on $S^m e_\alpha$. The slogan is 'Cells now - maps later'.

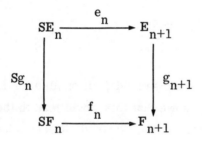

Of course, the functions $g : E' \to F$ have to be collected into equivalence classes. For this purpose we need to define homotopy, so we need cylinders. Recall that $I/\emptyset$ is the unit interval with a disjoint base-point added. If E is a spectrum, we make the obvious definition of the spectrum $E \wedge (I/\emptyset)$; its spaces are the spaces $E_n \wedge (I/\emptyset)$, and its maps are the maps

$$SE_n \wedge (I/\emptyset) \xrightarrow{e_n \wedge 1} E_{n+1} \wedge (I/\emptyset) .$$

Take two cofinal subspectra E', E'' in E. We say that two functions $g' : E' \to F$, $g'' : E'' \to F$ are <u>homotopic</u> if there is a cofinal subspectrum E''' contained in E' and E'', and a function

$$h : E''' \wedge (I/\emptyset) \to F ,$$

so that the restrictions of h to the two ends of the cylinder are the restrictions of g' and g'' to E'''. Homotopy is an equivalence relation, and the set of equivalence classes is written $[E, F]$.

With these definitions, the functor $[E, F]$ of E can be used to define a generalised cohomology theory, as required. If we apply this theory to a finite CW-complex X, it agrees with that given by the previous definition.

We turn now to the study of products in generalised homology and cohomology theories. In order to construct such products, G. W. Whitehead introduced the notion of a 'pairing of spectra'. A pairing of spectra from E and F to G, in the sense of G. W. Whitehead, consists of maps

$$\mu_{n,m} : E_n \wedge F_m \to G_{n+m}$$

which satisfy suitable axioms with respect to the maps e_n, f_m and g_q. For example, the maps

$$MU_n \wedge MU_m \to MU_{n+m}$$

(see Paper no. 23) define a pairing from MU and MU to MU.

However, G. W. Whitehead did not say (because it is not true) that every product in generalised cohomology theory arises from a pairing in this sense. For this reason, and for clarity, it seems best to explain matters by supposing that we can introduce a smash-product into our category of spectra, so that if E and F are spectra, then $E \wedge F$ is a spectrum in good standing, and has the properties you would expect. If so, we can replace a 'pairing of spectra' by a morphism from $E \wedge F$ to G. Unfortunately, no complete treatment of the necessary details has yet appeared in print. Until it does, the reader should treat the account which follows as a heuristic explanation which provides the easiest way of seeing what goes on, but which may need demythologising by substituting details from G. W. Whitehead or from some other author.

The definitions of the generalised homology and cohomology of a spectrum X now read as follows.

$$E^n(X) = [S^{-n} \wedge X, E]$$

$$E_n(X) = [S^n, E \wedge X] .$$

(We do not actually need smash products of spectra to extend the definition of E-homology from complexes to spectra; but we record these definitions here because these are the ones we shall use in discussing products.) Until further notice, every object in sight is a spectrum.

We will introduce the following four products.

$$E^p(X) \otimes F^q(Y) \to (E \wedge F)^{p+q}(X \wedge Y)$$

$$E_p(X) \otimes F_q(Y) \to (E \wedge F)_{p+q}(X \wedge Y)$$

$$E^p(X \wedge Y) \otimes F_q(Y) \to (E \wedge F)^{p-q}(X)$$

$$E^p(X) \otimes F_q(X \wedge Y) \to (E \wedge F)_{q-p}(Y)$$

The notation we use is

$$x^p \otimes y^q \to x^p \bar{\wedge} y^q$$

$$x_p \otimes y_q \to x_p \underline{\wedge} y_q$$

$$w^p \otimes y_q \to w^p / y_q$$

$$x^p \otimes w_q \to x^p \backslash w_q \, ,$$

for the four products in order. A few remarks on the notation may help to explain the conventions used here. The notation $\bar{\wedge}$, $\underline{\wedge}$ for the 'external smash products' incorporates a bar above or below to show whether we are dealing with contravariant functors or covariant ones. The two slant products may be thought of as analogous to fractions; this is particularly helpful when we come to the associativity laws. The notation is chosen so that a 'fraction' has the same variance as its numerator and the opposite variance to its denominator; the dimension of a 'fraction' is that of its numerator minus that of its denominator. The cohomology variable appears on the left and the homology variable on the right (as in the classical Kronecker product, when one considers cochains as functions defined on chains). The numerator always lies in a homology or cohomology group of $X \wedge Y$; the denominator lies in a cohomology group of X if it acts on the left of $X \wedge Y$, in a homology group of Y if it acts on the right of $X \wedge Y$. Finally, the left-hand variable in our products always lies in E-homology or E-cohomology, the right-hand variable in F, and the result in $E \wedge F$.

Of course, if we are given a map of spectra

$$\mu : E \wedge F \to G$$

(which is our analogue of a pairing of spectra) then we can apply it to a product lying in (for example) $(E \wedge F)^{p+q}(X \wedge Y)$ and obtain a result lying in $G^{p+q}(X \wedge Y)$. Similarly for the other three cases.

The definitions of the products are forced: they go as follows.

(i) Suppose given

$$S^{-p} \wedge X \xrightarrow{x} E, \quad S^{-q} \wedge Y \xrightarrow{y} F.$$

Form

$$S^{-p} \wedge S^{-q} \wedge X \wedge Y \xrightarrow{1 \wedge \tau \wedge 1} S^{-p} \wedge X \wedge S^{-q} \wedge Y \xrightarrow{x \wedge y} E \wedge F.$$

(Here, and in what follows, $\tau : U \wedge V \to V \wedge U$ is the usual map which interchanges the two factors.)

(13)

(ii) Suppose given

$$S^p \xrightarrow{x} E\wedge X, \qquad S^q \xrightarrow{y} F\wedge Y .$$

Form

$$S^p \wedge S^q \xrightarrow{x\wedge y} E\wedge X\wedge F\wedge Y \xrightarrow{1\wedge\tau\wedge 1} E\wedge F\wedge X\wedge Y .$$

(iii) Suppose given

$$S^{-p}\wedge X\wedge Y \xrightarrow{w} E , \qquad S^q \xrightarrow{y} F\wedge Y .$$

Form

$$\begin{array}{ccccc} S^{-p}\wedge S^q\wedge X & \xrightarrow{1\wedge y\wedge 1} & S^{-p}\wedge F\wedge Y\wedge X & & \\ & & \downarrow{\scriptstyle\rho} & & \\ & & S^{-p}\wedge X\wedge Y\wedge F & \xrightarrow{w\wedge 1} & E\wedge F . \end{array}$$

(Here ρ is the obvious permutation map.)

(iv) Suppose given

$$S^{-p}\wedge X \xrightarrow{x} E, \qquad S^q \xrightarrow{w} F\wedge X\wedge Y .$$

Form

$$\begin{array}{ccccc} S^{-p}\wedge S^q & \xrightarrow{1\wedge w} & S^{-p}\wedge F\wedge X\wedge Y & & \\ & & \downarrow{\scriptstyle 1\wedge\tau\wedge 1} & & \\ & & S^{-p}\wedge X\wedge F\wedge Y & \xrightarrow{x\wedge 1\wedge 1} & E\wedge F\wedge Y . \end{array}$$

(13)

Note that in (iii) and (iv) the order of composition agrees with the order of the variables in the product.

These products are natural for maps of E and F, in a sense which the student should make precise. The external smash products are also natural for maps of X and Y.

The behaviour of the slant products for maps of X and Y is given by the following formulae. We suppose given maps

$$\xi : X \to X', \quad \eta : Y \to Y'.$$

(i) Suppose given $w' \in E^p(X' \wedge Y')$, $y \in F_q(Y)$. Then

$$((\xi \wedge \eta)^* w') / y = \xi^*(w'/(\eta_* y)).$$

(ii) Suppose given $x' \in E^p(X')$, $w \in F_q(X \wedge Y)$. Then

$$x' \backslash ((\xi \wedge \eta)_* w) = \eta_*((\xi^* x') \backslash w).$$

The external smash products $\overline{\wedge}$ and $\underline{\wedge}$ satisfy anti-commutative laws which are fairly obvious. We can also formulate eight associativity statements, as follows.

(i) If $x \in E^p(X)$, $y \in F^q(Y)$ and $z \in G^r(Z)$ then

$$(x \overline{\wedge} y) \overline{\wedge} z = x \overline{\wedge} (y \overline{\wedge} z).$$

(ii) If $x \in E_p(X)$, $y \in F_q(Y)$ and $z \in G_r(Z)$ then

$$(x \underline{\wedge} y) \underline{\wedge} z = x \underline{\wedge} (y \underline{\wedge} z).$$

(iii) If $x \in E^p(X)$, $u \in F^q(Y \wedge Z)$ and $z \in G_r(Z)$ then

$$x \overline{\wedge} (u/z) = (x \overline{\wedge} u) / z .$$

(iv) If $x \in E^p(X)$, $w \in F_q(X \wedge Y)$ and $z \in G_r(Z)$ then

$$(x \backslash w) \underline{\wedge} z = x \backslash (w \underline{\wedge} z) .$$

(v) If $t \in E^p(X \wedge Y \wedge Z)$, $z \in F_q(Z)$ and $y \in G_r(Y)$ then

$$(t/z) / y = t / (\tau_*(z \underline{\wedge} y)) .$$

(Note that $z \underline{\wedge} y$ lies in $(F \wedge G)_{q+r}(Z \wedge Y)$ and $\tau_*(z \underline{\wedge} y)$ lies in $(F \wedge G)_{q+r}(Y \wedge Z)$.)

(iv) If $y \in E^p(Y)$, $x \in F^q(X)$ and $t \in G_r(X \wedge Y \wedge Z)$ then

$$y \backslash (x \backslash t) = (\tau^*(y \overline{\wedge} x)) \backslash t .$$

(Note that $y \overline{\wedge} x$ lies in $(E \wedge F)^{p+q}(Y \wedge X)$ and $\tau^*(y \overline{\wedge} x)$ lies in $(E \wedge F)^{p+q}(X \wedge Y)$.)

(vii) If $v \in E^p(X \wedge Z)$, $y \in F^q(Y)$ and $u \in G_r(Y \wedge Z)$ then

$$v/(y \backslash u) = [(1 \wedge \tau)^*(v \overline{\wedge} y)] / u .$$

(Note that $v \overline{\wedge} y$ lies in $(E \wedge F)^{p+q}(X \wedge Z \wedge Y)$ and $(1 \wedge \tau)^*(v \overline{\wedge} y)$ lies in $(E \wedge F)^{p+q}(X \wedge Y \wedge Z)$.)

(viii) If $w \in E^p(X \wedge Y)$, $y \in F_q(Y)$ and $v \in G_r(X \wedge Z)$ then

$$(w/y)\backslash v = w \backslash [(\tau \wedge 1)_*(y \underline{\wedge} v)] .$$

(Note that $y \underline{\wedge} v$ lies in $(F \wedge G)_{q+r}(Y \wedge X \wedge Z)$ and $(\tau \wedge 1)_*(y \underline{\wedge} v)$ lies in $(F \wedge G)_{q+r}(X \wedge Y \wedge Z)$.)

We note that the laws (iii) to (viii) are perfectly acceptable as rules for manipulating fractions, at least if we ignore the 'switch maps' τ necessary to keep X, Y and Z in the right order. For example, we remember (iii) as stating that

$$x(u/z) = (xu)/z .$$

The associativity laws can all be proved by diagram-chasing, provided that the smash-products in our category have the properties one expects.

One can also consider the case in which we incorporate pairings into the definition of our products; suppose we are given pairings

$$\alpha : E \wedge F \longrightarrow H$$
$$\beta : F \wedge G \longrightarrow K$$
$$\gamma : H \wedge G \longrightarrow L$$
$$\delta : E \wedge K \longrightarrow L .$$

Then the associativity laws need as part of their data the commutativity of the following diagram.

$$(13) \qquad \begin{array}{ccc} E\wedge F\wedge G & \xrightarrow{\alpha\wedge 1} & H\wedge G \\ \downarrow{\scriptstyle 1\wedge\beta} & & \downarrow{\scriptstyle \gamma} \\ E\wedge K & \xrightarrow{\delta} & L \end{array}$$

(The commutativity is supposed to hold in our category, that is, 'up to homotopy'.) With this data, the associativity laws in this form are easily deduced from those given above, by naturality.

Next we recall that in the case of CW-complexes the 0-sphere acts as a unit for the smash-product; so one of the properties we expect of the smash-product in our category of spectra is that the spectrum S^0 should act as a unit for the smash-product. Assuming this, we can identify $E^{-n}(S^0)$ with $E_n(S^0)$; each group is the coefficient group $\pi_n(E) = [S^n, E]$. With this identification, we have the following results.

(i) If $x \in E^p(X)$, $t \in \pi_q(F)$ then

$$x \overline{\wedge} t = x/t .$$

If $t = 1: S^0 \to S^0$ then

$$x \overline{\wedge} 1 = x/1 = x .$$

(ii) If $s \in \pi_p(E)$, $y \in F_q(Y)$ then

$$s \underline{\wedge} y = s \backslash y .$$

If $s = 1: S^0 \to S^0$ then

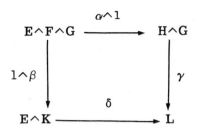

$1 \wedge y = 1 \backslash y = y$.

(iii) If $x \in E^p(X)$, $y \in F_q(X)$ then

$x/y = x \backslash y$ in $\pi_{q-p}(E \wedge F)$.

The third part, of course, gives us the definition of the Kronecker product for generalised theories.

Any other identities which are needed are deduced by combining the results which we have listed above. In particular, since the suspension isomorphisms can be defined by taking products with suitable classes on S^1, the properties of each product with respect to the suspension isomorphisms follow from the associative laws.

The products which incorporate a 'pairing' $E \wedge F \to G$ are particularly convenient when we can take $E = F = G$. We say that E is a ring-spectrum if we are given morphisms

$$\mu : E \wedge E \to E$$

$$i : S^0 \to E$$

which play the part of a product map and a unit map, and make the obvious diagrams commutative. We then say that F is a module-spectrum over E if we are given a morphism

$$\nu : E \wedge F \to F$$

which makes the obvious diagrams commutative.

So far we have been dealing with products in the homology and cohomology groups of spectra. It is now time to return to complexes. When we consider CW-complexes instead of spectra, and relative groups instead of reduced absolute groups, we use the following equivalence:

$$(X/A) \wedge (Y/B) \longrightarrow (X \times Y) \;/\; (A \times Y \cup X \times B) .$$

So, for example, we derive from the external smash-product (for spectra) an external cross product in the relative cohomology of pairs of complexes; this is a map

$$E^p(X, A) \otimes F^q(Y, B) \longrightarrow (E \wedge F)^{p+q}(X \times Y, \; A \times Y \cup X \times B) .$$

Or, of course, if we are using a pairing $E \wedge F \to G$ the result lies in $G^{p+q}(X \times Y, \; A \times Y \cup X \times B)$. It seems appropriate to write this product $\overline{\times}$ instead of $\overline{\wedge}$. Similarly for the other three products.

These products inherit from their predecessors properties of anticommutativity, associativity, and units; but the spectrum S^0 is now replaced by the point P, since $P/\phi = S^0$.

When we are dealing with relative groups, we have formulae for the behaviour of our products under the boundary or coboundary map; these are deduced from the analogous formulae giving the behaviour of our previous products under suspension isomorphisms. The proofs involve some tedious diagram-chasing. The results are strictly comparable to those results for ordinary homology and cohomology which are proved using the ordinary formula for the coboundary of the product of two cochains, or more generally, using the fact that the two Eilenberg-Zilber equivalences are chain maps. Compare Spanier 5. 3. 15 (p. 235), 5. 6. 6 (p. 250), 6. 1. 3 (p. 287).

One can now introduce internal products (cup and cap products) by using the diagonal map $\Delta: X \to X \times X$, as usual. More precisely, let X be a CW-complex with subcomplexes A and B;

then Δ maps $A \cup B$ into $A \times X \cup X \times B$, so we can define the cup-product

$$E^p(X, A) \otimes F^q(X, B) \longrightarrow (E \wedge F)^{p+q}(X, A \cup B)$$

by

$$x \cup y = \Delta^*(x \overline{\times} y) .$$

Similarly, we define the cap-product

$$E^p(X, A) \otimes F_q(X, A \cup B) \to (E \wedge F)_{q-p}(X, B)$$

by

$$x \cap y = x \backslash (\Delta_* y) .$$

Although we have so far considered our generalised homology and cohomology functors as defined on CW-pairs, we can extend the definition to more general spaces. For example, let X be any space; we may define the 'singular E-cohomology of X' to be the E-cohomology of any CW-complex weakly equivalent to X (such as the geometrical realisation of the total singular complex of X). (A weak equivalence is a map inducing isomorphisms of homotopy groups.) Similarly for homology groups, or relative groups of pairs.

Again, let K be a compact subset of a manifold M; then we can define the 'Cech E-cohomology of K' by

$$\check{E}^n(K) = \operatorname{InvLim} E^n(U)$$

where U runs over open subsets of M which contain K. (The cohomology group $E^n(U)$ is 'singular'.) Similarly for compact pairs K, L.

At this point the student will find that we have completed enough formal machinery to carry over almost word-for-word the account of duality in manifolds given in Spanier, Chapter 6 Section 2. (Preferably one avoids Spanier's reference to 4. 7. 13 by remarking that in the main proof we only need 6. 2. 12 for a simplex linearly embedded in R^n.)

It is not necessary to say very much about cohomology operations in E-cohomology. The study of stable cohomology operations in E-cohomology comes down to computing $E^*(E)$. This has been done for the spectrum $E = \underline{MU}$ (see Paper no. 23). Useful unstable operations are known for the case of K-theory, $E = \underline{BU}$ (see Paper no. 21).

For the universal coefficient theorem, and for some further reading, see my 'Lectures on Generalised Cohomology', Springer-Verlag, Lecture Notes in Mathematics No. 99, especially chapters 1 and 3.

14

The next piece, by Dold, offers a good exposition of the subject. It assumes a fair familiarity with the usual machinery of algebraic topology.

14

Matematisk Institut, Aarhus Universitet

RELATIONS BETWEEN ORDINARY AND EXTRAORDINARY HOMOLOGY

Albrecht Dold

Colloquium on Algebraic Topology, 1-10 August, 1962

1. Homology theories

Let **W** be the category whose objects are pairs (X, A) of finite CW-complexes (or of the same homotopy type) and whose morphisms are homotopy classes of maps. A (extraordinary) homology theory h on **W** with values in **A** (= abelian category) is a sequence of covariant functors $h_q : \mathbf{W} \to \mathbf{A}$, $q \in \underline{\underline{Z}}$, together with natural transformations $\partial = \partial_q : h_q(X, A) \to h_{q-1}(A)$ (connecting morphism) such that

(EC) $h(\text{inclusion}) : h(B, B \cap A) \cong h(A \cup B, A)$

for subcomplexes A, B ⊂ X, and such that the sequence

(LS) $h_{q+1}X \to h_{q+1}(X, A) \to h_q A \to h_q X \to h_q(X, A)$

is exact.

One then easily proves $h(X, A) \cong h(X/A, *)$ where X/A = X with A shrunk to a point *. The graded object $\check{h} = h$ (point) is called the coefficient object.

Examples. (i) If h is a homology theory and $t: \mathbf{A} \to \mathbf{A}^1$ an exact covariant functor, then $t \circ h$ is a homology theory.

(ii) (cf. Eilenberg-Steenrod) For every $G \in \mathbf{A}$ there is a unique h (up to equivalence) such that $\check{h}_q = 0$ for $q \neq 0$, and $\check{h}_0 = G$; for this h one writes $h_q(X, A) = H_q(X, A; G)$.

(iii) Given h we get a new homology theory $s^n h$ (= n-th suspension of h) by shifting the indices by n, i. e. $(s^n h)_q = h_{q-n}$.

(iv) A direct sum of homology theories is again a homology theory. In particular, for every graded $G = \{G_n\}$ we put $H_q(X, A; G) = \oplus_n H_{q-n}(X, A; G_n)$. This is the 'ordinary homology with coefficients G'.

(v) Given h and a fixed pair $(L, M) \in \mathbf{W}$ one defines a new homology theory $h \times (L, M)$ by

$$[h \times (L, M)]_q(X, A) = h_q(X \times L, X \times M \cup A \times L) ;$$

the new connecting morphism is the composite

$$h_q(X \times L, X \times M \cup A \times L) \to h_{q-1}(X \times M \cup A \times L) \to$$
$$h_{q-1}(X \times M \cup A \times L, X \times M) \overset{(EC)}{\cong} (A \times L, A \times M) ;$$

the new coefficients are $h \times (L, M)^{\cup} = h(L, M)$.

If L is a Moore space $L(\pi, n)$, and $M = *$ a point one writes $[h \times (L, M)]_{q+n}(X, A) = h_q(X, A; \pi)$, and one has an exact <u>universal coefficient sequence</u>

(14)

$$\text{(UC)}\quad 0 \to h_q(X, A) \otimes \pi \to h_q(X, A; \pi) \to \operatorname{Tor}(h_{q-1}(X, A), \pi) \to 0.$$

A special case of (UC) is the suspension isomorphism

$$(\sigma)\quad \sigma: h_q(X, A) \cong h_{q+1}(X \times S^1, X \times * \cup A \times S^1) \cong$$

$$h_{q+1}(S(X/A), *) ;$$

where S^1 = circle, $*$ = base point, S = suspension, $A \neq \emptyset$. One establishes (σ) in the usual way and uses it to prove (UC).

It follows from (σ) that h factors through the quotient category $\overline{\mathbf{W}}$ of $\mathbf{W}$ whose morphisms are stable homotopy classes of maps.

(vi) If $(L, *)$ is a topological space with base point, one defines a group valued homology theory Π^L, and a cohomology theory Π_L (= homology theory with values in the dual category) by

$$(\Pi^L)\quad \Pi^L_q(X, A) = \lim_{N \to \infty} \pi[S^{N+q}L, S^N(X/A)] ,$$

$$(\Pi_L)\quad \Pi^q_L(X, A) = \lim_{N \to \infty} \pi[S^N(X/A), S^{N+q}L]$$

(= stable homotopy classes of maps with base point). This generalises to spectra (see G. W. Whitehead).

Since every homology theory h factors through the stable category $\overline{\mathbf{W}}$ we can, for $(L, M) \in \mathbf{W}$, define a natural transformation

$$T: \Pi^{L/M}_q(X, A) \otimes h_n(L, M) \to h_{q+n}(X, A)$$

which for fixed $f \in \Pi_q^{L/M}(X, A)$ reduces to

$$T_f : h_n(L, M) = h_{N+q+n} S^{N+q}(L/M) \xrightarrow{h(f)} h_{N+q+n}(S^N(X/A)) =$$

$$h_{q+n}(X, A) .$$

In particular, if $(L, M) = (S^0, *)$ this becomes

$$\text{(T)} \qquad \pi_*^S(X, A) \otimes \check{h} \to h(X, A) ,$$

where π_i^S = i-th stable homotopy group. We shall see that T is an isomorphism if h is a vector space over $\underline{\underline{Q}}$ (= rationals); note already that in that case the left side of (T) is a homology theory by example (i), and T is a transformation of homology theories. --A dual construction and result holds for π_S^* and cohomology--.

(vii) The theory of vector bundles gives rise to the cohomology theories K_U^* and K_0^* .

2. h-fibrations and their spectral sequence

Let h be a homology theory and $\pi : E \to B$ a map in **W**. Define an exact couple (A, C) (cf. Massey) by

$$A_{pq} = h_{p+q}(\pi^{-1}B^{(p)}), \text{ and } C_{pq} = h_{p+q}(\pi^{-1}B^{(p)}, \pi^{-1}B^{(p-1)}) ,$$

where (p) denotes the p-skeleton, and where all pairs are assumed to lie in **W**. The corresponding spectral sequence converges to

$E^{\infty}_{p,n-p} = (F_p h_n E)/(F_{p-1} h_n E)$ where $F_p hE = \mathrm{im}[h(\pi^{-1}B^{(p)}) \to hE]$. The couple (A, C) (and hence the spectral sequence) are natural with respect to the variables π and h.

Assume now π is an <u>h-fibration,</u> i. e. is such that for every simplex $\Delta \subset B$ and every vertex $z \in \Delta$ the inclusion induces an isomorphism $h(\pi^{-1}v) \cong h(\pi^{-1}\Delta)$. (If one wants to avoid triangulability of B, one uses a corresponding definition for maps $E' \to \Delta$ induced from π by singular simplices $\Delta \to B$.) Then one finds for the E^1- and E^2-terms of the above spectral sequence

<u>Theorem 1</u> <u>(cf. Atiyah,</u> 2. 6). $E^1_{*q} \cong \underset{\sim}{C_*}(B, \tilde{h}_q F) =$ <u>cellular chain complex of</u> B <u>with coefficients</u> $\tilde{h}_q$, <u>hence</u>

$$E^2_{pq} \cong H_p(B, \tilde{h}_q F),$$

<u>where</u> $\tilde{h}_q$ <u>is the local coefficient system given by</u> $\{h_q(\pi^{-1}v)\}$ $v \in B^{(o)}$.

Remark. If h is also defined for infinite CW-complexes and commutes with direct sums, then the theorem is valid for all finite dimensional B. If h also commutes with direct limits, then B can be an arbitrary CW-complex. If, moreover, weak homotopy equivalence induce h-isomorphisms, then B can be an arbitrary space.

The geometric realization of a Kan-fibration is always an h-fibration, because then $\pi^{-1}v \quad \pi^{-1}\Delta$. Similarly, Serre-fibrations are h-fibrations.

3. The case $\pi = \mathrm{id}: B \to B$

In this case the fibre F is a point, the system $\tilde{h}$ is constant and equals $\check{h}$, hence a spectral sequence $H_p(B, \check{h}_q)$ $h_n B$ which is natural in (B, h).

Corollary 1. *If* $f: B \to B'$ *is a continuous map such that* $f_*: H(B, \check{h}) \cong H(B', \check{h})$ *(in particular, if integral homology is mapped isomorphically) then* $f_*: h(B) \cong h(B')$. *In particular,* $H(\ , \mathbb{Z})$-*fibrations are* h-*fibrations for all* h.

Corollary 2. *If* $\tau: h \to h'$ *is a natural transformation of homology theories such that* $\check{\tau}: \check{h} \cong \check{h}'$, *then* τ *is an equivalence.*

Corollary 3. *If* h *is a homology theory such that* $\check{h}$ *is a* $\mathbb{Q}$-*vectorspace, then*

$$T: \pi_*^S(X, A) \otimes \check{h} \cong h(X, A) \; ;$$

dually

$$\pi_S^*(X, A) \otimes \check{h}^* \cong h^*(X, A) \text{ for cohomology.}$$

It is enough to prove this for $(X, A) = (S^0, *)$. But then $\pi_q^S(X, A)$ is the stable homotopy of the sphere, hence finite except for $q = 0$ (and $\cong \mathbb{Z}$ there) after Serre, hence $\pi_*^S(S^0, *) \otimes \check{h} = \pi_0^S(S^0, *) \otimes \check{h} = \check{h} = h(S^0, *)$.

Corollary 4. *For every homology theory* h,

$$h(X, A) \otimes \mathbb{Q} \cong H(X, A; \check{h} \otimes \mathbb{Q}) \, ,$$

since both sides equal $\pi_*^S(X, A) \otimes \check{h} \otimes \underline{\underline{Q}}$ by Cor. 3. In particular, the differentials d^r of the spectral sequence become zero after tensoring with $\underline{\underline{Q}}$, $d^r \otimes \underline{\underline{Q}} = 0$.

Essentially, the same argument shows

Theorem 2. Let h, h' be a homology theories, $\phi: \check{h} \to \check{h}'$ a morphism (not necessarily degree-preserving or homogeneous) and assume $\check{h}'$ is a vector-space over $\underline{\underline{Q}}$. Then there exists a unique natural transformation $\Phi: h \to h'$ which extends ϕ and commutes with the connecting morphism (equivalently: commutes with the suspension isomorphism). If $\phi(h_i) \subset h'_i$, then Φ is a transformation of homology theories.

E. g. there is a unique $\Phi: K_U^* \to H^*(\ , \underline{\underline{Q}})$ which takes the generator of K_U^{2i} (point) $\cong \underline{\underline{Z}}$ into 1. It is called the Chern character, ch.

4. Multiplicative cohomology theories

A (group-valued) cohomology theory h is multiplicative if it is equipped with a natural map (a multiplication

$$h^i(X, A) \times h^j(Y, B) \xrightarrow{\times} h^{i+j}(X \times Y, X \times B \cup A \times Y),$$

all i, j

which is bilinear, associative, commutative (in the graded sense), and has a unit $1 \in h^0(S^0, *)$. Moreover, the following diagram has to be commutative ('comparibility with the connecting morphism')

(14)

$$
\begin{array}{ccc}
h^i(A) \times h^j(Y,\ B) & \xrightarrow{\quad \times \quad} & h^{i+j}(A \times Y,\ A \times B) \\
 & & \Big\| \ (EC) \\
\Big\downarrow \partial \times \mathrm{id} & & h^{i+j}(X \times B \cup A \times Y,\ X \times B) \\
 & & \Big\downarrow \partial \\
h^{i+1}(X,\ A) \times h^j(Y,\ B) & \xrightarrow{\quad \times \quad} & h^{i+j+1}(X \times Y,\ X \times B \cup A \times Y)
\end{array}
$$
.

Via the diagonal map $X \to X \times X$ the exterior product $\times$ becomes an interior product. In particular, $h(X,\ A)$ is a graded ring (with unit if $A = \emptyset$), and $\partial : h(A) \to h(X,\ A)$ is an $h(X)$-module homomorphism. The coefficients $\check{h}$ form a graded commutative ring with unit, and all maps of the theory are $\check{h}$-module homomorphism.

The axioms for a multiplication can, of course, also be formulated in terms of the interior product. Also, the conditions can be weakened in some places. Examples of multiplicative theories are: Ordinary cohomology H^* with coefficients in a ring, stable cohomotopy groups π_S^*, and K^*-theory.

Some of the implications of a multiplicative structure are as follows.

Proposition 1 (cf. Steenrod). *Let $\gamma^n \in h^n(S^n,\ *)$ be the element corresponding to $1 \in \check{h}^0$ under the n-fold suspension isomorphism $h^n(S^n,\ *) \cong h^0(S^0,\ *) \cong \check{h}^0$. Then*

$$\gamma^1 \times : h^i(X,\ A) \to h^{i+1}(S^1 \times X,\ S^1 \times A \cup * \times X) =$$

$$h^{i+1}(S(X/A).\ *)$$

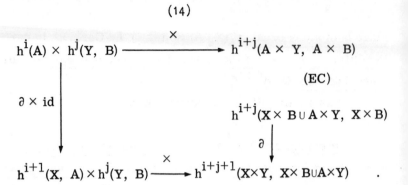

agrees with the suspension isomorphism σ, in particular, $\gamma^n = \gamma^1 \times \gamma^1 \times \ldots \times \gamma^1$.

Proposition 2. Let h, h' be multiplicative cohomology theories, h' a $\underline{\underline{Q}}$-module, $\phi:\check{h} \to \check{h}'$ a homomorphism, $\Phi:h \to h'$ its unique extension (see theorem 2). If ϕ is multiplicative (i. e. a homomorphism of rings with unit) then Φ is multiplicative.

This applies, for instance, to the Chern character. Proposition 1 allows, by the way, to replace the ∂-compatibility-condition for Φ (see theorem 2) by the condition '$\Phi(\gamma^1) = \gamma'^1$' (plus multiplicativity).

Proposition 3. Let $\pi:E \to B$ be an h-fibration, and assume there are elements $a_1, a_2, \ldots, a_r \in hE$ whose restriction to the fibre F form a base of the $\check{h}$-module hF. Then hE is a free hB-module (via $h\pi$) with base $a_1, a_2, \ldots, a_r$.

For the proof one simply notes that one has an E_1-isomorphism of exact couples

$$C \oplus C \oplus \ldots \oplus C \to C',$$

$$(z_1, z_2, \ldots, z_r) \to \sum \pi^*(z_i)\cdot a_i,$$

where C' resp. C is the exact couple of π resp. id_B.

This applies to $h = K^*_U$ if $\pi:E \to B$ is the projective bundle associated with a $Gl(n, \underline{\underline{C}})$-bundle, or if $E = B \times F$, and H(F) is free. Other example: Let $\pi:E \to B$ be an n-plane bundle (or micro-bundle), and let E^0 be the complement of the central section. Call the bundle h-orientable (compare G. W. Whitehead) if there exists an element $u \in h^n(E, E^0)$ whose restriction to the fibre is $\gamma^n \in h^n(F, F^0) = h^n(\underline{\underline{R}}^n, \underline{\underline{R}}^n - 0) = h^n(S^n, *)$. Then, applying Prop. 3 to the relative h-fibration (E, E^0) one gets the

Thom-isomorphism; $\Psi: h^i(B) \overset{\cong}{\to} h^{i+n}(E, E^0)$,

$\Psi(z) = u \cdot \pi^*(z)$.

In particular, there results for h-orientable n-plane bundles an exact Gysin-sequence

$$h^{i+n-1}E^0 \to h^i B \xrightarrow{h(j)(u)\cdot} h^{i+n}B \xrightarrow{h(\pi^0)} h^{i+n}E^0 \to h^{i+1}B .$$

($j: B \to E$ the central section).

For $h = H^*(\ , \underline{\underline{Z}})$ orientability has the usual meaning; for $h = K^*_U$ it is more restricted and holds for U(1)-Spin-bundles (i. e. if $w_1 = 0$, $\beta w_2 = 0$; Atiyah-Hirzebruch). π^*_s-orientability implies h-orientability for all h and is equivalent to being stably fibre homotopically trivial (compare G. W. Whitehead).

Theorem 3. If h is a multiplicative cohomology theory, then the spectral sequence of an h-fibration $\pi: E \to B$ is multiplicative, i. e. every E_r is a graded commutative ring, the differential d_r is a derivation, E_{r+1} is the homology ring of E_r, the multiplication in E_2 is the ordinary cup-product, and the one in E_∞ is induced from $h(E)$.

For K-theory Atiyah-Hirzebruch had already established the multiplicative properties of E_r, and they conjectured that the d_r are derivations. I have been informed by Hirzebruch that, since, Atiyah has also proved this conjecture.

Concluding, I wish to thank E. Dyer for many stimulating conversations.

REFERENCES

1. M. F. Atiyah, 'Characters and cohomology of finite groups', Publ. Mathem. de l'IHES 9, Paris (1961).
2. M. F. Atiyah and F. Hirzebruch, 'Vector bundles and homogeneous spaces', Proc. Symp. Pure Math. (Amer. Math. Soc.) 3 (1961), 7-38.
3. S. Eilenberg and N. Steenrod, 'Foundations of algebraic topology', Princeton Math. Series 15 (1952).
4. W. S. Massey, 'Exact couples in algebraic topology', Ann. of Math. 56 (1952), 363-396.
5. J.-P. Serre, 'Homologie singuliere des espaces fibres', Ann. of Math. 54 (1951), 425-505.
6. N. Steenrod, 'Cyclic reduced powers of cohomology classes', Proc. Nat. Acad. Sci. USA 39 (1953), 219-223.
7. G. W. Whitehead, 'Generalized homology theories', Trans. Amer. Math. Soc. 102 (1962), 227-283.

15, 16 & 17

The next paper, by Milnor, is the foundation paper on the cohomology of infinite complexes. It is fairly self-contained; the only prerequisite is some familiarity with axiomatic homology theory. To study the cohomology of infinite complexes, Milnor introduces an algebraic gadget, written Lim^1, the derived functor of the inverse-limit functor. For the applications, one needs to know conditions under which Lim^1 is zero; these are provided by the next two extracts. The extract by Atiyah introduces the Mittag-Leffler condition. It is implicit, but not explicit, in this work that the Mittag-Leffler condition is sufficient to ensure the vanishing of Lim^1; this is therefore an exercise for the reader. The extract by Anderson relates all this to the spectral sequence mentioned in §12.

15 ON AXIOMATIC HOMOLOGY THEORY

J. MILNOR

A homology theory will be called *additive* if the homology group of any topological sum of spaces is equal to the direct sum of the homology groups of the individual spaces.

To be more precise let H_* be a homology theory which satisfies the seven axioms of Eilenberg and Steenrod [1]. Let $\mathscr{A}$ be the admissible category on which H_* is defined. Then we require the following.

Additivity Axiom. If X is the disjoint union of open subsets X_α with inclusion maps $i_\alpha: X_\alpha \to X$, all belonging to the category $\mathscr{A}$, then the homomorphisms

$$i_{\alpha *}: H_n(X_\alpha) \to H_n(X)$$

must provide an injective representation of $H_n(X)$ as a direct sum.[1]

Similarly a cohomology theory H^* will be called *additive* if the homomorphisms

$$i_\alpha^*: H^n(X) \to H^n(X_\alpha)$$

provide a projective representation of $H^n(X)$ as a direct product.

It is easily verified that the singular homology and cohomology theories are additive. Also the Čech theories based on infinite coverings are additive. On the other hand James and Whitehead [4] have given examples of homology theories which are not additive.

Let $\mathscr{W}$ denote the category consisting of all pairs (X, A) such that both X and A have the homotopy type of a CW-complex; and all continuous maps between such pairs. (Compare [5].) The main object of this note is to show that there is essentially only one additive homology theory and one additive cohomology theory, with given coefficient group, on the category $\mathscr{W}$.

First consider a sequence $K_1 \subset K_2 \subset K_3 \subset \cdots$ of CW-complexes with union K. Each K_i should be a subcomplex of K. Let H_* be an additive homology theory on the category $\mathscr{W}$.

LEMMA 1. *The homology group $H_q(K)$ is canonically isomorphic to the direct limit of the sequence*

$$H_q(K_1) \to H_q(K_2) \to H_q(K_3) \to \cdots .$$

Received February 6, 1961.

[1] This axiom has force only if there are infinitely many X_α. (Compare pg. 33 of Eilenberg-Steenrod.) The corresponding assertion for pairs (X_α, A_α) can easily be proved, making use of the given axiom, together with the "five lemma."

The corresponding lemma for cohomology is not so easy to state. It is first necessary to define the "first derived functor" of the inverse limit functor.[2] The following construction was communicated to the author by Steenrod.

Let $A_1 \xleftarrow{p} A_2 \xleftarrow{p} A_3 \longleftarrow \cdots$ be an inverse sequence of abelian groups, briefly denoted by $\{A_i\}$. Let Π denote the direct product of the groups A_i, and define $d: \Pi \to \Pi$ by

$$d(a_1, a_2, \cdots) = (a_1 - pa_2, a_2 - pa_3, a_3 - pa_4, \cdots) .$$

The kernel of d is called the inverse limit of the sequence $\{A_i\}$ and will be denoted by $\mathfrak{L}\{A_i\}$.

DEFINITION. The cokernel $\Pi/d\Pi$ of d will be denoted by $\mathfrak{L}'\{A_i\}$; and $\mathfrak{L}'$ will be called the derived functor of $\mathfrak{L}$.

Now let $K_1 \subset K_2 \subset \cdots$ be CW-complexes with union K, and let H^* be an additive cohomology theory on the category $\mathscr{W}$.

LEMMA 2. *The natural homomorphism* $H^n(K) \to \mathfrak{L}\{H^n(K_i)\}$ *is onto, and has kernel isomorphic to* $\mathfrak{L}'\{H^{n-1}(K_i)\}$.

REMARK. The proofs of Lemmas 1 and 2 will make no use of the dimension axiom [1 pg. 12]. This is of interest since Atiyah and others have studied "generalized cohomology theories" in which the dimension axiom is not satisfied.

Proof of Lemma 1. Let $[0, \infty)$ denote the CW-complex consisting of the nonnegative real numbers, with the integer points as vertices. Let L denote the CW-complex

$$L = K_1 \times [0, 1] \cup K_2 \times [1, 2] \cup K_3 \times [2, 3] \cup \cdots ;$$

contained in $K \times [0, \infty)$. The projection map $L \to K$ induces isomorphisms of homotopy groups in all dimension, and therefore is a homotopy equivalence. (See Whitehead [6, Theorem 1]. Alternatively one could show directly that L is a deformation retract of $K \times [0, \infty)$.)

Let $L_1 \subset L$ denote the union of all of the $K_i \times [i-1, i]$ with i odd. Similarly let L_2 be the union of all $K_i \times [i-1, i]$ with i even. The additivity axiom, together with the homotopy axiom, clearly implies that

$$H_*(L_1) \approx H_*(K_1) \oplus H_*(K_3) \oplus H_*(K_5) \oplus \cdots$$

with a similar assertion for L_2, and similar assertions for cohomology. On the other hand $L_1 \cap L_2$ is the disjoint union of the $K_i \times [i]$, and

[2] This derived functor has been studied in the thesis of Z-Z. Yeh, Princeton University 1959; and by Jan-Eric Roos [8].

therefore

$$H_*(L_1 \cup L_2) \approx H_*(K_1) \oplus H_*(K_2) \oplus H_*(K_3) \oplus \cdots .$$

Note that the triad $(L; L_1, L_2)$ is proper. In fact each set $K_i \times [i-1, i]$ used in the construction can be thickened, by adding on $K_{i-1} \times [i-3/2, i-1]$, without altering its homotopy type. Hence this triad $(L; L_1, L_2)$ has a Mayer-Vietoris sequence. The homomorphism

$$\psi\colon H_*(L_1 \cap L_2) \to H_*(L_1) \oplus H_*(L_2)$$

in this sequence is readily computed, and turns out to be:

$$\psi(h_1, h_2, \cdots, 0, 0, \cdots)$$
$$= (h_1, ph_2 + h_3, ph_4 + h_5, \cdots) \oplus (-ph_1 - h_2, -ph_3 - h_4, \cdots) ;$$

where h_i denotes a generic element of $H_*(K_i)$, and $p\colon H_*(K_i) \to H_*(K_{i+1})$ denotes the inclusion homomorphism.

It will be convenient to precede ψ by the automorphism α of $H_*(L_1 \cap L_2)$ which multiplies each h_i by $(-1)^{i+1}$. After shuffling the terms on the right hand side of the formula above, we obtain

$$\psi\alpha(h_1, h_2, \cdots) = (h_1, h_2 - ph_1, h_3 - ph_2, h_4 - ph_3, \cdots) .$$

From this expression it becomes clear that ψ has kernel zero, and has cokernel isomorphic to the direct limit of the sequence $\{H_*(K_i)\}$. Now the Mayer-Vietoris sequence

$$0 \longrightarrow H_*(L_1 \cap L_2) \xrightarrow{\psi} H_*(L_1) \oplus H_*(L_2) \longrightarrow H_*(L) \longrightarrow 0$$

completes the proof of Lemma 1.

The proof of Lemma 2 is completely analogous. The only essential difference is that the dual homomorphism

$$H^*(L_1 \cap L_2) \xleftarrow{\psi} H^*(L_1) \oplus H^*(L_2)$$

is not onto, in general. Its cokernel gives rise to the term $\mathfrak{L}'\{H^{n-1}(K_i)\}$ in Lemma 2.

Now let K be a possibly infinite formal simplicial complex with subcomplex L, and let $|K|$ denote the underlying topological space in the weak (=fine) topology. (Compare [1 pg. 75]) Let H_* denote an additive homology theory with coefficient group $H_0(\text{Point}) = G$.

LEMMA 3. *There exists a natural isomorphism between $H_q(|K|, |L|)$ and the formally defined homology group $\mathbf{H}_q(K, L; G)$ of the simplicial pair.*

Proof. If K is a finite dimensional complex then the proof given on pages 76–100 of Eilenberg-Steenrod applies without essential change. Now let K be infinite dimensional with n-skeleton K^n. It follows from this remark that the inclusion homomorphism

$$H_q(|K^n|) \to H_q(|K^{n+1}|)$$

is an isomorphism for $n > q$. Applying Lemma 1, it follows that the inclusion

$$H_q(|K^n|) \to H_q(|K|)$$

is also an isomorphism. Therefore the inclusion

$$H_q(|K^n|, |L^n|) \to H_q(|K|, |L|)$$

is an isomorphism for $n > q$. Together with the first remark this completes the proof of Lemma 3.

The corresponding lemma for cohomology groups can be proved in the same way. The extra term in Lemma 2 does not complicate the proof since $\mathfrak{L}' = 0$ for an inverse sequence of isomorphisms.

Uniqueness Theorem. *Let H_* be an additive homology theory on the category $\mathscr{W}$ (see introduction) with coefficient group G. Then for each (X, A) in $\mathscr{W}$ there is a natural isomorphism between $H_q(X, A)$ and the qth singular homology group of (X, A) with coefficients in G.*

Proof. Let $|SX|$ denote the geometric realization of the total singular complex of X, as defined by Giever, Hu, or Whitehead. (References [2, 3, 7].) Recall that the second barycentric subdivision $S''X$ is a simplicial complex. Since X has the homotopy type of a CW-complex, the natural projection

$$|SX| = |S''X| \to X$$

is a homotopy equivalence. (Compare [7, Theorem 23]). Using the five lemma it follows that the induced homomorphism

$$H_*(|S''X|, |S''A|) \to H_*(X, A)$$

is an isomorphism. But the first group, by Lemma 3, is isomorphic to

$$\mathbf{H}_*(S''X, S''A; G) \approx \mathbf{H}_*(SX, SA; G) ,$$

which by definition is the singular homology group of the pair X, A.

It is easily verified that the resulting isomorphism

$$\mathbf{H}_*(SX, SA; G) \to H_*(X, A)$$

commutes with mappings and boundary homomorphisms. (Compare pp. 100–101 of [1] for precise statements.) This completes the proof of the

Uniqueness Theorem.

The corresponding theorem for cohomology groups can be proved in the same way.

References

1. S. Eilenberg and N. Steenrod, *Foundations of algebraic topology*, Princeton 1952.

2. J. B. Giever, *On the equivalence of two singular homology theories*, Annals of Math., **51** (1950), 178-191.

3. S. T. Hu, *On the realizability of homotopy groups and their operations*, Pacific J. Math., **1** (1951), 583-602.

4. I. James and J. H. C. Whitehead, *Homology with zero coefficients*, Quarterly J. Math., Oxford (1958), 317-320.

5. J. Milnor, *On spaces having the homotopy type of a CW-complex*, Trans. Amer. Math. Soc., **90** (1959), 272-280.

6. J. H. C. Whitehead, *Combinatorial homotopy* I, Bull. Amer. Math. Soc., **55** (1949), 213-245.

7. J. H. C. Whitehead, *A certain exact sequence*, Annals of Math., **52** (1950), 51-110.

8. J. E. Roos, *Sur les foncteurs dérivés de lim*, Comptes Rendus Acad. des Science (Paris), **252** (1961), 3702-3704.

PRINCETON UNIVERSITY

16

CHARACTERS AND COHOMOLOGY OF FINITE GROUPS

M. F. Atiyah

§ 3. Inverse limits and completions.

Let M be a filtered abelian group, i.e. we have a sequence of subgroups:

$$M = M_0 \supset M_1 \supset \ldots \supset M_n \supset \ldots$$

This filtration gives M the structure of a topological group, the subgroups M_n being taken as a fundamental system of neighbourhoods of o in M. We denote by $\hat{M}$ (or $M^\wedge$) the completion of M for this topology, i.e.

$$\hat{M} = \varprojlim M/M_n \text{ (inverse limit).} \tag{3.1}$$

We remark that the topology of M is not necessarily Hausdorff so that the natural map $M \to \hat{M}$ may have non-zero kernel. In fact we have:

$$\operatorname{Ker}(M \to \hat{M}) = \bigcap_{n=1}^{\infty} M_n. \tag{3.2}$$

If $\{{}_nA\}$ is an inverse system of abelian groups (indexed by the non-negative integers), the inverse limit $A = \varprojlim {}_nA$ has a natural filtration defined by (1)

$$A_n = \operatorname{Ker}\{A \to {}_{n-1}A\}$$

Moreover A is complete for the topology defined by this filtration, i.e. $A \cong \hat{A}$. Thus an inverse limit is in a natural way a *complete filtered group*. This applies in particular to the group $\hat{M}$ given by (3.1). It is easy to see that the subgroups of the filtration may be identified with the completions $\hat{M}_n$ of the subgroups M_n (for the induced topology).

If M is a finite group then the filtration necessarily terminates, i.e. $M_n = M_{n+1}$ for all $n \geqslant n_0$, and so $\hat{M} \cong M/M_{n_0}$. We record this for future reference.

Lemma (3.3). — *If* M *is a finite filtered group* $M \to \hat{M}$ *is an epimorphism.*

We also state the following elementary properties of inverse limits, the verifications being trivial.

Lemma (3.4). — *Let* $\{{}_{\alpha,\beta}A\}$ *be an inverse system indexed by pairs* $(\alpha, \beta) \in I \times J$, *where* I, J *are two directed sets. Then*

$$\varprojlim_{\alpha} \varprojlim_{\beta} {}_{\alpha,\beta}A \cong \varprojlim_{(\alpha,\beta)} {}_{\alpha,\beta}A \cong \varprojlim_{\beta} \varprojlim_{\alpha} {}_{\alpha,\beta}A$$

Lemma (3.5). — *If* $0 \to \{{}_\alpha A\} \to \{{}_\alpha B\} \to \{{}_\alpha C\} \to 0$ *is an exact sequence of inverse systems* (α *belonging to some directed set*), *then*

$$0 \to \varprojlim {}_\alpha A \to \varprojlim {}_\alpha B \to \varprojlim {}_\alpha C$$

is exact.

In order for $\varprojlim$ to be right exact we need a condition. Following Dieudonné-Grothendieck [8] we adopt the following definition. An inverse system $\{{}_\alpha A\}$ is said to satisfy the *Mittag-Leffler* condition (ML) if, for each α, there exists $\beta \geqslant \alpha$ such that

$$\operatorname{Im}({}_\beta A \to {}_\alpha A) = \operatorname{Im}({}_\gamma A \to {}_\alpha A)$$

(1) We put ${}_{-1}A = 0$ so that $A_0 = A$.

for all $\gamma \geqslant \beta$. Moreover we shall assume from now on that all inverse systems are over *countable* directed sets. The following properties of (ML) are proved in [8, chapter 0 (complements)].

(3.6) *If* $\{_\alpha A\} \to \{_\alpha B\} \to 0$ *is exact and* $\{_\alpha A\}$ *satisfies* (ML), *so does* $\{_\alpha B\}$.

(3.7) *If* $0 \to \{_\alpha A\} \to \{_\alpha B\} \to \{_\alpha C\} \to 0$ *is exact, and if* $\{_\alpha A\}$ *and* $\{_\alpha C\}$ *each satisfy* (ML), *so does* $\{_\alpha B\}$.

(3.8) *If* $0 \to \{_\alpha A\} \to \{_\alpha B\} \to \{_\alpha C\} \to 0$ *is exact, and if* $\{_\alpha A\}$ *satisfies* (ML), *then*

$$0 \to \varprojlim {}_\alpha A \to \varprojlim {}_\alpha B \to \varprojlim {}_\alpha C \to 0$$

is exact.

(3.9) *Let* $\{_\alpha C^*\}$ *be an inverse system of complexes, with differentials of degree r. Suppose that, for each p,* $\{_\alpha C^p\}$ *and* $\{H^p(_\alpha C^*)\}$ *satisfy* (ML), *then* $\varprojlim H^p(_\alpha C^*) \cong H^p(\varprojlim {}_\alpha C^*)$.

Remark. — In [8] the differentials in (3.9) are supposed to have degree 1, but this does not affect the argument.

17
EXTRACT FROM THESIS

D. W. Anderson

The first thing which we must do to formulate our theorem is to describe the topology on $KU^*(B_G)$. If we let k^* stand for any cohomology theory defined on the category of CW-complexes, and if X is a CW-complex with n-skeleton X^n, $k^*(X)$ can be given the structure of a topological group if we take as the fundamental system of neighbourhoods of zero the groups, kernel $(k^*(X) \to k^*(X^n))$. The resulting topology will be called the inverse limit topology on $k^*(X)$. We denote the inverse limit functor by $\lim^0$. There is one non-zero right derived functor $\lim^1$ when we work in the category of abelian groups, and $\lim^0$ is left exact. The relationship between $k^*(X)$ and $\lim^0(k^*(X^n))$ has been described by Milnor for a class of theories k^* which satisfy one more axiom than the Eilenberg-Steenrod axioms. This axiom is called 'additivity' by Milnor, and says that when applied to topological sums, k^* gives the direct product of the k^*'s of the individual components. Any representable theory clearly satisfies this condition, and any additive theory is representable [4]. Milnor's result [6] states that for an additive theory k^* and a CW-complex X, there is for all p an exact sequence:

$$0 \to \lim{}^1(k^{p-1}(X^n)) \to k^p(X) \to \lim{}^0(k^p(X^n)) \to 0 .$$

For a representable theory, of course, the surjectiveness of the map $k^*(X) \to \lim^0(k^*(X^n))$ follows from the homotopy extension property for CW-pairs. The following is implicit in §3, 4 [2]:

Proposition 4.1. A sufficient condition that $\lim^1(k^*(X^n)) = 0$ is that in the spectral sequence [5] which connects $H^*(X; k^*(\text{point}))$ with $k^*(X)$, for all (p, q) there exists an r such that $E_r^{p,q} = E_\infty^{p,q}$.

If both $H^*(X; Z)$ and $k^*(\text{point})$ are of finite type, it is clear from this proposition that a sufficient condition that $\lim^1(k^*(X^n)) = 0$ is that every element of $E_2^{p,q} = H^p(X; k^q(\text{point}))$ have some multiple which is an infinite cycle in the spectral sequence. If $k^* = KU^*$, we know from [3] that this will happen if $ch^p: KU^*(X) \otimes Q \to H^p(X; Q)$ is onto for every p. If we take X to be the classifying space of a compact Lie group, G, then $ch^p: RU(G) \otimes Q \to H^p(B_G; Q)$ is surjective for all p (see [3] for the case when G is connected, and use this to conclude the same result for any compact Lie group). Thus, we see that for a compact Lie group G, $KU^*(B_G) = \lim^0(KU^*(B_G^n))$.

18

The scope of the next paper, by Dyer, is indicated in §12. It offers a good introduction to a central topic of interest in this field. It assumes a fair familiarity with the usual machinery of algebraic topology.

18

Matematisk Institut, Aarhus Universitet

RELATIONS BETWEEN COHOMOLOGY THEORIES

Eldon Dyer†

Colloquium on Algebraic Topology, 1-10 August 1962

This lecture is principally an exposition of a folk theorem of a Riemann-Roch type for general cohomology theories known to Adams, Atiyah, Hirzebruch . . . .

First we need to make a few remarks about Poincare duality in such theories. We could follow the approach of G. W. Whitehead in [3], but more convenient for our purposes is Dold's isomorphism theorem for h-orientable bundles. To avoid a discussion of micro-bundles, we limit our discussion here to differentiable manifolds.

Poincaré duality

Let h^* be a multiplicative cohomology theory in the sense of Dold. Let $\pi: E \to B$ be a vector bundle λ ($B \simeq$ connected, finite CW complex). Let E' be the complement of the 0-section and $i:(F, F') \to (E, E')$ be the inclusion of a fibre.

Theorem (Dold). *If there is a class* $u \in h^n(E, E')$ *such that* i^*u *generates* $h^*(F, F') \cong h^*(S_n, \text{pt.})$ *as* $h^*(\text{pt.})$-*module, then the homomorphism*

† The author is an A. P. Sloan Foundation Fellow.

$$\phi : h^i(B) \to h^{i+n}(E, E')$$

defined by $\phi(\xi) = \pi^*(\xi) \cup u$ is an isomorphism.

Of course, (E, E') has the homotopy type of $(T(\lambda), \text{pt.})$, $T(\lambda)$ the Thom space of the bundle λ.

Let M be a closed, differentiable manifold and ν a normal bundle of M. If Dold's isomorphism holds for this bundle, i. e., if a suitable class $u_M \in h^*(T(\nu), \text{pt.})$ exists, we may say M is h*-orientable and satisfies Poincaré duality in the h*-theory. In fact, Dold's isomorphism may be regarded as the statement of Poincare duality in light of the Milnor-Spanier result that $T(\nu)$ is S-dual to $M/\emptyset$. (Note corollaries 7.8 and 7.10 of Whitehead [3].) Let $\mathscr{H}$ denote the class of h*-oriented manifolds.

We may regard M as the solid unit tube in ν, $\mathring{M}$ as its boundary and $(T(\nu), \text{pt.})$ as $(M, \mathring{M})$. Then $h^*(M, \mathring{M})$ is an $h^*(M)$-module and the action

$$h^*(M) \otimes h^*(M, \mathring{M}) \to h^*(M, \mathring{M}), \quad \text{or equivalently}$$

$$h^*(M) \otimes h^*(T(\nu), \text{pt.}) \to h^*(T(\nu), \text{pt.})$$

is the action of $h^*(M)$ on $h^*(M, \mathring{M})$ through π^* used in Dold's theorem. We shall regard this action as a cap-product.

Let $f: X \to Y$ be a continuous map of manifolds in $\mathscr{H}$. This induces a dual map $\hat{f}: \hat{Y} \to \hat{X}$, where $\hat{Y}$ and $\hat{X}$ are S-duals as given by Thom-spaces of normal bundles. We have a diagram

$$\begin{array}{ccccc} h^*(X) & \otimes & h^*(\hat{X}) & \to & h^*(\hat{X}) \\ \uparrow f^* & & \downarrow \hat{f}^* & & \downarrow \hat{f}^* \\ h^*(Y) & \otimes & h^*(\hat{Y}) & \to & h^*(\hat{Y}) \end{array},$$

in which

$$\hat{f}^*(f^*(y) \cup x) = y \cup \hat{f}^*(X) .$$

This is the usual formula for cap-products.

Also, there is an 'inverse-homomorphism' $f_! : h^*(X) \to h^*(Y)$ given by the composition

$$h^*(X) \xrightarrow{\phi} \tilde{h}^*(\hat{X}) \xrightarrow{\hat{f}^*} \tilde{h}^*(\hat{Y}) \xrightarrow{\phi^{-1}} h^*(Y) .$$

We notice that $f_!$ is functorial and that

$$f_!(f^*(y) \cup x) = y \cup f_!(x) ,$$

again as in the usual case.

Riemann-Roch Theorem. Let h^* and k^* be multiplicative cohomology theories. Let $\tau : h^* \to k^*$ be a natural transformation such that

(1) for each X, $\tau : h^*(X) \to k^*(X)$ is multiplicative, and

(2) if $\alpha \in h^1(S_1, \text{pt.})$ and $\beta \in k^1(S_1, \text{pt.})$ are suspensions of the units in $h^0(S_0, \text{pt.})$ and $k^0(S_0, \text{pt.})$, then $\tau(\alpha) = \beta$.

Such a transformation we shall call multiplicative.

For $X \in \mathscr{H} \cap \mathscr{K}$ let $\tau(X) \in k^*(X)$ be the image of $1 \in h^*(X)$ under the composition

$$h^*(X) \xrightarrow{\phi_h} \tilde{h}^*(\hat{X}) \xrightarrow{\tau} \tilde{k}^*(\hat{X}) \xrightarrow{\phi_k^{-1}} k^*(X); \text{ i. e. ,}$$

$$\tau(X) = \phi_k^{-1}\ \tau\ \phi_h\ (1)\ .$$

Proposition. For $x \in h^*(X)$, $\tau(x) \cup \tau(X) = \phi_k^{-1}\ \tau\ \phi_h(x)$.

Proof. Let $\phi_h(1) = U_h$ and $\phi_k(1) = U_k$. Then

$$\begin{aligned}\phi_k(\tau(x) \cup \phi_k^{-1}\ \tau\ \phi_h(1)) &= \pi_k^*\ \tau(x) \cup \pi_k^*\ \phi_k^{-1}\ \tau\ \phi_h(1) \cup U_k \\ &= \tau\ \pi_h^*(x) \cup \tau\ \phi_h(1) \\ &= \tau\ \phi_h(x)\ .\end{aligned}$$

Theorem. Let X and Y lie in $\mathscr{H} \cap \mathscr{K}$, $\tau{:}h^* \to k^*$ be a multiplicative transformation of cohomology theories, and $f{:}X \to Y$ be a continuous map. Then for $x \in h^*(X)$,

$$f_!^k(\tau(x) \cup \tau(X)) = \tau(f_!^h(x)) \cup \tau(Y)\ .$$

Proof.
$$\begin{aligned}\phi_k^{-1}\ \hat{f}^*[\phi_k(\tau(x) \cup \tau(X))] &= \phi_k^{-1}\ \hat{f}^*\ \tau\ \phi_h(x) \\ &= \phi_k^{-1}\ \tau\ \hat{f}^*\ \phi_h(x) \\ &= \phi_k^{-1}\ \tau\ \phi_h[\phi_h^{-1}\ \hat{f}^*\ \phi_h(x)] \\ &= \tau[\phi_h^{-1}\ \hat{f}^*\ \phi_h(x)] \cup \tau(Y) \\ &= \tau\ f_!^h(x) \cup \tau(Y)\ .\end{aligned}$$

Examples. We give two examples taken from papers of Atiyah and Hirzebruch, [1] and [2].

(A) Let $h^*(\ ) = k^*(\ ) = H^*(\ ; Z_p)$ and let λ be a cohomology automorphism (multiplicative; identity on $H^1(S_1, \text{pt.})$). In [1]

Atiyah and Hirzebruch have defined $Wu(\lambda, X)$ to be the image of 1 in the composition

$$H^*(X) \xrightarrow{\phi} H^*(X^\tau) \xrightarrow{\lambda} H^*(X^\tau) \xrightarrow{\phi^{-1}} H^*(X) \xrightarrow{\lambda^{-1}} H^*(X) ,$$

where X^τ is the Thom space of the tangent bundle τ of X. For λ we have $\lambda(X) = Wu(\lambda^{-1}, X)$. The theorem then states that for $f: X \to Y$,

$$f_!\,(\lambda(x) \cup Wu(\lambda^{-1}(X))) = \lambda\, f_!\,(x) \cup Wu(\lambda^{-1}, Y) .$$

In this case $f_!$ is the ordinary inverse homomorphism. This is Theorem 3.2 of [1].

An illustration explaining the notation $Wu(\ ,X)$ is given in [1]. Let $p = 2$ and $\lambda^{-1} = Sq = Sq^0 + Sq^1 + \dots$. Take Y to be a point. Then $f_!\,(x) = x[x]$ and the equation becomes

$$(\lambda(x) \cup Wu(\lambda^{-1}, X))[X] = x[X] .$$

$Wu(Sq, X) = Sq^{-1}W$, where W is the total Stiefel-Whitney class of X. Let $U = Sq^{-1}x$ and $V = Sq^{-1}W$. The above equation becomes

$$(U \cup V)\,[X] = (Sq\, U)\,[X] ,$$

which is the formula of Wu characterizing W.

Atiyah and Hirzebruch give other applications of this theorem and a computation of $\lambda(X)$ for $\lambda = Sq^{-1}$ or P^{-1} in terms of Todd polynomials and $\hat{a}$.

(B) As a second example we let $h^* = K^*_C$, $k^* = H^*(\ ;Q)$ and $\tau = ch$, the Chern character.

An orientable manifold X is a C_1-manifold if for some imbedding of X in a sphere, there is a class $C_1 \in H^2(X; Z)$ whose mod 2 reduction is the second Stiefel-Whitney class of the normal bundle ν of X in the imbedding. It follows by representation theory (Sec. 5. 3 and 5. 4 of [2]) that for such a manifold there is an element

$$\phi(1) \in K^*_C(X^\nu) \quad \text{such that}$$

$$ch\ \phi(1) = \phi(e^{c_1/2}\hat{a}(\nu)^{-1}) = \phi(e^{c_1/2}\hat{a}(X)) .$$

As the first non-zero term of ch is 1 in the dimension of the fibre, $\phi(1)$ pulls back correctly to show a C_1-manifold is orientable in the K^*_C-theory. Thus we have the

Theorem. Let $f: X \to Y$ be a continuous map of C_1-manifolds. There is a homomorphism

$$f_! : K^*_C(X) \to K^*_C(Y)$$

such that for $x \in K^*_C(X)$,

$$f^H_!(ch(x) \cup e^{c_1(X)/2}\hat{a}(X)) = ch(f_!(x)) \cup e^{c_1(Y)/2}\hat{a}(Y) .$$

$f_!$ is functorial for maps of C_1-manifolds and

$$f_!(f^*_K(y) \cup x) = y \cup f_!(x) .$$

(18)

This theorem has had several interesting applications in stable homotopy theory and in differential topology.

REFERENCES

1. M. F. Atiyah and F. Hirzebruch, 'Cohomologie-Operationen und charakteristische Klassen', Math. Z. 77 (1961), 149-187.
2. F. Hirzebruch, 'A Riemann-Roch theorem for differentiable manifolds', Sem. Bourbaki 58/59, #177.
3. G. W. Whitehead, 'Generalized homology theories', Trans. Amer. Math. Soc. 102 (1962), 227-283.

19, 20, 21 & 22

The next paper, by Atiyah and Hirzebruch, is the foundation paper on K-theory. It requires no more prerequisites than may be expected at this stage, but the writing may be found condensed in places. The following exposition, by Hirzebruch, can be recommended for its clarity. The next note, by myself, is merely a summary of results which were later published in full. For all these three papers (but especially the last) the reader can consult Husemoller for further details; or see Atiyah.

The fourth extract in this group is intended to give an idea of one particular application of K-theory. The prerequisite is a certain sympathy for homotopy-theory and homological algebra.

19

VECTOR BUNDLES AND HOMOGENEOUS SPACES

BY

M. F. ATIYAH AND F. HIRZEBRUCH

Dedicated to Professor Marston Morse

Introduction. In [**1**] we introduced for a space X the "ring of complex vector bundles" $K(X)$. The Bott periodicity of the infinite unitary group [**8**; **9**; **10**] implied that K satisfied the "Künneth formula"

$$K(X \times S^2) \simeq K(X) \otimes K(S^2)$$

which was fundamental for the proof of the differentiable Riemann-Roch theorems [**1**; **16**].

Using the Bott periodicity we construct in §1 a "periodic cohomology theory": For every integer n, the abelian group $K^n(X)$ is defined, $K^0(X)$ is $K(X)$ and $K^{n+2}(X)$ is isomorphic with $K^n(X)$, the group $K^1(X)$ is the kernel of the homomorphism $K^0(X \times S^1) \to K^0(X)$ induced from the embedding $X \to X \times S^1$. This cohomology theory satisfies all the axioms of Eilenberg-Steenrod [**14**] except the "dimension axiom." For the space consisting of a single point, K^n is infinite cyclic for even n and vanishes for odd n. The axioms without the dimension axiom do not characterize the theory, even if the values of K^n are given for a point. There is a spectral sequence relating the ordinary cohomology theory with our periodic theory (§2).

In §§3–5 we try to get information on K^0 and K^1 for classifying spaces and certain homogeneous spaces. An important tool is the differentiable Riemann-Roch theorem which we recall in the beginning of §3. The final goal would be to answer all those questions for the K-theory on homogeneous spaces which for the ordinary cohomology theory have been treated so successfully by A. Borel (see for example [**3**]). We can give only partial results in this direction. The new cohomology theory can be applied to various topological questions and may give better results than the ordinary cohomology theory, even if the latter one is enriched by cohomology operations (see [**2**] and M. F. Atiyah and J. A. Todd, *On complex Stiefel manifolds*, to appear in Proc. Cambridge Philos. Soc.). This justifies the new theory.

In spite of its length, the present paper is by no means a final exposition. The proofs are often sketchy and the definitions and results could be generalized in certain cases. For example, using real vector bundles and the Bott periodicity of the infinite orthogonal group, we can define a periodic cohomology theory with period 8. This is not more difficult than in the unitary case. Furthermore, the definition of $K(X)$ in 1.1 can be given for any topological space. For convenience, we have restricted the theory to the special class $\mathfrak{A}$ (see 1.1). We have then the homotopy classification theorem (1.3)

$$(1) \qquad K(X) \cong [X, Z \times B_U], \qquad (X \in \mathfrak{A}).$$

For this actually $\mathfrak{A}$ could be chosen much larger. But in general (1) would be wrong. The restriction to $\mathfrak{A}$ simplifies the presentation of certain consequences drawn from the spectral sequence. For any topological space, we can take the right side of (1) as a definition of a functor $k(X)$. If $Z \times B_U$ is endowed with a natural structure of a commutative ring (up to homotopy), then $k(X)$ has a natural (commutative) ring structure for any space X and the rings $K(X)$ and $k(X)$ are isomorphic if $X \in \mathfrak{A}$. Such a "ring" structure on $Z \times B_U$ has been defined by Milnor (not published). In view of Milnor's construction it would perhaps be more natural to study the functor $k(X)$, but since Milnor's result is not yet at our disposal we have studied $K(X)$ where sum and product structure is automatically given by the Whitney sum and the tensor product of vector bundles.

For the classifying spaces B_G we have defined $\mathcal{K}(B_G)$ as an inverse limit indicating by the curly letter that we mean neither $K(B_G)$ nor $k(B_G)$. We conjecture that $\mathcal{K}(B_G)$ is isomorphic to $k(B_G)$ for any compact Lie group G. But we shall deal with this question elsewhere. We prove for a compact connected Lie group G that $\mathcal{K}(B_G)$ is isomorphic with the completed representation ring $\hat{R}(G)$ (see 4.8).

1. A cohomology theory derived from the unitary groups.

1.1. Let $\mathfrak{A}$ be the class of those spaces which can carry the structure of a finite CW-complex. For $X \in \mathfrak{A}$ we have defined in [1] an abelian group $K(X)$. There we gave the definition only for a connected X, but we may define $K(X)$ in general as the direct sum of the groups $K(X_i)$ where the X_i are the connectedness components of X. For the sake of completeness we recall the definition of $K(X)$ and give it directly for a space $X \in \mathfrak{A}$ not necessarily connected.

We adopt the usual definition of a complex vector bundle over X except that we allow the bundle to have fibres of different dimensions over the various connectedness components of X. We can now verbally repeat the definition of [1]:

Let $F(X)$ be the free abelian group generated by the set of all isomorphism classes of complex vector bundles over X. To every triple $t = (\xi, \xi', \xi'')$ of vector bundles with $\xi \cong \xi' \oplus \xi''$ we assign the element $[t] = [\xi] - [\xi'] - [\xi'']$ of $F(X)$, where $[\xi]$ denotes the isomorphism class of ξ. The group $K(X)$ is defined as the quotient of $F(X)$ by the subgroup generated by all the elements of the form $[t]$.

The tensor product of vector bundles defines a commutative ring structure in $K(X)$; the unit 1 is given by the trivial bundle of dimension 1.

K is a contravariant functor: for a continuous map $f : Y \to X$ $(Y, X \in \mathfrak{A})$ we have the natural ring homomorphism $f^! : K(X) \to K(Y)$ induced by the lifting of bundles under f. We denote it by $f^!$ to distinguish it from the analogous homomorphism f^* in ordinary cohomology theory.

1.2. Let $\tilde{\mathfrak{A}}$ be the class whose objects are the pairs (X, x_0) with $X \in \mathfrak{A}$ and $x_0 \in X$. Usually we shall write an object of $\tilde{\mathfrak{A}}$ simply by indicating the space X. Very often the base point x_0 of X is naturally given by the context. For $X \in \tilde{\mathfrak{A}}$ we define the *reduced group* $\tilde{K}(X)$ as follows: the ring $K(\{x_0\})$ is canonically isomorphic with Z (the ring of integers). The imbedding $i : \{x_0\} \to X$ induces the ring homomorphism

$$i^! : K(X) \to K(\{x_0\}) = Z.$$

We define $\tilde{K}(X)$ to be the kernel of $i^!$. It is an ideal of $K(X)$. Whenever a symbol like $\tilde{K}(X)$ occurs it is to be understood that X is a space with base point, i.e., an object of $\tilde{\mathfrak{A}}$.

We now consider the class $\mathfrak{B}$ consisting of pairs (X, Y) where X can be given the structure of a finite CW-complex in such a way that Y becomes a subcomplex. For $(X, Y) \in \mathfrak{B}$ we define

$$K(X, Y) = \tilde{K}(X/Y).$$

Here X/Y is obtained from X by collapsing Y to a point which becomes then the base point of X/Y. By [19], $X/Y \in \tilde{\mathfrak{A}}$. Note that $\tilde{K}(X) = K(X, x_0)$ for $X \in \tilde{\mathfrak{A}}$. If Y is empty ($Y = \emptyset$), then $X/\emptyset = X^+$ (where X^+ is the topological sum of X with an extra point which becomes base point of X^+) and $K(X, \emptyset) = \tilde{K}(X^+) = K(X)$.

For $X, Y \in \tilde{\mathfrak{A}}$ the objects $X \vee Y$ and $X \wedge Y$ of $\tilde{\mathfrak{A}}$ are defined. (In the literature, $X \wedge Y$ is also denoted by $X \# Y$). $X \vee Y$ is obtained from the topological sum of X and Y by identifying the base point of X with the base point of Y to one point which becomes the base point of $X \vee Y$. The space $X \wedge Y$ is $X \times Y$ with the union of the axis $x_0 \times Y$ and $X \times y_0$ collapsed to a point which becomes the base point of $X \wedge Y$. We have the natural maps

$$X \vee Y \to X \times Y \to X \wedge Y$$

and may write

$$X \wedge Y = X \times Y/X \vee Y. \tag{1}$$

The operations $\vee$ and $\wedge$ are associative and commutative and $\wedge$ is distributive over $\vee$. This means, for example, that there is a *canonical homeomorphism* between $X \wedge Y$ and $Y \wedge X$.

If $S^n \in \tilde{\mathfrak{A}}$ is the standard n-sphere with base point, we write

$$S^n(X) = S^n \wedge X, \qquad (X \in \tilde{\mathfrak{A}}). \tag{2}$$

This is the nth suspension of X. Since

$$S^n = S^1 \wedge S^1 \wedge \cdots \wedge S^1 \qquad (n \text{ times})$$

it follows that $S^n(X)$ is the n times iterated suspension of X.

DEFINITION. *For any integer* $n \geq 0$ *we put* $K^{-n}(X, Y) = \tilde{K}(S^n(X/Y))$, $((X, Y) \in \mathfrak{B})$. *For* $X \in \mathfrak{A}$ *we put* $K^{-n}(X) = K^{-n}(X, \emptyset) = \tilde{K}(S^n(X^+))$. *For* $X \in \tilde{\mathfrak{A}}$ *with base point* x_0 *we put* $\tilde{K}^{-n}(X) = K^{-n}(X, x_0) = \tilde{K}(S^n(X))$.

For $n = 0$ we have the groups already defined:

$$K^0(X, Y) = K(X, Y), \quad K^0(X) = K(X), \quad \tilde{K}^0(X) = \tilde{K}(X).$$

Of course, the K^{-n} are also contravariant functors.

1.3. We write $[A, B]$ for the set of homotopy classes of maps of the space A into the space B and correspondingly $[A, U; B, V]$ for the homotopy classes of maps of the pair (A, U) into the pair (B, V). If the spaces A and B have base points, then we write $[A, B]_0$ for the set of homotopy classes of maps preserving base points.

Let B_U be the classifying space of the infinite unitary group [10] and $Z \times B_U$ the cartesian product of it with the group of integers (Z having the discrete topology). In $Z \times B_U$ we choose a base point lying in $0 \times B_U$. The classification theorem for unitary bundles [18, §19.3] gives rise to the following natural bijective maps (compare also [16, §1.7, 2.1]):

$$K(X) \cong [X, Z \times B_U], \qquad (X \in \mathfrak{A}).$$

$$\tilde{K}(X) \cong [X, Z \times B_U]_0, \qquad (X \in \tilde{\mathfrak{A}}).$$

$$\begin{aligned} K^{-n}(X, Y) &\cong [S^n(X/Y), Z \times B_U]_0, \qquad ((X, Y) \in \mathfrak{B}), \\ &\cong [X/Y, \Omega^n(Z \times B_U)]_0, \\ &\cong [X, Y; \Omega^n(Z \times B_U), \text{point}], \\ &\cong [X, Y; \Omega^{n-1}U, \text{point}], \qquad n > 0. \end{aligned}$$

We recall that $Z \times B_U$ is weakly homotopy equivalent to an H-space, namely to ΩU (Bott, see [8]). Thus all the above sets of homotopy classes are endowed with a natural group structure. The above bijections are in fact all group isomorphisms. Since U is weakly homotopy equivalent to $\Omega(Z \times B_U)$, the space $\Omega^2(Z \times B_U)$ is weakly homotopy equivalent to $Z \times B_U$ and we have an isomorphism

$$K^{-(n+2)}(X, Y) \cong K^{-n}(X, Y), \qquad n \geq 0. \tag{3}$$

We shall give later an explicit description of an isomorphism between these two groups.

If x_0 denotes the space consisting of a single point, then

$$K^{-n}(x_0) = \pi_n(Z \times B_U), \qquad n \geq 0,$$

and thus [9]

$$K^{-n}(x_0) \cong Z \text{ for } n \text{ even and } K^{-n}(x_0) = 0 \text{ for } n \text{ odd.}$$

1.4. PROPOSITION. *If* $(X, Y) \in \mathfrak{B}$ *we have exact sequences*

(i) $\cdots \to K^{-(n+1)}(Y) \xrightarrow{\delta} K^{-n}(X, Y) \to K^{-n}(X)$

$$\to K^{-n}(Y) \to \cdots \to K^0(X, Y) \to K^0(X) \to K^0(Y),$$

(ii) $\cdots \to \tilde{K}^{-(n+1)}(Y) \xrightarrow{\delta} K^{-n}(X, Y) \to \tilde{K}^{-n}(X)$

$$\to \tilde{K}^{-n}(Y) \to \cdots \to K^0(X, Y) \to \tilde{K}^0(X) \to \tilde{K}^0(Y).$$

For (ii) *we assume* $X, Y \in \tilde{\mathfrak{A}}$ *with* $x_0 = y_0 \in Y$.

Proof. We use the paper of Puppe [17]. If Y and X are arbitrary spaces with base point and $f : Y \to X$ a map preserving base points, then there is a sequence of spaces and maps (with base points)

$$Y \xrightarrow{f} X \xrightarrow{Pf} C_f \xrightarrow{Qf} S^1Y \to S^1X \to S^1C_f \to S^2Y \to S^2X \to \cdots$$

such that the following is true: if V is any space with base point, then the functor $[\ ,\ V]_0$ gives an exact sequence of sets. Here we note that exactness is a property of sets with preferred elements—the group structure is irrelevant. The preferred element is always given by the constant map onto the base point. We recall the construction of C_f. First we take the cone

$$CY = Y \times I/Y \times 1 \cup y_0 \times I.$$

Then we take the topological sum $CY + X$ in which we identify $(y, 0) \in CY$ with $f(y)$ for each $y \in Y$. The space C_f contains X as subspace. C_f/X is (canonically homeomorphic with) the first suspension of Y. This gives rise to the maps $Y \to^f X \to^{Pf} C_f \to^{Qf} S^1Y$. All the other maps in Puppe's sequence are suspensions of these. If Y is a subspace of X and f the injection, then we have a natural homeomorphism $X/Y \cong C_f/CY$. If (X, Y) belongs to $\mathfrak{B}$ then it satisfies the homotopy extension condition and according to Puppe the map $C_f \to C_f/CY$ followed by the above mentioned homeomorphism is a homotopy equivalence h. The composition $h \circ Pf$ is the natural projection $X \to X/Y$. Taking this into account Puppe's theorem applied to $V = Z \times B_U$ gives the exact sequence (ii) and all homomorphisms in this sequence are canonically defined by Puppe's maps. $K^{-n}(X, Y) \to \tilde{K}^{-n}(X)$ is induced from $X \to X/Y$. The sequence (i) is obtained by replacing in (ii) Y and X by Y^+ and X^+ respectively.

Remark. If $Y = \{x_0\}$ then the sequence (i) breaks off in split exact sequences

$$0 \to \tilde{K}^{-n}(X) \to K^{-n}(X) \to K^{-n}(x_0) \to 0.$$

Hence

$$K^{-n}(X) \cong \tilde{K}^{-n}(X) \oplus \pi_n(Z \times B_U), \qquad \text{(see 1.3).}$$

The exact sequence (i) is obtained from (ii) by adding to $\tilde{K}^{-n}(X)$ and also to $\tilde{K}^{-n}(Y)$ the direct summand $\pi_n(Z \times B_U)$.

1.5. We have mentioned in 1.1 that $K(X) = K^0(X)$ is a commutative ring. We wish to define more generally products also involving the groups K^{-n} $(n \geq 0)$.

Suppose $X, Y \in \tilde{\mathfrak{A}}$. Then $X \vee Y = X \times y_0 \cup x_0 \times Y$ is a subspace of $X \times Y$. We apply 1.4 (ii) to the pair $(X \times Y, X \vee Y)$. The exact sequence breaks off in this case into split exact sequences.

(4) $$0 \to \tilde{K}^{-n}(X \wedge Y) \to \tilde{K}^{-n}(X \times Y) \to \tilde{K}^{-n}(X \vee Y) \to 0, \qquad (n \geqq 0),$$

and we have a canonical decomposition

(5) $$\tilde{K}^{-n}(X \times Y) \cong \tilde{K}^{-n}(X \wedge Y) \oplus \tilde{K}^{-n}(X) \oplus \tilde{K}^{-n}(Y).$$

For the proof of (4) we observe that $\tilde{K}^{-n}(X \times Y) \to \tilde{K}^{-n}(X \vee Y)$ is surjective and that this homomorphism may be regarded as the projection onto a direct summand. For this we make use of

$$\begin{aligned} \tilde{K}^{-n}(X \vee Y) = \tilde{K}(S^n(X \vee Y)) = \tilde{K}(S^nX \vee S^nY) &= \tilde{K}(S^nX) \oplus \tilde{K}(S^nY) \\ &= \tilde{K}^{-n}(X) \oplus \tilde{K}^{-n}(Y). \end{aligned}$$

We have the following natural group homomorphisms which are all induced by the tensor product of vector bundles

(6) $$K(X) \otimes K(Y) \to K(X \times Y), \qquad (X, Y \in \mathfrak{A}),$$

(7) $$\tilde{K}(X) \otimes \tilde{K}(Y) \to \tilde{K}(X \wedge Y), \qquad (X, Y \in \tilde{\mathfrak{A}}),$$

(8) $$K(X, X_0) \otimes K(Y, Y_0) \to K(X \times Y, X_0 \times Y \cup X \times Y_0),$$

for (X, X_0) and $(Y, Y_0) \in \mathfrak{B}$.

It is clear how (6) is defined. If $a \in \tilde{K}(X)$ and $b \in \tilde{K}(Y)$, then the product is in the kernel of $\tilde{K}(X \times Y) \to \tilde{K}(X \times y_0 \cup x_0 \times Y)$. By (4) and (5) the product is well defined as element of $\tilde{K}(X \wedge Y)$, $(n = 0)$. If we replace in (7) X by X/X_0 and Y by Y/Y_0 we get the definition of (8). More generally we have a group homomorphism

(9) $$K^{-m}(X, X_0) \otimes K^{-n}(Y, Y_0) \to K^{-(m+n)}(X \times Y, X_0 \times Y \cup X \times Y_0),$$

for $(X, X_0), (Y, Y_0) \in \mathfrak{B}$ and $m \geq 0, n \geqq 0$.

We get this from (7) and the fact that

$$\begin{aligned} S^m(X/X_0) \wedge S^n(Y/Y_0) &= S^{m+n}(X/X_0 \wedge Y/Y_0) \\ &= S^{m+n}(X \times Y/X_0 \times Y \cup X \times Y_0). \end{aligned}$$

The equality sign means that there is a natural homeomorphism between these spaces. If one uses the natural identification of $X \times Y$ with $Y \times X$, one gets from (9) a product

(9′) $$K^{-n}(Y, Y_0) \otimes K^{-m}(X, X_0) \to K^{-(m+n)}(X \times Y, X_0 \times Y \cup X \times Y_0).$$

LEMMA. *If $a \in K^{-m}(X, X_0)$ and $b \in K^{-n}(Y, Y_0)$, then $ab = (-1)^{mn}ba$ where ab is the image of $a \otimes b$ under (9) and ba the image of $b \otimes a$ under (9′).*

PROOF. The sign comes from the use of the various "natural identifications" between different spaces. $S^m \wedge (X/X_0) \wedge S^n \wedge (Y/Y_0)$ and $S^n \wedge (Y/Y_0) \wedge S^m \wedge (X/X_0)$ are identified just by the permutation. However, for the definition of (9) we employ the identification

$$\alpha_{m,n} : S^m \wedge S^n \to S^{m+n}$$

which comes from a map $S^m \times S^n \to S^{m+n}$ of degree $+1$ (all spheres and also the cartesian product in this order have the standard orientations). If β is the permutation $S^m \times S^n \to S^n \times S^m$, then $\alpha_{n,m} \circ \beta \circ \alpha_{m,n}^{-1}$ has degree $(-1)^{mn}$. This shows that ab and ba correspond to elements of

$$G = [S^{m+n}(X \times Y/X_0 \times Y \cup X \times Y_0), Z \times B_U]_0$$

which are related with each other by a map of S^{m+n} onto itself of degree $(-1)^{mn}$. Since the group structure of G can also be defined by the suspension coordinate like a homotopy group, the lemma follows.

1.6. Using the diagonal map as in the definition of the cup product we get:

PROPOSITION. *Let $X \in \mathfrak{A}$. Then $\sum_{n \geq 0} K^{-n}(X)$ is a graded anti-commutative ring. Let $(X, Y) \in \mathfrak{B}$. Then there is a "graded homomorphism"*

$$\Big(\sum_{n \geq 0} K^{-n}(X)\Big) \otimes \Big(\sum_{m \geq 0} K^{-m}(X, Y)\Big) \to \sum_{i \geq 0} K^{-i}(X, Y),$$

making $\sum_{m \geq 0} K^{-m}(X, Y)$ a graded module over $\sum_{n \geq 0} K^{-n}(X)$.

The products have functorial properties. For example, if $(X, X_0) \to^f (X', X_0')$ and $(Y, Y_0) \to^g (Y', Y_0')$ are maps with the pairs all belonging to $\mathfrak{B}$, then we have the commutative diagram

$$\begin{array}{ccc} K^{-m}(X', X_0') \otimes K^{-n}(Y', Y_0') & \to & K^{-(m+n)}(X' \times Y', X_0' \times Y' \cup X' \times Y_0') \\ \downarrow f^! \otimes g^! & & \downarrow (f \times g)^! \\ K^{-m}(X, X_0) \otimes K^{-n}(Y, Y_0) & \to & K^{-(m+n)}(X \times Y, X_0 \times Y \cup X \times Y_0)\,. \end{array} \tag{10}$$

Furthermore, for $f : Y \to X$, the induced homomorphism $f^! : \sum_{m \geq 0} K^{-m}(X) \to \sum_{m \geq 0} K^{-m}(Y)$ is a ring homomorphism, etc.

1.7. *The Bott isomorphism.* The existence of the Bott isomorphism (see 1.3 (3)) is the central and deep point of the cohomology theory we are developing. We give now the explicit description of this isomorphism.

Let x_0 be the space consisting of a single point. Then (1.3) $K^{-2}(x_0)$ is infinite cyclic. By definition $K^{-2}(x_0) = \tilde{K}(S^2)$. Let η be the complex line bundle over S^2 whose first Chern class equals the canonical generator of $H^2(S^2, Z)$. Then η represents an element $[\eta] \in K(S^2)$ and $[\eta] - 1$ is a generator of $\tilde{K}(S^2) = K^{-2}(x_0)$ which we denote by g. If $a \in K^{-m}(X, X_0)$, then $ag \in K^{-(m+2)}(X, X_0)$. Here we use 1.5 (9) with $Y = x_0$ and Y_0 empty.

THEOREM. *The map $a \to ag$ is an isomorphism of $K^{-m}(X, X_0)$ onto $K^{-(m+2)}(X, X_0)$. In particular, $\sum_{n \geq 0} K^{-n}(x_0)$ is the polynomial ring $Z[g]$.*

For a proof of this central theorem we refer to [10]. For any $(X, X_0) \in \mathfrak{B}$ the graded group $\sum_{n\geq 0} K^{-n}(X, X_0)$ is a module over $Z[g]$. Multiplication with g^k gives an isomorphism of $K^{-n}(X, X_0)$ onto $K^{-(n+2k)}(X, X_0)$. This holds in particular if X_0 is empty or reduces to the base point, i.e., $\sum_{n\geq 0} K^{-n}(X)$ and $\sum_{n\geq 0} \tilde{K}^{-n}(X)$ are both modules over $Z[g]$. Let β denote the multiplication by g. The next lemma follows from 1.6 (10).

Lemma. *If (X, Y) and (X', Y') belong to $\mathfrak{B}$ and if $f : (X, Y) \to (X', Y')$ is a continuous map, then $f^!\beta = \beta f^!$ where $f^! : K^{-n}(X', Y') \to K^{-n}(X, Y)$, $(n \geq 0)$, is the induced homomorphism, in other words: $f^!$ is a homomorphism of $Z[g]$-modules.*

Lemma. *If $(X, Y) \in \mathfrak{B}$, then β gives a homomorphism of exact sequences* (1.4 (ii)), *i.e., we have the commutative diagram $(n \geq 0)$*

$$\begin{array}{ccccccc}
\tilde{K}^{-(n+1)}(Y) & \xrightarrow{\delta} & K^{-n}(X, Y) & \to & \tilde{K}^{-n}(X) & \to & \tilde{K}^{-n}(Y) \\
\downarrow \beta & & \downarrow \beta & & \downarrow \beta & & \downarrow \beta \\
\tilde{K}^{-(n+3)}(Y) & \xrightarrow{\delta} & K^{-(n+2)}(X, Y) & \to & \tilde{K}^{-(n+2)}(X) & \to & \tilde{K}^{-(n+2)}(Y).
\end{array}$$

The corresponding statement holds for the exact sequence (1.4 (i)).

Proof. This follows from the preceding lemma. We take into account that the homomorphism δ is also induced by a map, namely by $C_f \to S^1Y$.

1.8. The group $K^{-2n}(X, Y)$ can be identified with $K^0(X, Y)$ and $K^{-(2n+1)}(X, Y)$ with $K^{-1}(X)$ by the Bott isomorphisms:

$$\beta^n : K^0(X, Y) \to K^{-2n}(X, Y),$$

$$\beta^n : K^{-1}(X, Y) \to K^{-(2n+1)}(X, Y).$$

This allows us to define $K^n(X, Y)$ for any integer n by

$$K^n(X, Y) = K^0(X, Y) \quad \text{if } n \text{ is even},$$

$$K^n(X, Y) = K^{-1}(X, Y) \quad \text{if } n \text{ is odd}.$$

The groups $K^n(X, Y)$ satisfy the usual axioms of a cohomology theory [14] (in the category $\mathfrak{B}$ with all continuous maps of one pair into another one being admissable) except that $K^n(x_0)$ does not vanish for $n \neq 0$ (1.3). The existence of an exact sequence

$$(11) \quad \cdots \to K^n(Y) \xrightarrow{\delta} K^{n+1}(X, Y) \to K^{n+1}(X) \to K^{n+1}(Y) \to \cdots (-\infty < n < \infty)$$

follows from 1.4 and the second lemma of 1.7.

Let (X, Y, Z) be a triple with $X \supset Y \supset Z$ and all the pairs (X, Y), (X, Z), (Y, Z) belonging to $\mathfrak{B}$. Then we have an exact sequence

$$(11^*) \quad \cdots \to K^n(Y, Z) \xrightarrow{\delta} K^{n+1}(X, Y) \to K^{n+1}(X, Z) \to K^{n+1}(Y, Z) \to \cdots, \quad (-\infty < n < \infty),$$

where the δ of (11*) is the composition $K^n(Y, Z) \to K^n(Y) \to^{\delta} K^{n+1}(X, Y)$.

The exactness of (11*) would follow from 1.4(ii) applied to the pair $(X/Z, Y/Z)$ if this belonged to $\mathfrak{B}$. But (11*) is also a consequence of the cohomology axioms. (Excision-, homotopy-, and dimension axioms are not needed for this formal deduction of (11*); compare [**14**, Chapter I, §10].)

1.9. In 1.8 we have completed the construction of a cohomology theory satisfying all axioms except the "dimension axiom." Since these "cohomology groups" are periodic $(K^n(X, Y) = K^{n+2}(X, Y))$ it is convenient to define

$$K^*(X, Y) = K^0(X, Y) \oplus K^1(X, Y), \qquad (X, Y) \in \mathfrak{B},$$

and similarly for $K^*(X)$ and $\tilde{K}^*(X)$. $K^*(X)$ is then an anti-commutative ring, graded by Z_2, i.e., $K^0(X)$ is a subring and

$$K^0(X) \cdot K^1(X) \subset K^1(X), \qquad K^1(X) \cdot K^1(X) \subset K^0(X).$$

Moreover $K^*(X, Y)$ is a Z_2-graded module over $K^*(X)$. Since δ respects the periodicity, we have the exact triangle

$$\text{(12)} \qquad \begin{array}{ccc} K^*(Y) & \xrightarrow{\delta} & K^*(X, Y) \\ & \nwarrow \quad \swarrow & \\ & K^*(X) & \end{array}$$

which resolves in an exact hexagon

$$\begin{array}{ccccc} & K^1(X, Y) & \rightarrow & K^1(X) & \\ {}^{\delta}\nearrow & & & & \searrow \\ K^0(Y) & & & & K^1(Y) \\ \nwarrow & & & & \swarrow_{\delta} \\ & K^0(X) & \leftarrow & K^0(X, Y) & \end{array}$$

and which has, so to speak, the exact sequence (11) as "universal covering."

For a triple X, Y, Z (see 1.8) we have the exact triangle

$$\text{(12*)} \qquad \begin{array}{ccc} K^*(Y, Z) & \rightarrow & K^*(X, Y) \\ & \nwarrow \quad \swarrow & \\ & K^*(X, Z) & \end{array}$$

and the corresponding hexagon.

1.10. *The Chern character.* For each complex vector bundle ξ over the space $X \in \mathfrak{A}$ the Chern character $ch(\xi)$ is defined as an element of the rational cohomology ring $H^*(X, Q)$, [**5**, §9.1]. If $H^{**}(X, Q)$ denotes the direct sum of the even dimensional cohomology groups (which is a commutative subring of $H^*(X, Q)$), then $ch(\xi) \in H^{**}(X, Q)$. The definition of $ch(\xi)$ uses only the total Chern class $c(\xi)$. The classes $ch(\xi)$ and $c(\xi)$, both regarded as elements of $H^{**}(X, Q)$, determine each other. The Chern character induces a ring homomorphism [**15**, §12.1 (5)]

$$\text{(13)} \qquad ch: K(X) = K^0(X) \rightarrow H^{**}(X, Q) \subset H^*(X, Q)$$

with

$$ch(\tilde{K}(X)) \subset \tilde{H}^*(X, Q) = \text{Kernel} \quad (H^*(X, Q) \rightarrow H^*(\{x_0\}, Q)).$$

We are now going to define a group homomorphism

$$(14) \qquad ch\colon K^{-n}(X, Y) \to H^*(X, Y; Q), \quad ((X, Y) \in \mathfrak{B}, n \geq 0).$$

By definition, $K^{-n}(X, Y) = \tilde{K}(S^n(X/Y))$. We have the suspension isomorphism

$$\sigma^n\colon \tilde{H}^*(A, Q) \to \tilde{H}^*(S^n(A), Q), \qquad A \in \mathfrak{A},$$

which raises degrees by n and is defined by tensoring $a \in \tilde{H}^*(A, Q)$ (from the left) with the canonical generator of $H^n(S^n, Z)$. If $\xi \in K^{-n}(X, Y)$, let ξ' be the "corresponding element" of $\tilde{K}(S^n(X/Y))$. Then $ch(\xi') \in \tilde{H}^*(S^n(X/Y), Q)$ and $(\sigma^n)^{-1}\, ch(\xi') \in \tilde{H}^*(X/Y; Q)$. We have the canonical isomorphism

$$\alpha : \tilde{H}^*(X/Y, Q) \to H^*(X, Y; Q)$$

and we define

$$ch(\xi) = \alpha((\sigma^n)^{-1} ch(\xi')).$$

In 1.7 and 1.8 we described the Bott isomorphism. Since $ch([\eta] - 1)$ is the canonical generator of $H^2(S^2, Z)$ and since ch preserves products, it follows easily, that $ch(\beta(\xi)) = ch(\xi)$ for $\xi \in K^{-n}(X, Y)$. Therefore we can define $ch(\xi)$ for $\xi \in K^n(X, Y)$, n any integer. Using the notation of 1.9 we have now defined the Chern character as a homomorphism

$$ch\colon K^*(X, Y) \to H^*(X, Y; Q).$$

ch maps $K^0(X, Y)$ into $H^{ev}(X, Y; Q)$ and $K^1(X, Y)$ into $H^{od}(X, Y; Q)$ which denotes the direct sum of the odd-dimensional cohomology groups. The following theorem is easy to check.

Theorem. *The Chern character is a "natural transformation" of the "cohomology theory" described in* 1.9 *into the ordinary cohomology theory with rational coefficients for which one only considers the Z_2-grading $H^* = H^{ev} \oplus H^{od}$. In particular, ch preserves products, commutes with maps, $ch \circ f^! = f^* \circ ch$, and one has commutative diagrams*

$$\begin{array}{ccc} K^0(Y) & \xrightarrow{\delta} & K^1(X, Y) \\ {\scriptstyle ch}\downarrow & & \downarrow{\scriptstyle ch} \\ H^{ev}(Y, Q) & \xrightarrow{\delta} & H^{od}(X, Y; Q), \end{array} \qquad \begin{array}{ccc} K^1(Y) & \xrightarrow{\delta} & K^0(X, Y) \\ {\scriptstyle ch}\downarrow & & \downarrow{\scriptstyle ch} \\ H^{od}(Y, Q) & \xrightarrow{\delta} & H^{ev}(X, Y; Q). \end{array}$$

The commutativity of these diagrams can be deduced from the fact that the δ of both theories is induced from the map $C_f \to S^1Y$ (compare 1.4). One has to be careful with the signs. We hope to have chosen the various definitions such that commutativity (not only commutativity up to sign) holds in these diagrams.

2. The spectral sequence. Let X be a finite simplicial complex. We shall establish a spectral sequence relating the integral cohomology ring of X with $K^*(X)$.

2.1. Let X^n be the n-skeleton of X. We use the K-theory defined in 1.8. We filter $K^n(X)$ by defining

$$K_p^n(X) = \text{Kernel}\, [K^n(X) \to K^n(X^{p-1})].$$

THEOREM. *Let X be a finite simplicial complex. Let x_0 be the space consisting of a single point, so that $K^q(x_0) \cong Z$ if q is even and $K^q(x_0) = 0$ if q is odd. There exists a spectral sequence $E_r^{p,q}$ $(r \geqq 1, -\infty < p, q < \infty)$ with*

(1) $$E_1^{p,q} \cong C^p(X, K^q(x_0)),$$

d_1 being the ordinary coboundary operator.

(2) $$E_2^{p,q} \cong H^p(X, K^q(x_0)),$$

(3) $$E_\infty^{p,q} \cong G_p K^{p+q}(X) = K_p^{p+q}(X)/K_{p+1}^{p+q}(X).$$

The differential $d_r : E_r^{p,q} \to E_r^{p+r,q-r+1}$ vanishes for r even since $E_r^{p,q} = 0$ for all odd values of q.

PROOF. We use the method of [12, Chapter XV, §7] and define the graded group

$$H(p, q) = \sum_{-\infty<n<\infty} H^n(p, q) = \sum_{-\infty<n<\infty} K^n(X^{q-1}, X^{p-1}), \qquad q \geqq p.$$

These $H(p, q)$ satisfy the axiom (SP.1)–(SP.5) of [12, loc. cit.]. For axiom (SP.4) see 1.8 (11*).

$$E_1^{p,q} = K^{p+q}(X^p, X^{p-1}) = \sum_i K^{p+q}(\sigma_i^p, \dot\sigma_i^p),$$

where σ_i^p runs through all p-simplices. But $\sigma_i^p/\dot\sigma_i^p = S^p$. Therefore $K^{p+q}(\sigma_i^p, \dot\sigma_i^p) \cong \tilde K^{p+q}(S^p) \cong \tilde K^q(S^0) \cong K^q(x_0)$. This proves (1). To get (2) one has to check that d_1 is the ordinary coboundary operator.

2.2. REMARK. The preceding spectral sequence can be generalized to a fibre bundle (Y, X, F) with projection $\pi : Y \to X$. If this fibre bundle satisfies certain conditions, then there is a spectral sequence with $E_1^{p,q} \cong C^p(X, K^q(F))$ and $E_2^{p,q} \cong H^p(X, K^q(F))$ (local coefficients). Furthermore $E_\infty^{p,q} \cong G_p K^{p+q}(Y)$ with respect to a certain filtration of $K^{p+q}(Y)$. This spectral sequence specializes to the one of the theorem for $Y = X$ and π the identity.

2.3. The whole spectral sequence of 2.1 is compatible with the Bott periodicity. This makes it possible to forget about the grading and to use the notation of 1.9.

THEOREM. *Let X be a finite simplicial complex. Let $K_p^*(X)$ be the kernel of $K^*(X) \to K^*(K^{p-1})$. There exists a spectral sequence $E_r^p(X)$, $r \geqq 1$, with*

$$E_1^p(X) \cong C^p(X, Z),$$

$$E_2^p(X) \cong H^p(X, Z),$$

$$E_\infty^p(X) \cong G_p K^*(X) = K_p^*(X)/K_{p+1}^*(X).$$

The differentials d_r vanish for even r.

This spectral sequence could also be obtained directly by the method of [12, Chapter XV, §7] by putting $H(p, q) = K^*(X^{q-1}, X^{p-1})$.

Remark. It is easy to show (by the notion of 1-equivalence, [12, p. 336]) that the $E_r^p(X)$ together with the differentials d_r are homotopy type invariants of X for $r \geqq 2$. Also $K_q^*(X)$ is a homotopy type invariant. It can be invariantly defined as follows: an element ξ of $K^*(X)$ belongs to $K_q^*(X)$ if and only if for any finite simplicial complex Y of dimension $\leqq q - 1$ and any continuous map $f : Y \to X$ we have $f^!\xi = 0$. Thus the spectral sequence $\{E_r^p(X), r \geqq 2\}$ is well-defined for any space X of the homotopy type of a finite simplicial complex. By a theorem of J. H. C. Whitehead [19, p. 239, Theorem 13] any finite CW-complex is of the homotopy type of a finite simplicial complex. Hence the spectral sequence $\{E_r^p(X), r \geqq 2\}$ is well-defined for spaces of the class $\mathfrak{A}$ (see 1.1).

The differentials d_r are certain (higher order) cohomology operations. $d_3 : E_3^p \cong H^p(X, Z) \to E_3^{p+3} \cong H^{p+3}(X, Z)$ is the Steenrod operation Sq^3.

2.4. Let X be a finite simplicial complex. We propose to study the spectral sequence of 2.3 in its relation with the Chern character. Let $'E_r^p$ be the spectral sequence with

$$'E_1^p = C^p(X, Q), \qquad d_1 \text{ the ordinary coboundary operator,}$$

$$'E_r^p = H^p(X, Q) \quad \text{for} \quad r \geqq 2, \qquad 'd_r = 0 \quad \text{for} \quad r \geq 2.$$

This trivial spectral sequence is obtained by the method of [12, Chapter XV, §7] by putting $'H(r, s) = H^*(X^{s-1}, X^{r-1}; Q)$ for $s \geqq r$. The spectral sequence of 2.3 comes from $H(r, s) = K^*(X^{s-1}, X^{r-1})$. The Chern character gives a homomorphism

$$ch: H(r, s) \to 'H(r, s),$$

and since ch is a natural transformation from the K^*-theory to the rational cohomology theory, we get a homomorphism ch from the spectral sequence $\{E_r^p\}$ of 2.3 into the spectral sequence $\{'E_r^p\}$. Using ch we can prove:

Theorem. *Suppose $X \in \mathfrak{A}$ (see 1.1). The spectral sequence $\{E_r^p(X)\}$ collapses (i.e., $d_r = 0$ for $r \geqq 2$ and thus $E_2^p(X) \cong E_\infty^p(X)$) if one of the following conditions is satisfied:*

(i) *$H^*(X, Z)$ has no torsion,*

(ii) *$H^*(X, Z) = 0$ for all odd integers i.*

Proof. We may assume that X is a finite simplicial complex. $ch : E_r^p \to {'E_r^p}$ is always injective for $r = 1$, since then it is just the coefficient homomorphism $C^p(X, Z) \to C^p(X, Q)$. For $r = 2$ it is the homomorphism $H^p(X, Z) \to H^p(X, Q)$ which is injective if X has no torsion. Since the $'d_r$ vanish for $r \geqq 2$ it follows by induction on r that the d_r also vanish for $r \geqq 2$ if $E_2^p \to {'E_2^p}$ is injective. This proves the theorem under the assumption (i). If (ii) holds, then $d_r (r \geqq 3$, odd) vanishes since it maps $E_r^p(X)$ in $E_r^{p+r}(X)$, and one of these groups is 0. The d_r (r even) vanish anyhow.

THEOREM. *Suppose $X \in \mathfrak{A}$ (see 1.1). The spectral sequence $\{E_r^p(X) \otimes Q\}$ collapses (i.e., $d_r \otimes Q = 0$ for $r \geqq 2$).*

$$ch: K^*(X) \otimes Q \to H^*(X, Q)$$

is bijective and maps $K^0(X) \otimes Q$ onto $H^{ev}(X, Q)$ and $K^1(X) \otimes Q$ onto $H^{od}(X, Q)$.

PROOF. We may assume that X is a finite simplicial complex. The spectral sequence $\{E_r^p(X) \otimes Q\}$ is obtained by putting $''H(p, q) = K^*(X^{q-1}, X^{p-1}) \otimes Q$. The Chern character gives a homomorphism of this spectral sequence into the spectral sequence $\{'E_r^p(X)\}$ which is bijective for $r = 1$. This implies the theorem (compare [**12**, Chapter XV, Theorem 3.2]).

COROLLARY. *Suppose $X \in \mathfrak{A}$ (see 1.1). If $K^*(X)$ has no torsion, then*

$$ch: K^*(X) \to H^*(X, Q)$$

is injective.

2.5. The preceding results on the spectral sequence imply:

COROLLARY. *Let X be a space belonging to $\mathfrak{A}$ (see 1.1). Then $K^*(X)$ is additively a finitely generated abelian group. The rank of $K^0(X)$ equals the sum of the even dimensional Betti numbers of X, whereas the rank of $K^1(X)$ is the sum of the odd dimensional Betti numbers of X.*

For any $\xi \in K^*(X)$ let $ch_n(\xi)$ be the n-dimensional component of $ch(\xi)$.

COROLLARY. *Suppose $X \in \mathfrak{A}$ and that $H^*(X, Z)$ has no torsion. Then*

(i) *$\xi \in K_p^*(X)$ if and only if $ch_r(\xi) = 0$ for $r < p$, in particular*

$$ch: K^*(X) \to H^*(X, Q)$$

is injective and $K^*(X)$ is without torsion, i.e., free abelian.

(ii) *If $\xi \in K_p^*(X)$, then $ch_p(\xi) \in H^p(X, Q)$ comes from an integral class which is uniquely determined and equal to the image of ξ in $K_p^*(X)/K_{p+1}^*(X) \cong H^p(X, Z)$. To every integral p-dimensional class x, there exists $\xi \in K_p^*(X)$ with $ch_p(\xi) = x$, i.e., $ch(\xi) = x +$ higher terms.*

(iii) *Let A be a subgroup of $K^*(X)$. If for every $x \in H^p(X, Z)$, $p \geqq 0$, there exists $\xi \in A$ with $ch(\xi) = x +$ higher terms, then $A = K^*(X)$.*

2.6. So far we have not studied the behaviour of the spectral sequence (2.3) with respect to the product structure of $K^*(X)$. We have only been able to get a partial result which we summarize without proof in the following theorem.

THEOREM. *Suppose $X \in \mathfrak{A}$. We consider the spectral sequence $E_r^p(X)$ ($r \geqq 2$) with the operators d_r. Let Z_r^p be the kernel and B_r^p the image of d_r. There exist pairings $\prod_r : E_r^p(X) \otimes E_r^q(X) \to E_r^{p+q}(X)$ with*

$$Z_r^p(X) \otimes Z_r^q(X) \to Z_r^{p+q}(X),$$
$$(4) \qquad Z_r^p(X) \otimes B_r^q(X) \to B_r^{p+q}(X) \quad \text{and} \quad B_r^p(X) \otimes Z_r^q(X) \to B_r^{p+q}(X),$$

and such that $\prod_{r+1}$ *is induced from* $\prod_r$ *in virtue of* (4). *Moreover,* $\prod_2$ *is the cup-product and* $\prod_\infty$ *is the product in* $GK^*(X)$ *induced by the ring structure of* $K^*(X)$ *for which*

$$K^*_p(X) \cdot K^*_{p'}(X) \subset K^*_{p+p'}(X). \tag{5}$$

We conjecture that d_r is an anti-derivation. This would imply (4). We shall only need (5) in the sequel. (5) admits a straightforward proof.

By (5) the mth power of an element of $K^*_1(X)$ belongs to $K^*_m(X)$. If m is sufficiently large then $K^*_m(X)$ is zero, hence any element of $K^*_1(X)$ is nilpotent. Clearly, $\xi \in K^*_1(X)$ if and only if $ch_0(\xi) = 0$. This special case of 2.5 (i) holds for any $X \in \mathfrak{A}$. We conclude:

PROPOSITION. *An element* ξ *of* $K^*(X)$ *is nilpotent if and only if* $ch_0(\xi) = 0$. *An element* η *of* $K^*(X)$ *is invertible if and only if* $ch_0(\eta) = \pm 1$.

PROOF. It remains to show that η is invertible if $ch_0(\eta) = \pm 1$. In this case, $\pm\eta = 1 - \xi$ with $ch_0(\xi) = 0$ and thus ξ nilpotent. Then $\eta^{-1} = \pm(1 + \xi + \xi^2 + \cdots + \xi^{m-1})$ if $\xi^m = 0$.

3. The differentiable Riemann-Roch theorem and some applications.

3.1. We recall the Riemann-Roch theorem given in [1] in a slightly more general formulation. Let X, Y be compact oriented differentiable manifolds. By the triangulation theorem of Cairns, X and Y belong to the class $\mathfrak{A}$ of 1.1. A continuous map $f : Y \to X$ will be called a c_1-map if we are given an element $c_1(f) \in H^2(Y, Z)$ such that $c_1(f) \equiv w_2(Y) - f^*w_2(X) \bmod 2$ where $w_2(Y)$ and $w_2(X)$ are the second Stiefel-Whitney classes of Y and X respectively ($w_2 \in H^2(\ , Z_2)$). As in [5; 1], if ξ is a real vector bundle with finite-dimensional base B_ξ we define

$$\hat{\mathfrak{A}}(\xi) = \prod_i (x_i/2)/(\sinh(x_i/2)) \in H^*(B_\xi, Q)$$

where the Pontrjagin classes of ξ are the elementary symmetric functions in the x_i^2. If ξ is the tangent bundle of the differentiable manifold X we write $\hat{\mathfrak{A}}(X)$ instead of $\hat{\mathfrak{A}}(\xi)$.

THEOREM. *Let* Y *and* X *be as before. Let* $f : Y \to X$ *be a* c_1*-map. Then there exists a homomorphism*

$$g : K^*(Y) \to K^*(X)$$

such that

(i) $$ch(g(y)) \cdot \hat{\mathfrak{A}}(X) = f_*(ch(y)e^{c_1(f)/2} \cdot \hat{\mathfrak{A}}(Y)), \qquad y \in K^*(Y),$$

where f_* *is the Gysin homomorphism (Poincaré dual of the homology homomorphism).*

(ii) g *maps* $K^0(Y)$ *into* $K^0(X)$ *and* $K^1(Y)$ *into* $K^1(X)$ *if* $\dim Y \equiv \dim X \pmod 2$.

g *maps* $K^0(Y)$ *into* $K^1(X)$ *and* $K^1(Y)$ *into* $K^0(X)$ *if* $\dim Y \not\equiv \dim X \pmod 2$.

(iii) *g is related to the homomorphism $f^!: K^*(X) \to K^*(Y)$ by the formula*

$$g(f^!(x)\cdot y) = x\cdot g(y), \qquad x \in K^*(X), \qquad y \in K^*(Y).$$

If we define $\hat{\mathfrak{A}}(f) = \hat{\mathfrak{A}}(Y)\cdot f^*(\hat{\mathfrak{A}}(X)^{-1})$, then (i) *may be written as*

(i′) $$ch(g(y)) = f_*(ch(y)\, e^{c_1(f)/2}\cdot\hat{\mathfrak{A}}(f)).$$

This theorem is slightly more general than Theorem 1 of [1] which was formulated for $K^0(X)$. Here we have stated it for $K^*(X)$ which makes the assumption $\dim Y \equiv \dim X \pmod 2$ superfluous. The proof does not have to be changed once one has developed the cohomology theory of §1. Moreover we assert here the existence of the homomorphism g satisfying (iii). This brings no additional difficulty. One just has to follow up the proof of Theorem 1 of [1] (see also [16]). Something new would be involved if we tried to choose g in a natural way (call it then $f_!$) and prove certain functorial properties of it. We shall take up this question in a more detailed exposition. Nevertheless we permit ourselves to call the g of the theorem $f_!$. But we are not allowed then to use for $Z \to^{\bar{f}} Y \to^{f} X$ the formula $(f\circ \bar{f})_! = f_! \circ \bar{f}_!$. (The composition of two c_1-maps is a c_1-map in a natural way.) The formula (i′) shows that $ch(g(y))$ is uniquely determined for a c_1-map f. Therefore (2.4, 2.5), $g = f_!$ is given without ambiguity if $K^*(X)$ or $H^*(X, Z)$ has no torsion.

It follows easily from (i′) that

$$ch((f\circ\bar{f})_! z) = ch(f_!(\bar{f}_! z)) \quad \text{for} \quad z \in K^*(Z).$$

By (2.4, 2.5)

$$(f\circ\bar{f})_! z = f_!(\bar{f}_! z) \quad \text{if} \quad K^*(X) \quad \text{or} \quad H^*(X, Z) \quad \text{has no torsion.}$$

3.2. Let Y be a compact oriented differentiable manifold. It is called a c_1-manifold if we are given an element $c_1(Y) \in H^2(Y, Z)$ whose restriction mod 2 is $w_2(Y)$. For a c_1-manifold Y the Todd genus $T(Y)$ is defined. It is equal to the value of the top-dimensional component of $e^{c_1(Y)/2}\cdot\hat{\mathfrak{A}}(Y)$ on the fundamental cycle of Y. By definition, $T(Y)$ is a rational number. It is an integer as follows by applying Theorem 3.1 to the map of Y onto a point. Compare [1], see also [6]. If Y is almost-complex and $c_1(Y)$ the first Chern class, then $T(Y)$ is the usual Todd genus which is equal to the arithmetic genus if Y is a projective algebraic manifold [15].

3.3. Let $\xi = (E_\xi, B_\xi, F_\xi, \pi_\xi)$ be a differentiable fibre bundle in the sense of [5, §7.4]. Assume that E_ξ, B_ξ, F_ξ are compact oriented differentiable manifolds. As in [5] we let $\hat{\xi}$ be the bundle along the fibres. This is a real vector bundle over E_ξ whose second Stiefel-Whitney class $w_2(\hat{\xi})$ equals $w_2(E_\xi) - \pi^*w_2(B_\xi)$. Assume that $\pi = \pi_\xi$ is a c_1-map. Then $c_1(\pi) \equiv w_2(\hat{\xi}) \bmod 2$. If $i: F_\xi \to E_\xi$ is the injection of a fibre in the total space then

$$i^*c_1(\pi) \equiv w_2(F_\xi) \mod 2.$$

Therefore if we put $c_1(F_\xi) = i^*c_1(\pi)$, the manifold F_ξ becomes a c_1-manifold and we can speak of the Todd genus $T(F_\xi)$. Assume that ξ is endowed with a complex structure, i.e., we are given a complex vector bundle η over E_ξ which considered as real vector bundle is ξ. Then F_ξ is almost complex in a natural way. Furthermore π is a c_1-map with $c_1(\pi) = c_1(\eta)$. The Todd genus $T(F_\xi)$ is then the same whether we consider F_ξ as c_1-manifold with $c_1(F_\xi) = i^*c_1(\eta)$ or as almost complex manifold.

3.4. THEOREM. *Let ξ be a differentiable fibre bundle as in 3.3. Let $\pi = \pi_\xi$ be a c_1-map. If the Todd genus $T(F_\xi) = \pm 1$ then the homomorphism*

$$\pi^! : K^*(B_\xi) \to K^*(E_\xi)$$

is injective. Moreover $\pi^!$ identifies $K^*(B_\xi)$ with a direct summand of $K^*(E_\xi)$. The endomorphism $\pi_! \circ \pi^!$ of $K^*(B_\xi)$ is the multiplication with a fixed invertible element of $K^*(B_\xi)$.

PROOF. We shall use Theorem 3.1 for π with $Y = E_\xi$ and $X = B_\xi$. First we observe that

$$\mathfrak{A}(\xi) = \mathfrak{A}(E_\xi)\cdot(\pi^*\mathfrak{A}(B_\xi))^{-1} = \mathfrak{A}(\pi).$$

Therefore with $g = \pi_!$ we have by 3.1 (i′)

$$ch(\pi_!(y)) = f_*(ch(y)\cdot e^{c_1(\pi)/2}\cdot\mathfrak{A}(\xi)), \qquad y \in K^*(E_\xi).$$

Now put $y = 1$, the unit of $K^*(E_\xi)$. Then $ch(y) = 1$ and it follows easily that the zero-dimensional component of $ch(\pi_! 1)$ equals $T(F_\xi)$. Since $T(F_\xi) = \pm 1$, $\pi_! 1$ is an invertible element in $K^*(B_\xi)$ (see 2.6) whose inverse we denote by a. Now let h be the homomorphism $K^*(E_\xi) \to K^*(B_\xi)$ equal to $\pi_!$ followed by multiplication with a; then (iii) of 3.1 gives

$$h(\pi^!(x)) = x \quad \text{for all} \quad x \in K^*(B_\xi)$$

which proves the theorem.

The preceding theorem admits various generalisations. For example, if the Todd genus $T(F_\xi) = m \neq 0$, $(m \in Z)$, then $\pi^!$ is injective on the direct sum of those p-primary components of $K^*(B_\xi)$ with $p = 0$ or a prime not dividing m. This type of theorem is analogous to 3.2 of [4].

3.5. Let G be a compact connected Lie group and T a maximal torus of G. Let ξ be a principal G-bundle whose base space B_ξ is a compact oriented differentiable manifold. Consider the associated bundle with G/T as fibre. Its total space is E_ξ/T, its base space is B_ξ. With these assumptions we have:

PROPOSITION. *Let π be the projection $E_\xi/T \to B_\xi$. Then*

$$\pi^! : K^*(B_\xi) \to K^*(E_\xi/T)$$

is injective. $\pi^! K^*(B_\xi)$ is a direct summand of $K^*(E_\xi/T)$.

PROOF. We may assume that ξ is differentiable. The bundle along the fibres of E_ξ/T admits a complex structure such that G/T has Todd genus 1 (see

[5, §§7.4, 22.3]). The complex structure along the fibres and the orientation of B_ξ define an orientation for the compact differentiable manifold E_ξ/T. The proposition follows from 3.4.

THEOREM. *We make the preceding assumptions. Let U be a closed connected subgroup of G of maximal rank, i.e., we may assume $U \supset T$. Then E_ξ/U is the total space of the bundle associated to ξ and with G/U as fibre. Let σ be the projection $E_\xi/U \to B_\xi$. Then*

$$\sigma^! : K^*(B_\xi) \to K^*(E_\xi/U)$$

is injective. $\sigma^! K^*(B_\xi)$ is a direct summand of $K^*(E_\xi/U)$.

PROOF. We have the diagram

$$E_\xi/T \xrightarrow{\rho} E_\xi/U \xrightarrow{\sigma} B_\xi, \qquad \sigma \circ \rho = \pi, \qquad \pi^! = \rho^! \circ \sigma^!.$$

By the above proposition $\pi^!$ is injective which implies $\sigma^!$ is injective. Also the last statement of the theorem follows immediately.

REMARK. We have proved this theorem under the assumption that B_ξ is a compact oriented differentiable manifold. A small generalization of the Riemann-Roch Theorem 3.1 makes it possible to drop the assumption on orientability. It is probably also true when B_ξ is any finite CW-complex.

The preceding theorem holds in particular for bundles with an even dimensional sphere as fibre and the special orthogonal group as structure group. If $\pi : Y \to X$ is such a bundle (X compact oriented differentiable), then $\pi^! : K^*(X) \to K^*(Y)$ is injective. The corresponding theorem for integral cohomology holds if X has no 2-torsion (more generally, π^* is injective on the direct sum of the p-primary components of $H^*(X, Z)$ with $p = 0$ or p an odd prime).

3.6. THEOREM. *Let G be a compact connected Lie group, U a closed connected subgroup of G of maximal rank. Then $K^1(G/U) = 0$ and $K^0(G/U)$ is a free abelian group with rank equal to the quotient of the order of the Weyl group of G by the order of the Weyl group of U.*

PROOF. The theorem is true if $U = T$ (maximal torus of G). In this case G/T has no torsion in integral cohomology and its odd dimensional cohomology groups vanish [7]. The theorem follows then from 2.5 if one takes into account that the order of $W(G)$ (Weyl group of G) is the Euler number of G/T which equals $\dim_Q H^{**}(G/T, Q)$. For the general case, we assume that $U \supset T$ and consider the map $\pi : G/T \to G/U$. Then $\pi^!$ is injective by 3.5. It follows that $K^1(G/U) = 0$ and that $K^0(G/U)$ has no torsion. It is well-known [3] that the odd-dimensional Betti numbers of G/U vanish and that the Euler number of G/U equals ord $W(G)$/ord $W(U)$. Thus $\dim_Q H^{**}(G/U, Q) =$ ord $W(G)$/ord $W(U)$ which completes the proof in virtue of 2.5.

REMARK. As in the case of G/T, Theorem 3.6 follows immediately from 2.5 if $H^*(G/U, Z)$ has no torsion.

4. The classifying space of a compact connected Lie group.

4.1. *Completions of modules.* We shall summarize here some known results of commutative algebra which we learned from J. P. Serre. These results are needed in the sequel. For references see Zariski and Samuel, *Commutative algebra*, Van Nostrand, and [**13**, Exposé 18 (Godement)].

Let A be a Noetherian ring, $\mathfrak{a}$ an ideal of A. We give every finitely generated A-module M the topology defined by the submodules $\mathfrak{a}^n M$. The completion of M for this "$\mathfrak{a}$-adic topology" is by definition

$$\hat{M} = \varprojlim M/\mathfrak{a}^n M \qquad \text{(inverse limit).}$$

(i) *Let N be a submodule of M. Then the $\mathfrak{a}$-adic topology of N coincides with the topology induced on N by the $\mathfrak{a}$-adic topology of M.*

This is a consequence of the lemma of Artin-Rees which says that there exists a positive integer h such that $(\mathfrak{a}^n M) \cap N = \mathfrak{a}^{n-h}((\mathfrak{a}^h M) \cap N)$ for $n \geqq h$; see [**13**, Exposé 2, Théorème 2].

(ii) *Let $0 \to N \to M \to P \to 0$ be an exact sequence of (finitely generated) A-modules; then*

$$0 \to \hat{N} \to \hat{M} \to \hat{P} \to 0$$

is exact. Thus "completion" is an exact functor [**12**, Chapter II, §4].

Proof. We have the exact sequence

$$0 \to N/(\mathfrak{a}^n M \cap N) \to M/\mathfrak{a}^n M \to P/\mathfrak{a}^n P \to 0.$$

By (i), $\hat{N}$ is the inverse limit of the first inverse system in this sequence. Since $N/(\mathfrak{a}^{n+k} M \cap N) \to N/(\mathfrak{a}^n M \cap N)$ is onto for all n and all $k \geqq 0$, this inverse system satisfies the "Mittag-Leffler condition." According to the forthcoming book of Dieudonné-Grothendieck (Complements to Chapter 0) the assertion (ii) follows. There is, of course, a direct proof along the lines of [**11**, §3].

(iii) *Let B be a commutative ring, G a finite group of automorphisms of B and let $A = B^G$ be the subring of those elements of B which are invariant under all automorphisms of G. Assume B is, as an A_0-algebra, finitely generated over a Noetherian subring A_0 of A. Then B and A are Noetherian and B is a finitely generated A-module.*

Proof. Since A_0 is Noetherian, B (as a quotient ring of a polynomial ring over A_0) is also Noetherian. If $x \in B$, then $\prod_{\sigma \in G} (x - \sigma(x)) = 0$. Thus x is integral over A. Let $x_1, \cdots, x_n$ be generators of B over A_0. Then we have equations

$$x_i^q + a_{i1}x_i^{q-1} + \cdots + a_{iq} = 0, \qquad a_{ij} \in A, \qquad q = \text{order of } G.$$

Thus B is generated as an A-module by the monomials $x_1^{m_1} \cdots x_n^{m_n}$ $(m_i \leqq q - 1)$, hence is a finitely generated A-module. Let A' be the subring of A generated over A_0 by the a_{ij}. The ring A' is Noetherian since it is a finitely generated

A_0-algebra. B is even a finitely generated A'-module. Thus also A is a finitely generated A'-module. If $c_1, \cdots, c_k$ are generators of A as module over A', then the c_r and the a_{ij} generate A as A_0-algebra. Hence A is a Noetherian ring.

(iv) *We make the assumptions of* (iii). *Let* $\mathfrak{b}$ *be an ideal of* B *which is stable under* G ($\sigma(\mathfrak{b}) = \mathfrak{b}$ *for* $\sigma \varepsilon G$). *Put* $\mathfrak{a} = \mathfrak{b} \cap A$. *Let* $\mathfrak{b}' = \mathfrak{a} \cdot B$ *be the ideal of* B *generated by* $\mathfrak{a}$. *Then there exists a positive integer* n *such that* $\mathfrak{b}^n \subset \mathfrak{b}' \subset \mathfrak{b}$. *Thus* $\mathfrak{b}$ *and* $\mathfrak{b}'$ *define the same topology on* B.

Proof. In a noetherian ring, to prove that a power of the ideal $\mathfrak{b}$ is contained in $\mathfrak{b}'$, it is enough to show that all prime ideals $\mathfrak{p}$ containing $\mathfrak{b}'$ also contain $\mathfrak{b}$ (see for example [**13**, Exposé 2]). Let $\mathfrak{p}$ be a prime containing $\mathfrak{b}'$ and let $x \varepsilon \mathfrak{b}$. Then $x' = \prod_{\sigma \varepsilon G} \sigma(x) \varepsilon A \cap \mathfrak{b} = \mathfrak{a} \subset \mathfrak{b}'$. Hence $x' \varepsilon \mathfrak{p}$. Hence there is a σ with $\sigma(x) \varepsilon \mathfrak{p}$ and thus $x \varepsilon \sigma^{-1}(\mathfrak{p})$. Hence $\mathfrak{b}$ is contained in the union of the prime ideals $\sigma(\mathfrak{p})$, $\sigma \varepsilon G$. But it is an easy lemma (see Northcott, *Ideal theory*, Cambridge Tracts, pp. 12–13), true in any ring, that if an ideal $\mathfrak{b}$ is contained in the union of a finite number of prime ideals, it is contained in one of them. Thus in our case, $\mathfrak{b} \subset \sigma(\mathfrak{p})$ for some $\sigma \varepsilon G$. But $\mathfrak{b} = \sigma^{-1}(\mathfrak{b})$ by assumption. Thus $\mathfrak{b} \subset \mathfrak{p}$ as contended.

We consider A and B both as A-modules and complete them with respect to the $\mathfrak{a}$-adic topology. We have a map $\hat{A} \to \hat{B}$ which is injective by (ii). In view of (iv) $\hat{B}$ is also the completion of B with respect to the $\mathfrak{b}$-adic topology of B. The group G operates naturally on $\hat{B}$. Let $(\hat{B})^G$ be the ring of invariants.

(v) *Under the preceding assumptions the map* $\hat{A} \to \hat{B}$ *maps* $\hat{A}$ *(bijectively) onto* $(\hat{B})^G$. *Thus* $(B^G)^\wedge = (\hat{B})^G$.

Proof. Let $B(G)$ be the ring of all maps from G into B. This is a direct sum of g copies of B where g is the order of G. We consider the exact sequence

$$0 \to B^G \to B \xrightarrow{\alpha} B(G)$$

where $\alpha(b)$, $b \varepsilon B$, is the map which attaches to $\sigma \varepsilon G$ the element $b - \sigma(b) \varepsilon B$. All rings in this exact sequence have to be considered as A-modules ($A = B^G$). We complete them with respect to the $\mathfrak{a}$-adic topology. "Completion" is an exact functor, hence $(B(G))^\wedge = \hat{B}(G)$ and the resulting sequence

$$0 \to (B^G)^\wedge \to \hat{B} \xrightarrow{\hat{\alpha}} \hat{B}(G)$$

is exact which proves (v).

4.2. *The representation ring of a compact Lie group*. Let G be a compact Lie group. Let $(\rho_1, \rho_2, \cdots)$ be the (equivalence classes of) irreducible complex representations of G. Let $R(G)$ be the free abelian group generated by the ρ_i. The tensor product of representations makes $R(G)$ into a ring which we shall call the representation ring of G. The complex representations of G may be identified with the elements $\sum n_i\rho_i$ of $R(G)$ where the n_i are non-negative integers.

Let $\epsilon : R(G) \to Z$ be the "augmentation homomorphism" obtained by attaching to each representation of G its dimension. Let $I(G)$ be the kernel of ϵ; it will be called the augmentation ideal of $R(G)$. We define the completed representation ring with respect to the $I(G)$-adic topology:

$$\hat{R}(G) = \varprojlim R(G)/I(G)^n \qquad \text{(inverse limit)}.$$

Let G, H be compact Lie groups and $G \to H$ a homomorphism; then we have an induced homomorphism $R(H) \to R(G)$ which maps $I(H)$ in $I(G)$ and is therefore continuous with respect to the $I(H)$-adic topology of $R(H)$ and the $I(G)$-adic topology of $R(G)$. It induces therefore a homomorphism $\hat{R}(H) \to \hat{R}(G)$. Suppose now $G = H$. Then any automorphism of G induces automorphisms of $R(G)$ and of $\hat{R}(G)$. An inner automorphism induces the identity. If G is connected and T a maximal torus of G, then the Weyl group $W(G)$ is a group of automorphisms of T and thus operates also on $R(T)$ and $\hat{R}(T)$.

4.3. *The completed representation ring of a torus.* Let T be a torus. We write it as the group of k-tuples of reals mod 1. Every irreducible representation of T is 1-dimensional and given by a homomorphism

$$(x_1, \cdots, x_k) \to \exp(2\pi i(a_1x_1 + \cdots + a_kx_k)), \qquad a_i \in Z,$$

of T into $U(1)$. The ring $R(T)$ may be identified with that subring of the ring of formal power series $C[[x_1, \cdots, x_k]]$ which is generated over Z by

$$\exp(2\pi i x_1), \exp(-2\pi i x_1), \cdots, \exp(2\pi i x_k), \exp(-2\pi i x_k).$$

Hence *$R(T)$ is Noetherian.*

Let $z_1, \cdots, z_k$ be indeterminates. We give the polynomial ring $Z[z_1, \cdots, z_k]$ the $(z_1, \cdots, z_k)$-adic topology, and define a ring homomorphism

$$\phi : Z[z_1, \cdots, z_k] \to R(T)$$

by setting $\phi(z_i) = \exp(2\pi i x_i) - 1$. Then $\phi(z_i) \in I(T)$, thus ϕ is continuous and induces a homomorphism $\hat{\phi}$ of the completed rings.

PROPOSITION. *The homomorphism*

$$\hat{\phi} : Z[[z_1, \cdots, z_k]] \to \hat{R}(T)$$

is bijective. ($Z[[z_1, \cdots, z_k]]$ *is the ring of formal power series.*)

PROOF. Put $A = Z[z_1, \cdots, z_k]$. Under ϕ we may identify A with a subring of $R(T)$. For the latter ring we may write

$$R(T) = Z[z_1, \cdots, z_k, (1 + z_1)^{-1}; \cdots, (1 + z_k)^{-1}].$$

We have then $I(T) = (z_1, \cdots, z_k, (1 + z_1)^{-1} - 1, \cdots, (1 + z_k)^{-1} - 1)$. Thus the ideal $I(T)^n$ of $R(T)$ contains only formal power series with lowest term of degree $\geq n$. Thus $I(T)^n \cap A$ contains only polynomials with lowest term of degree $\geq n$. Therefore, $I(T)^n \cap A \subset (z_1, \cdots, z_k)^n$. Clearly, $(z_1, \cdots, z_k)^n \subset I(T)^n \cap A$. Thus

$$I(T)^n \cap A = (z_1, \cdots, z_k)^n = (I(T) \cap A)^n.$$

This shows that the $(z_1, \cdots, z_k)$-adic topology of A coincides with the topology induced from the embedding of A in $R(T)$. Thus $\hat{\phi}$ is injective. Since $\hat{\phi}(\hat{A})$ contains $R(T)$, the map $\hat{\phi}$ is surjective.

NOTE. We have just considered $R(T)$ as subring of $\hat{R}(T)$. This is all right, since $R(T)$ *is Hausdorff* in its $I(T)$-adic topology. In fact, $\bigcap_n I(T)^n = 0$, since an element of this intersection would be a power series whose lowest term has an arbitrarily high degree.

4.4. *The completed representation ring of a compact connected Lie group.* Let G be a compact *connected* Lie group and T a maximal torus of G. The Weyl group $W(G)$ operates on $R(T)$; see 4.2. We have a ring homomorphism $R(G) \to R(T)$ (by the restriction map) which is injective. $R(G)$ maps (bijectively) onto the ring of invariants of $R(T)$ under the action of $W(G)$. This classical result follows from the fact that the highest weight of an irreducible representation has multiplicity one (compare [5, §3.4]). We denote this ring of invariants by $R(T)^{W(G)}$ and identify $R(G)$ with it. We have the situation of 4.1 (iii). A_0 is here the ring Z of integers. Thus we know that $R(G)$ is Noetherian and that $R(T)$ is a finitely generated module over $R(G)$.

$W(G)$ operates naturally on $\hat{R}(T)$ and we have an induced map $\hat{R}(G) \to \hat{R}(T)$ (see 4.2).

THEOREM. *Let G be a compact connected Lie group, T a maximal torus of G. Then $\hat{R}(G) \to \hat{R}(T)$ maps $\hat{R}(G)$ bijectively onto $(\hat{R}(T))^{W(G)}$, the ring of invariants of $W(G)$ in $\hat{R}(T)$.*

PROOF. We are exactly in the situation of 4.1. Here $R(T)$ plays the role of B, $R(G)$ of A, $W(G)$ of G, and Z of A_0. The ideal $\mathfrak{b}$ corresponds to $I(T)$, the ideal $\mathfrak{a}$ to $I(T) \cap R(G) = I(G)$.

NOTE. $R(G)$ is Hausdorff, since $\bigcap_n I(G)^n \subset \bigcap_n I(T)^n = 0$. The homomorphism $R(G) \to \hat{R}(G)$ is injective. This is in general not true if G is not connected (Atiyah, *Characters and cohomology*, in preparation).

4.5. Let X be a space belonging to the class $\mathfrak{A}$ of 1.1. Let ξ be a principal G-bundle over X where G is a compact Lie group. ξ induces a ring homomorphism

$$\alpha_\xi : R(G) \to K^0(X) \subset K^*(X)$$

in the following way. Consider a representation of G viewed as a homomorphism $\rho : G \to U(m)$. Then $\rho(\xi)$ is a principal $U(m)$-bundle and defines an element $\alpha_\xi(\rho)$ of $K^0(X)$. Since the (equivalence classes of) irreducible representations are free generators of the additive group $R(G)$ the homomorphism α_ξ is well-defined.

If we have a map $f : Y \to X$ ($Y, X \in \mathfrak{A}$), if ξ is a principal G-bundle over X and $\eta = f^*\xi$ the principal G-bundle over Y induced from ξ by f, then we have the commutative diagram

$$\text{(1)} \qquad \begin{array}{ccc} & \overset{\alpha_\xi}{\nearrow} & K^*(X) \\ R(G) & & \big\downarrow f^! \\ & \underset{\alpha_\eta}{\searrow} & K^*(Y). \end{array}$$

If Y consists of a single point, then $K^*(Y) \cong Z$ and α_η is just the augmentation $\epsilon : R(G) \to Z$. This shows that the ideal $I(G)$ is mapped by α_ξ into $K_1^*(X)$ (see 2.3 Remark). By 2.6 (5) there exists an n_0 such that $\alpha_\xi(I(G)^n) = 0$ for $n \geqq n_0$. Since $\hat{R}(G)$ is the inverse limit of the $R(G)/(I(G))^n$ with $n \geqq n_0$, we have a natural ring homomorphism

$$\hat{\alpha}_\xi : \hat{R}(G) \to K^*(X).$$

Obviously, α_ξ is $R(G) \to \hat{R}(G)$ followed by $\hat{\alpha}_\xi$.

If we have as before a map $f : Y \to X$, then we have the commutative diagram

$$\text{(2)} \qquad \begin{array}{ccc} & & K^*(X) \\ & \overset{\hat{\alpha}_\xi}{\nearrow} & \\ \hat{R}(G) & & \Big\downarrow f^! \\ & \underset{\hat{\alpha}_\eta}{\searrow} & \\ & & K^*(Y) \end{array}, \qquad \eta = f^*\xi.$$

4 6. *Classifying spaces.* Let F be a contravariant functor on the class $\mathfrak{A}$ (see 1.1), i.e., F attaches to each $X \in \mathfrak{A}$ an algebraic object of a given type, say an abelian group for convenience, and for each continuous map $f : Y \to X$ ($Y, X \in \mathfrak{A}$) there is given a homomorphism $f^* : F(X) \to F(Y)$ satisfying the functorial properties and the homotopy axiom ($f^* = g^*$ if the maps $f, g : Y \to X$ are homotopic).

Let G be a compact Lie group, B_G its (infinite) classifying space determined up to homotopy type. We shall define $\mathfrak{F}(B_G)$ to be an algebraic object of the same type as all the $F(X)$, $X \in \mathfrak{A}$. The definition will be such that an element of $\mathfrak{F}(B_G)$ is completely given by the group G. The classifying space B_G is not needed for the definition, but we write $\mathfrak{F}(B_G)$ rather than $\mathfrak{F}(G)$ to avoid the confusion with $F(G)$.

Definition. *An element a of $\mathfrak{F}(B_G)$ is an operator which attaches to each X and each principal G-bundle ξ over X an element $a(\xi) \in F(X)$ depending only on the equivalence class of ξ such that the following holds: for a map $f : Y \to X$ ($Y, X \in \mathfrak{A}$), a principal G-bundle ξ over X and the principal G-bundle $f^*\xi$ over Y induced from ξ by f, we have $a(f^*\xi) = f^*(a(\xi))$. Using the notation of* [15, §3] *this means that the diagram*

$$\begin{array}{cccc} f^* : & H^1(X, G_c) & \to & H^1(Y, G_c) \\ & a \Big\downarrow & & a \Big\downarrow \\ f^* : & F(X) & \to & F(Y) \end{array}$$

is commutative.

If U, G are compact Lie groups and $\rho : U \to G$ a homomorphism, then we have the induced homomorphism

$$\rho^* : \mathfrak{F}(B_G) \to \mathfrak{F}(B_U).$$

For $a \in \mathfrak{F}(B_G)$, $\rho^*a : H^1(X, U_c) \to F(X)$ is the composition $H^1(X, U_c) \to^\rho H^1(X, G_c) \to^a F(X)$.

If $U = G$ and ρ is an inner automorphism of G, then ρ^* is the identity since $\rho : H^1(X, G_c) \to H^1(X, G_c)$ is the identity.

According to the classification theorem [**18**, §19] we can choose a principal G-bundle ξ_n which is classifying up to n, i.e., $\pi_i(E_{\xi_n}) = 0$ for $i \leqq n$, and whose base space $B_{\xi_n} = B_n$ belongs to $\mathfrak{A}$. Let $n_1 < n_2 < n_3 < \cdots$ be a sequence of positive integers such that $\dim B_{n_i} \leqq n_{i+1}$. Then ξ_{n_i} is induced from $\xi_{n_{i+1}}$ by a map $B_{n_i} \to B_{n_{i+1}}$ uniquely determined up to homotopy. Thus we have a homomorphism $F(B_{n_{i+1}}) \to F(B_{n_i})$. This enables us to write $\mathfrak{F}(B_G)$ as an inverse limit

$$\text{(3)} \qquad \mathfrak{F}(B_G) \cong \lim_{\leftarrow} F(B_{n_i}).$$

This isomorphism is canonical. In particular, we have:

(4) *An element a of $\mathfrak{F}(B_G)$ vanishes if and only if there exists for every n_0 an integer $n \geqq n_0$ and a principal G-bundle ξ_n classifying up to n such that $a(\xi_n) = 0$.*

If we take for F the ordinary cohomology theory with coefficients in some abelian group, then $\mathfrak{F}(B_G)$ becomes $H^{**}(B_G, A)$; see [5, §6.1]. If we take for F the K^*-theory of 1.9 then we define the ring

$$\mathcal{K}^*(B_G) = \mathcal{K}^0(B_G) \oplus \mathcal{K}^1(B_G) = \mathfrak{F}(B_G)$$

$\mathcal{K}^0(B_G)$ is the $\mathcal{K}(B_G)$ mentioned in the introduction. In this theory we write $\rho^!$ instead of ρ^*. The Chern character $ch : \mathcal{K}^*(B_G) \to H^{**}(B_G, Q)$ is clearly defined.

4.7. Because of the diagrams (1) and (2) of 4.5 we have canonical ring homomorphisms

$$\alpha : R(G) \to \mathcal{K}^*(B_G), \qquad \hat{\alpha} : \hat{R}(G) \to \mathcal{K}^*(B_G).$$

α equals $R(G) \to \hat{R}(G)$ followed by $\hat{\alpha}$. Of course, α and $\hat{\alpha}$ map into $\mathcal{K}^0(B_G)$. We sometimes write more explicitly α_G instead of α and $\hat{\alpha}_G$ instead of $\hat{\alpha}$.

Let G and H be compact Lie groups and $\rho : G \to H$ a homomorphism; then we have a commutative diagram

$$\begin{array}{ccccccc} R(G) & \to & \hat{R}(G) & \xrightarrow{\hat{\alpha}} & \mathcal{K}^*(B_G) & \xrightarrow{ch} & H^{**}(B_G, Q) \\ \uparrow & & \uparrow & & \uparrow \rho^! & & \uparrow \rho^{**} \\ R(H) & \to & \hat{R}(H) & \xrightarrow{\hat{\alpha}} & \mathcal{K}^*(B_H) & \xrightarrow{ch} & H^{**}(B_H, Q). \end{array}$$

4.8 We state now the main theorem of §4 and give a corollary. The proof of the theorem will be given in the following sections.

THEOREM. *Let G be a compact connected Lie group. Then $\hat{\alpha}$ is an isomorphism of $\hat{R}(G)$ onto $\mathcal{K}^*(B_G)$.*

COROLLARY. *Let G be a compact connected Lie group. Then $\mathcal{K}^1(B_G) = 0$. Moreover, $\mathcal{K}^*(B_G) = \mathcal{K}^0(B_G)$ has no torsion and no zero divisors.*

We have seen in 4.4 that $\hat{R}(G)$ is a subring of $\hat{R}(T)$ which is a ring of formal power series over Z. Thus the corollary follows from the theorem.

REMARK. We conjecture the theorem to hold for any compact Lie group. It holds if G is finite (Atiyah, loc. cit. in 4.4 Note).

4.9. We prove Theorem 4.8 first for the case where G is a torus T which we describe as in 4.3 as the group of k-tuples of reals mod 1. Let P_n be the complex projective space of complex dimension n. Over P_n we take the $U(1)$-bundle η_n whose first Chern class is the canonical generator g of $H^2(P_n, Z) \cong Z$; see [15, §4.2]. η_n is induced from η_{n+1} by the embedding $P_n \to P_{n+1}$. Let B_{2n} be the cartesian product of k copies of P_n. Over B_{2n} we have the T-bundle ξ_{2n} which is the Whitney sum of the $\pi_i^*(\eta_n)$, $1 \leq i \leq k$, where π_i is the projection of B_{2n} on its ith factor. ξ_{2n} is classifying up to dimension $2n$. We have the embedding $B_{2n} \to B_{2n+2}$ which induces ξ_{2n} from ξ_{2n+2} and which gives rise to the homomorphism $K^*(B_{2n+2}) \to K^*(B_{2n})$. It follows from 4.6 (3) that

$$\mathcal{K}^*(B_T) \cong \varprojlim K^*(B_{2n}),$$

the inverse limit being taken with respect to the maps $K^*(B_{2n+2}) \to K^*(B_{2n})$ just defined.

Let us denote by x_i the first Chern class of $\pi_i^*(\eta_n)$, i.e., $x_i = \pi_i^*(g)$. Then

$$H^*(B_{2n}, Z) = Z[x_1, \cdots, x_k]/I_{n+1} \tag{5}$$

where I_{n+1} is the ideal $(x_1^{n+1}, \cdots, x_k^{n+1})$.

We consider the map $ch \circ \alpha_{\xi_{2n}} \circ \phi$ of the polynomial ring $Z[z_1, \cdots, z_k]$ into $H^*(B_{2n}, Q)$, see 4.3 and 4.5. It maps z_i onto $e^{x_i} - 1$. Since $e^{x_i} - 1 = x_i +$ higher terms, it follows from 2.5 (iii) that $\alpha_{\xi_{2n}} \circ \phi$ maps $Z[z_1, \cdots, z_k]$ onto $K^*(B_{2n}) = K^0(B_{2n})$, the kernel being the ideal $J_{n+1} = (z_1^{n+1}, \cdots, z_k^{n+1})$ as follows from (5). Thus

$$K^*(B_{2n}) \cong Z[z_1, \cdots, z_k]/J_{n+1}$$

and

$$\mathcal{K}^*(B_T) \cong \varprojlim Z[z_1, \cdots, z_k]/J_{n+1}. \tag{6}$$

If we identify $\hat{R}(T)$ with $Z[[z_1, \cdots, z_k]]$ (Proposition 4.3) and $\mathcal{K}^*(B_T)$ with the above inverse limit (6), then $\hat{\alpha} : \hat{R}(T) \to \mathcal{K}^*(B_T)$ is just the natural map

$$Z[[z_1, \cdots, z_k]] \to \varprojlim Z[z_1, \cdots, z_k]/J_{n+1}.$$

To prove that this map is bijective, one has to check that the $(z_1, \cdots, z_k)$-adic topology of $Z[z_1, \cdots, z_k]$ and the topology defined by the sequence J_n of ideals coincide. But this is easy to do.

4.10. PROPOSITION. *Let G be a compact connected Lie group, T a maximal torus of G and $\rho : T \to G$ the embedding. Then the map $\rho^! : \mathcal{K}^*(B_G) \to \mathcal{K}^*(B_T)$ (see 4.6) is injective.*

Proof. We first observe that there exist principal G-bundles which are classifying up to n (n arbitrary) *and which have a compact oriented differentiable manifold as base.* This is true for $G = U(m)$, since then we have the complex Grassmannians as "universal" base spaces. An arbitrary G may be embedded in $U(m)$ for m sufficiently large. G has thus "universal" base spaces which are fibred with $U(m)/G$ as typical fibre and complex Grassmannians. The bundle along the fibres is orientable, since it is an extension of a principal G-bundle and G is connected [**5**, §7.5]. Hence we have constructed universal base spaces for G with the desired properties (compare [**18**, §19.6]).

Let a be an element of $\mathcal{K}^*(B_G)$ for which $\rho^!(a) = 0$. Then we must show that $a = 0$. By 4.6 (4) and the above observation on classifying bundles, it suffices to prove that $a(\xi) = 0$ where ξ is any principal G-bundle over an arbitrary compact oriented differentiable manifold X. Using the notation and the proposition of 3.5 with $B_\xi = X$ it suffices to prove that $\pi^! a(\xi) = 0$. But $\pi^! a(\xi) = a(\pi^*\xi)$, the lifted bundle $\pi^*\xi$ equals $\rho(\eta)$ where η is a principal T-bundle. Now $a(\rho(\eta)) = (\rho^! a)(\eta) = 0$.

4.11. **Proof of Theorem 4.8.** We have the commutative diagram

$$\begin{array}{ccc} \hat{R}(T) & \xrightarrow{\hat\alpha_T} & \mathcal{K}^*(B_T) \\ {\scriptstyle i}\uparrow & & \uparrow{\scriptstyle \rho^!} \\ \hat{R}(G) & \xrightarrow[\hat\alpha_G]{} & \mathcal{K}^*(B_G). \end{array}$$

The vertical maps are injective, the upper horizontal one is bijective (4.4, 4.10, 4.9). Thus $\hat\alpha_G$ is injective. The Weyl group $W(G)$ as group of automorphisms of T operates on $\mathcal{K}^*(B_T)$ (see definition of $\rho^!$ in 4.6). Since these automorphisms come from inner automorphisms of G, every element of $\rho^!\mathcal{K}^*(B_G)$ is invariant under $W(G)$. The operation of $W(G)$ on $\hat{R}(T)$ and $\mathcal{K}^*(B_T)$ is the same if one identifies the two rings under $\hat\alpha_T$; this follows from the diagram in 4.7. Therefore by 4.4

$$\hat\alpha_T^{-1}(\rho^!\mathcal{K}^*(B_G)) \subset i\hat{R}(G),$$

$$\rho^!\mathcal{K}^*(B_G) \subset \hat\alpha_T i(\hat{R}(G)) = \rho^!\hat\alpha_G\hat{R}(G).$$

Since $\rho^!$ is injective, $\mathcal{K}^*(B_G) \subset \hat\alpha_G\hat{R}(G)$ which completes the proof.

References

1. M. F. Atiyah and F. Hirzebruch, *Riemann-Roch theorems for differentiable manifolds*, Bull. Amer. Math Soc. vol. 65 (1959) pp. 276–281.

2. ———, *Quelques théorèmes de non-plongement pour les variétés différentiables*, Bull. Soc. Math. France vol. 87 (1959) pp. 383–396.

3. A. Borel, *Sur la cohomologie des espaces fibrés principaux et des espaces homogènes de groupes de Lie compact*, Ann. of Math. vol. 57 (1953) pp. 115–207.

4. ———, *Sur la torsion des groupes de Lie*, J. Math. Pures. Appl. vol. 35 (1956) pp. 127–139.

5. A. Borel and F. Hirzebruch, *Characteristic classes and homogeneous spaces*. I, II, Amer. J. Math. vol. 80 (1958) pp. 458–538 and vol. 81 (1959) pp. 315–382.

6. ———, *Characteristic classes and homogeneous spaces*. III, Amer. J. Math., to appear.

7. R. Bott, *An application of the Morse theory to the topology of Lie groups*, Bull. Soc. Math. France vol. 84 (1956) pp. 251–281.

8. ———, *The space of loops on a Lie group*, Michigan Math. J. vol. 5 (1958) pp. 35–61.

9. ———, *The stable homotopy of the classical groups*, Ann. of Math. vol. 70 (1959) pp. 313–337.

10. ———, *Quelques remarques sur les théorèmes de périodicité*, Bull. Soc. Math. France vol. 87 (1959) pp. 293–310.

11. N. Bourbaki, *Topologie générale*, Chapter 9, 2nd ed., Paris, Hermann, 1958.

12. H. Cartan and S. Eilenberg, *Homological algebra*, Princeton University Press, 1956.

13. H. Cartan and C. Chevalley, *Géométrie algébrique*, Séminaire Ecole Norm. Sup., Paris, 1956.

14. S. Eilenberg and N. Steenrod, *Foundations of algebraic topology*, Princeton University Press, 1952.

15. F. Hirzebruch, *Neue topologische Methoden in der algebraischen Geometrie*, Berlin-Göttingen-Heidelberg, Springer-Verlag, 1956.

16. ———, *A Riemann-Roch theorem for differentiable manifolds*, Séminaire Bourbaki, Février, 1959.

17. D. Puppe, *Homotopiemengen und ihre induzierten Abbildungen*. I, Math. Z. vol. 69 (1958) pp. 299–344.

18. N. Steenrod, *The topology of fibre bundles*, Princeton University Press, 1951.

19. J. H. C. Whitehead, *Combinatorial homotopy*. I, Bull. Amer. Math. Soc. vol. 55 (1949) pp. 213–245.

20

Lecture notes of the American Math. Soc. Summer Topology Institute, Seattle 1963

LECTURES ON K-THEORY

F. Hirzebruch

(Notes prepared by Paul Baum)

In these two lectures on K-theory I shall:

(i) Give the elementary proof (due to Atiyah and Bott) of the Bott periodicity theorem.

(ii) Develop the basic machinery of K-theory (using Dold's lectures on half-exact functors) and show how Adams and Dyer have applied it to obtain Adams' result on the nonexistence of elements of Hopf invariant one.

I shall under (i) only give the surjectivity of the Bott homomorphism whereas the injectivity is obtained under (ii) from 'general nonsense'. This shortens the exposition. For the full elementary proof see the notes of Atiyah and Bott distributed during this Summer Institute.

1. Vector bundles on X and vector bundles on $X \times S^2$

Let X be a compact topological space. We shall prove the Bott periodicity theorem by examining the relationship between complex vector bundles on X and complex vector bundles on $X \times S^2$. A complex vector bundle may have different fibre dimensions over the various connectedness components of its base space. We consider S^2 as $\mathbf{C} \cup \infty$ and S^1 as $\{z \mid |z| = 1\}$.

If E is a complex vector bundle on X, E_x denotes the fibre of E at $x \in X$. A clutching function for E is a function p which continuously assigns to each $(x, z) \in X \times S^1$ an automorphism $p(x, z)$ of E_x. An endomorphism of E is a function a which continuously assigns to each $x \in X$ an endomorphism $a(x)$ of E_x. a is an automorphism of E if $a(x)$ is non-singular for all $x \in X$. A clutching function p is linear if there exist endomorphisms a, b of E such that $p(x, z) = a(x)z + b(x)$. A clutching function p is polynomial if there exist endomorphisms $a_0, a_1, \ldots, a_n$ of E such that $p(x, z) = a_n(x)z^n + a_{n-1}(x)z^{n-1} + \ldots + a_1(x)z + a_0(x)$.

Given E and a clutching function p for E a complex vector bundle (E, p) over $X \times S^2$ can be constructed as follows:

Set

$$D^+ = \{z \mid |z| \leq 1\}, \quad D^- = \{z \mid |z| \geq 1\}$$

$$E^+ = E \text{ lifted to } X \times D^+, \quad E^- = E \text{ lifted to } X \times D^- .$$

Then (E, p) is formed from the disjoint union of E^+ and E^- by identifying the point $(x, z, v) \in E^+$ with the point $(x, z, p(x, z)v) \in E^-$ for $x \in X$ and $z \in S^1$.

The operation (E, p) has these elementary properties:

(i) Any vector bundle F over $X \times S^2$ is isomorphic to an (E, p). Put $E = i^*F$ where $i: X \to X \times S^2$ with $i(x) = x \times 1$.

(ii) (E, p) is isomorphic to (E, p') if p and p' are homotopic as clutching functions, i.e. if there is a continuous family p_t,

$0 \leq t \leq 1$, of clutching functions such that $p_0 = p$, $p_1 = p'$.

(iii) If a is an automorphism of E then $(E, p) = (E, ap) = (E, pa)$. Here and in the following we allow ourselves to use for isomorphic the equality sign.

(iv) If X is a point and E is the trivial line bundle over X, then (E, z^{-1}) is the Hopf bundle H over S^2. For any integer m, $(E, z^m) = H^{-m}$. The line bundle H belongs to the divisors of degree 1 of the algebraic curve S^2.

(v) $(E, z^m p) = (E, p) \otimes \pi_2^*(H^{-m})$, where $\pi_2 : X \times S^2 \to S^2$ is the projection.

Notation. If E is a bundle on X and F is a bundle on S^2 then $E \hat{\otimes} F$ denotes $\pi_1^*(E) \otimes \pi_2^*(F)$, where $\pi_1 : X \times S^2 \to X$, $\pi_2 : X \times S^2 \to S^2$ are the projections.

Lemma 1.1. Let X be a compact topological space, E a vector bundle on X, and p a linear clutching function for E. Then bundles E_1, E_2 can be chosen on X such that

$$(E, p) = E_1 \hat{\otimes} H^{-1} + E_2 \otimes 1 .$$

Proof. Set $p(x, z) = a(x)z + b(x)$. Since X is compact a complex number z_0 with $|z_0| < 1$ can be chosen such that $a(x) + b(x) \cdot \bar{z}_0$ is non-singular for all $x \in X$.

For all $z \in S^1$, $t \in [0, 1]$, we have $\left| \dfrac{z + tz_0}{1 + zt\bar{z}_0} \right| = 1$, so

setting $p_t(x, z) = a(x) \cdot \dfrac{z + tz_0}{1 + zt\bar{z}_0} + b(x)$ a homotopy of clutching functions between p and p_1 is obtained where $p_1(x, z) =$

$$a(x) \cdot \frac{z + z_0}{1 + z\bar{z}_0} + b(x) .$$

Set $p_2(x, z) = (1 + z\bar{z}_0)p_1(x, z) = [a(x) + b(x)\bar{z}_0]z + a(x) \cdot z_0 + b(x)$. The number z_0 was chosen so that $a(x) + b(x)\bar{z}_0$ is non-singular, so set $p_3(x, z) = [a(x) + b(x)\bar{z}_0]^{-1}p_2(x, z) = z + [a(x) + b(x)\bar{z}_0]^{-1}[a(x) \cdot z_0 + b(x)] = z + b'(x)$. From elementary properties (ii) and (iii) above it follows that $(E, p) = (E, p_3)$. Since $p_3(x, z)$ is non-singular for $|z| = 1$, the endomorphism $b'(x)$ has no eigenvalues of absolute value one. For each $x \in X$ and each $\lambda \in S^1$ let $E_x^\lambda = \{v \in E_x | \quad n \text{ with } (\lambda - b'(x))^n v = 0\}$. Set $E_x^+ = \bigoplus_{|\lambda| < 1} E_x^\lambda$, $E_x^- = \bigoplus_{|\lambda| > 1} E_x^\lambda$. Let E_i, $i = 1, 2$, be the sub-bundles of E whose fibres at x are $E_x^+ \; E_x^-$ respectively. Then $E = E_1 \oplus E_2$ and each E_i is mapped into itself by $p_3(x, z) = z + b'(x)$.

So $(E, p_3) = (E_1, p_3|E_1) \oplus (E_2, p_3|E_2)$. $p_3|E_1$ is homotopic to z with the homotopy given by $z + tb'$, $0 \le t \le 1$. This is a homotopy of clutching functions since $b'|E_1$ has no eigenvalue of absolute value ≥ 1. $p_3|E_2$ is homotopic to $b'|E_2$ with the homotopy given by $tz + b'$, $0 \le t \le 1$. b' is non-singular on E_2 so $(E_2, b') = (E_2, 1)$. Thus $(E, p) = (E, p_3) = (E_1, z) + (E_2, 1) = E_1 \hat{\otimes} H^{-1} + E_2 \hat{\otimes} 1$.

Lemma 1. 2. *Let X be a compact topological space, E a vector bundle on X, p a polynomial clutching function for E. Then there are vector bundles E_1, E_2, E_3 on X such that*

$$(E, p) + E_3 \hat{\otimes} 1 = E_1 \hat{\otimes} H^{-1} + E_2 \hat{\otimes} 1.$$

(20)

Proof. Set $p(x, z) = a_n(x)z^n + a_{n-1}(x)z^{n-1} + \ldots + a_0(x)$. Let $L^n p$ be the clutching function for $(n+1)E$ represented by the $(n+1) \times (n+1)$ matrix

$$\begin{bmatrix} 1 & -z & 0 & 0 & \ldots & 0 \\ 0 & 1 & -z & 0 & \ldots & 0 \\ 0 & 0 & 1 & -z & \ldots & 0 \\ \cdot & \cdot & \cdot & \cdot & & \cdot \\ \cdot & \cdot & \cdot & \cdot & & \cdot \\ \cdot & \cdot & \cdot & \cdot & & \cdot \\ 0 & 0 & 0 & \ldots & 1 & -z \\ a_n & a_{n-1} & a_{n-2} & \ldots & a_1 & a_0 \end{bmatrix}$$

$L^n p$ is a linear clutching function for $(n+1)E$. By lemma 1.1 there exist vector bundles E_1, E_2 on X such that

$((n+1)E, L^n p) = E_1 \hat{\otimes} H^{-1} + E_2 \hat{\otimes} 1$.

The matrix for $L^n p$ can be transformed by elementary row and column operations (i.e. adding a multiple of one row (or column) to another row (or column)) to the matrix

$$\begin{bmatrix} 1 & 0 & 0 & \ldots & 0 \\ 0 & 1 & 0 & \ldots & 0 \\ 0 & 0 & 1 & \ldots & 0 \\ \cdot & \cdot & \cdot & & \\ \cdot & \cdot & \cdot & & \\ \cdot & \cdot & \cdot & & \\ 0 & 0 & 0 & \ldots & a_n z^n + \ldots + a_0 \end{bmatrix}$$

Each row (or column) operation can be achieved as a homotopy of clutching functions by adding $t\lambda \cdot$ (row) to another row and letting t range continuously from zero to one. Thus by (ii)

$$((n+1)E,\ L^n p) = (nE,\ 1) + (E,\ p) = nE \hat{\otimes} 1 + (E,\ p)$$

and

$$(E,\ p) + nE \hat{\otimes} 1 = E_1 \hat{\otimes} H^{-1} + E_2 \hat{\otimes} 1 .$$

Proposition 1. 3. Let X be a compact topological space. If F is any vector bundle over $X \times S^2$ then bundles E_1, E_2, E_3 over X and integers m_1, m_2, m_3 can be chosen such that

$$F \oplus E_3 \hat{\otimes} H^{m_3} = E_1 \hat{\otimes} H^{m_1} + E_2 \hat{\otimes} H^{m_3} .$$

Proof. Choose a vector bundle E over X and a clutching function p for E such that $(E,\ p) = F$. For each integer n define an endomorphism a_n of E by

$$a_n(x) = \frac{1}{2\pi i} \int_{S'} \frac{p(x,\ z)dz}{z^{n+1}}$$

Set $p_k(x) = \sum_{n=-k}^{n=k} a_n(x)z^n$. The sequence $\frac{1}{n}(p_0 + p_1 + \dots + p_n)$ converges uniformly to p (Fejer's theorem). Taking a close enough approximation to p we conclude that p is homotopic to a clutching function of the form $z^{-m}q$ where q is a polynomial clutching function. By lemma 1. 2 vector bundles E_3, E_1, E_2 on X can be chosen such that

$$(E,\ q) + E_3 \hat{\otimes} 1 = E_1 \hat{\otimes} H^{-1} + E_2 \hat{\otimes} 1 .$$

But

$$(E, p) = (E, z^{-m}q) = (E, q) \hat{\otimes} H^m \tag{20}$$

so

$$(E, p) + E_3 \hat{\otimes} H^m = E_1 \hat{\otimes} H^{m-1} + E_2 \hat{\otimes} H^m .$$

2. Definition of K(X)

$\underline{W}$ denotes the category whose objects are finite CW complexes and whose morphisms are homotopy classes of continuous maps. For $X \in \underline{W}$ set:

$B(X) =$ commutative semi-group of isomorphism classes of (complex) vector bundles over X with addition given by the Whitney sum

$F(X) =$ free abelian group generated by the isomorphism classes of vector bundles on X

$R(X) =$ subgroup of $F(X)$ generated by all elements of the form $\xi_1 \oplus \xi_2 - \xi_1 - \xi_2$ where $\oplus$ indicates the Whitney sum.

Then $K(X) = F(X)/R(X)$. The operation in $K(X)$ is denoted by $+$.

As sets $B(X) \subset F(X)$. Composing this inclusion with the projection $F(X) \to K(X)$, a homomorphism $i:B(X) \to K(X)$ of abelian semi-groups is obtained. $\oplus$ carries over into $+$. Thus it will cause no confusion that we have written $+$ for the Whitney sum in the preceding section. If h is any homomorphism of abelian semi-groups mapping $B(X)$ into an abelian group A, then there is a unique homomorphism $\tilde{h}$ of abelian groups mapping $K(X)$ into A

such that $h = \tilde{h} \circ i$

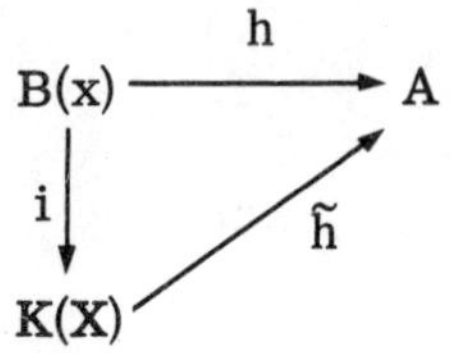

If $X, Y \in \underline{W}$ and $f: X \to Y$ is a morphism in $\underline{W}$ then by associating with each bundle E over Y the induced bundle f^*E over X a map $B(Y) \to B(X)$ is constructed. By the universal mapping property for i there is a unique homomorphism $K(f): K(Y) \to K(X)$ such that the diagram

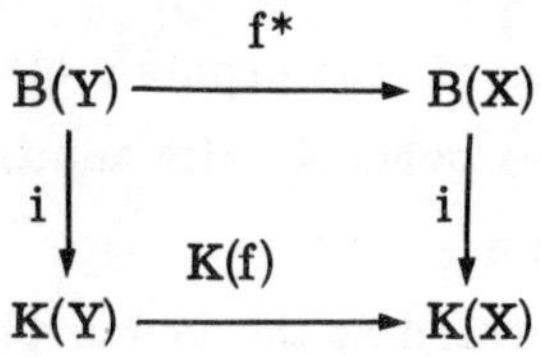

is commutative. ($K(f)$ is denoted by $f^!$.) Thus K is a functor from $\underline{W}$ to the category of abelian groups. The operation of forming the tensor product of two bundles makes K into a functor from $\underline{W}$ to the category of commutative rings with unit.

If $\xi \in B(X)$ then $\xi \to \dim(\text{fibre } \xi)$ defines a homomorphism $B(X) \to H^0(X; Z)$. The induced ring homomorphism $K(X) \to H^0(X; Z)$ is called the rank.

Given an element $\xi \in B(X)$ let

$$c(\xi) = 1 + c_1(\xi) + c_2(\xi) + \ldots \in \bigoplus_i H^{2i}(X; Z) = H^{\text{even}}(X; Z)$$

be the total Chern class of ξ. Then there is an induced homomorphism of abelian groups $c:K(X) \to G^{\text{even}}(X; Z)$ where $G^{\text{even}}(X; Z)$ is the set of all sums $1 + a_2 + a_4 + \dots$, with $a_{2i} \in H^{2i}(X; Z)$ and with group operation given by the cup product in $H^*(X; Z)$.

For $\xi \in B(X)$ define

$$\mathrm{ch}(\xi) = \mathrm{rank}\ \xi + \sum_{k=1}^{\infty} \frac{s_k(c_1(\xi), \dots, c_k(\xi))}{k!}$$

where s_k is the universal polynomial which expresses the sum $x_1^k + \dots x_N^k$ $(N \geq k)$ in terms of the elementary symmetric function of the x_i. We have $s_1(\xi) = c_1(\xi)$, $s_2(\xi) = c_1^2(\xi) - 2c_2(\xi)$, $s_n(\xi) = \pm\, nc_n(\xi)$ + composite terms.

$\mathrm{ch}:K(X) \to H^{\text{even}}(X; Q)$ is a natural ring homomorphism and is called the Chern character.

Since X has a unique map onto a point there is a homomorphism $Z = K(\text{point}) \to K(X)$ whose cokernel we call $\tilde{K}(X)$. There are (unnatural) splittings $K(X) = Z \oplus \tilde{K}(X)$ depending on the choice of a base point: If $x_0 \in X$, then $\tilde{K}(X) = \text{Kernel}\ (K(X) \to K(x_0) = Z)$. As an abelian group $\tilde{K}(X)$ can be naturally identified with the group of stable vector bundles on X. If X is connected, we have

$$\tilde{K}(X) = [X, B_{U(n)}] \text{ for } n \gg \dim X.$$

$\tilde{K}$ is a half-exact functor in the sense of Dold, i.e. if $A \to X \to X/A$ is a sequence in $\underline{W}$ obtained by collapsing the subcomplex A of X to a point, then the sequence

$$\tilde{K}(X/A) \to \tilde{K}(X) \to \tilde{K}(A) \text{ is exact.}$$

Given objects X, Y in $\underline{W}$, choose base points in X and Y and form the sequence

$$X \vee Y \xrightarrow{\alpha} X \times Y \xrightarrow{\beta} X \# Y$$

Then the sequence

$$0 \to \tilde{K}(X \# Y) \xrightarrow{\beta!} \tilde{K}(X \times Y) \xrightarrow{\alpha!} \tilde{K}(X \vee Y) \to 0$$

is split exact as is true for all half-exact functors. The homomorphism $\gamma : \tilde{K}(X) \oplus \tilde{K}(Y) \to \tilde{K}(X \times Y)$ given by the two projections gives rise to the splitting once we have identified $\tilde{K}(X \vee Y)$ with $\tilde{K}(X) \oplus \tilde{K}(Y)$. In order to avoid reference to base points we shall think of $\tilde{K}(X \# Y)$ as the cokernel of γ. If $\xi \in K(X)$ and $\eta \in K(Y)$ then the external product $\xi \hat{\otimes} \eta = \pi_1^! \xi \otimes \pi_2^! \eta \in K(X \times Y)$ is defined. It induces an external product for the cokernels $\tilde{K}$:

$$\hat{\otimes} : \tilde{K}(X) \otimes \tilde{K}(Y) \to \tilde{K}(X \# Y)$$

The Chern character gives a natural transformation of half-exact functors

$$\mathrm{ch} : \tilde{K} \to \tilde{H}^{\mathrm{even}}(\ ; Q)$$

compatible with the multiplicative structures.

Complex vector bundles E over S^2 are classified by two integers: (i) dim fibre E, (ii) $c_1(E)[S^2]$. From this it follows that $K(S^2) = Z + Z$. An additive basis for $K(S^2)$ is given by 1, H - 1. H denotes, as in section 1, the canonical line bundle over S^2. $\mathrm{ch} : K(S^2) \to H^*(S^2; Q)$ maps $K(S^2)$ isomorphically onto

$H^*(S^2; Z) \subset H^*(S^2; Q)$. As bundles $H \otimes H + 1 = H + H$. In $K(S^2)$ $H^2 + 1 = 2H$, $(H - 1)^2 = 0$ and $H^m = (1 + (H - 1))^m = 1 + m(H - 1)$.

Notation. The map from $\tilde{K}(X)$ to $\tilde{K}(X \# S^2)$ which takes an element $\xi \in K(X)$ to $\xi \hat{\otimes} (H - 1) \in K(X \# S^2)$ will be denoted by $\hat{\otimes}(H - 1):K(X) \to K(X \# S^2)$. It is the Bott homomorphism. Here $H - 1$, the generator of $K(S^2)$ is actually the image of $H - 1$ and of H under $K(S^2) \to \tilde{K}(S^2)$).

Proposition 2.1. $\hat{\otimes}(H - 1):\tilde{K}(X) \to \tilde{K}(X \# S^2)$ *is surjective.*

Proof. From proposition 1.3 and the remark that $H^m = 1 + m(H - 1)$ it follows that for any element $\xi \in K(X \times S^2)$ elements $\xi_1, \xi_2 \in K(X)$ can be chosen such that

$$\xi = \pi_1^!(\xi_2) + \pi_1^!(\xi_2) \otimes \pi_2^!(H - 1) .$$

Going over to the cokernels $\tilde{K}$ we get the result.

3. Proof of Bott periodicity

Theorem 3.1. (a) *If* X *is a finite CW-complex then* $\hat{\otimes}(H - 1):\tilde{K}(X) \to \tilde{K}(X \# S^2)$ *is bijective.*

(b) *For each non negative integer* k, ch *maps* $\tilde{K}(S^{2k})$ *isomorphically onto* $H^{2k}(S^{2k}; Z) \subset H^{2k}(S^{2k}; Q)$.

Proof. $\hat{\otimes}(H - 1):\tilde{K}(X) \to \tilde{K}(X \# S^2)$ is a natural transformation of half-exact functors so it suffices to show that this is bijective for the special case when X is a sphere (see Dold's lecture). Consider $\hat{\otimes}(H - 1):\tilde{K}(S^n) \to \tilde{K}(S^{n+2})$.

Since U(m) is connected $\tilde{K}(S^1) = 0$. By proposition 2.1 $\hat{\otimes}(H - 1):\tilde{K}(S^n) \to \tilde{K}(S^{n+2})$ is surjective, so $\tilde{K}(S^n) = 0$ for all odd n by induction.

If n is even $(n = 2k)$ consider the commutative diagram

$$\begin{array}{ccc} \tilde{K}(S^{2k}) & \xrightarrow{\hat{\otimes}(H - 1)} & \tilde{K}(S^{2k+2}) \\ \downarrow{\scriptstyle ch} & & \downarrow{\scriptstyle ch} \\ \tilde{H}^*(S^{2k}; Q) & \xrightarrow[U]{} & \tilde{H}^*(S^{2k+2}; Q) \end{array}$$

where the lower horizontal arrow stands for the external cup product with the canonical generator $U \in H^2(S^2; Z)$. The lower horizontal arrow is bijective. The upper horizontal arrow is surjective.

Proceed by induction. The theorem is obviously true for $n = 0$. The induction hypothesis is that the left vertical arrow is an isomorphism of $\tilde{K}(S^{2k})$ onto $\tilde{H}^*(S^{2k}; Z) \subset \tilde{H}^*(S^{2k}; Q)$. Since $\hat{\otimes}(H - 1)$ is surjective and U maps $\tilde{H}^*(S^{2k+2}; Z)$ it follows from the inductive hypothesis that ch maps $\tilde{K}(S^{2k+2})$ into $\tilde{H}^*(S^{2k+2}; Z)$. Thus the diagram may be replaced by the diagram:

$$\begin{array}{ccc} \tilde{K}(S^{2k}) & \xrightarrow{\hat{\otimes}(H - 1)} & \tilde{K}(S^{2k+2}) \\ \downarrow & & \downarrow \\ \tilde{H}^*(S^{2k}; Z) & \xrightarrow[U]{} & \tilde{H}^*(S^{2k+2}; Z) \end{array}$$

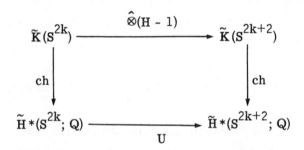

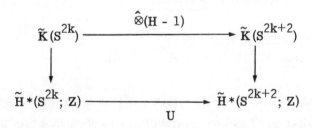

Since $ch \circ \hat{\otimes}(H - 1) = (\hat{\cup} U) \circ ch$ and $\hat{\cup} U \circ ch$ is injective, it follows that $\hat{\otimes}(H - 1)$ is injective. $\hat{\otimes}(H - 1)$ is, therefore, bijective and so is $ch:\tilde{K}(S^{2k+2}) \to \tilde{H}^*(S^{2k+2}; Z)$. This completes the induction.

Corollary 3.2. *Let* v *be a generator of* $H^{2n}(S^{2n}; Z)$. *Then for any complex vector bundle* ξ *over* S^{2n}, *the Chern class* $c_n(\xi)$ *is an integral multiple of* $(n - 1)!\, v$ *and for every integer* $r \equiv 0 \bmod (n - 1)!$ *there is one and only one* $\xi \in \tilde{K}(S^{2n})$ *with* $c_n(\xi) = r \cdot v$.

Proof. This follows from assertion (b) of theorem 3.1 and the formula for $ch\ \xi$ in terms of the Chern classes of ξ.

Corollary 3.4. *For* $n \gg i$, $\pi_i(BU(n)) = 0$ *if* i *is odd and* $\pi_i(BU(n)) = Z$ *if* i *is even* $(i \geq 2)$.

Proof. For $n \gg \dim X$, $\tilde{K}(X) = [X, BU(n)]$. So for $n \gg i > 0$ we have $\pi_i(BU(n)) = \tilde{K}(S^i)$. $\tilde{K}(S^i) = 0$ if i is odd, and $\tilde{K}(S^i) = Z$ if i is even.

Corollary 3.5. $ch:\tilde{K}(X) \otimes Q \to \tilde{H}^{even}(X; Q)$ *is a natural equivalence of functors.*

Proof. $ch:\tilde{K}(X) \otimes Q \to \tilde{H}^{even}(X; Q)$ is a natural transformation of half-exact functors, so it suffices to show that $ch:\tilde{K}(X) \otimes Q \to \tilde{H}^{even}(X; Q)$ is bijective for the special case when X is a sphere. This follows from assertion (b) of theorem 3.1.

Corollary 3.6. *Let* X *be a space* $\in W$ *with no torsion in* $H_*(X; Z)$. *Then* $ch:\tilde{K}(X) \to \tilde{H}^{even}(X; Q)$ *is injective and for each*

$a \in H^{2k}(X; Q)$ <u>there exists</u> $\xi \in K(X)$ <u>with</u>

$$ch\,\xi = a + a_{2k+2} + a_{2k+4} + \ldots \quad (a_{2k+2q} \in H^{2k+2q}(X; Q))$$

<u>if and only if</u> $a \in H^{2k}(X; Z)$.

Proof. $ch:\tilde{K}(X) \to \tilde{H}^{even}(X; Q)$ is a natural transformation of half-exact functors. Take the spectral sequence of both sides (see Dold, proposition 6.2). At the E_2-level we have essentially $\tilde{H}^{even}(X; Z) \to \tilde{H}^{even}(X; Q)$ with the map induced by the inclusion $Z \subset Q$. (See Theorem 3.1, b).) Since $H_*(X; Z)$ is torsion-free this map is injective. The spectral sequence for $\tilde{H}^{even}$ obviously collapses (all differentials vanish). Thus the same holds for K. We have in both cases $E_2 = E_\infty$. $a \in E_\infty^{2p,-2p}(\tilde{H}^{even}) = H^{2p}(X, Q)$ is in $H^{2p}(X, Z)$ if and only if it is in the image of $E_\infty^{2p,-2p}(K) = F_{2p}(K)/F_{2p+1}(K)$. The injectivity on the E_∞-level implies also $K(X) \to H^{even}(X; Q)$ is injective.

4. Elements of Hopf invariant one

In order to apply K-theory to obtain Adams' result on the non-existence of Hopf invariant one we need two theorems of Adams and one number-theoretic fact. We state these without proof:

Notation. For each $b \in H^*(X; Z)$ the reduction of b mod 2 is denoted by $(b)_2$. Sq^{-1} denotes $(1 + Sq^1 + Sq^2 + \ldots)^{-1}$. If $r \in Q$, $[r]$ denotes the largest integer n such that $n \leq r$. If q is an integer > 0 set $\mu(q) = \prod_{p\ \text{prime}} p^{[\frac{q}{p-1}]}$. B_n denotes the n-th Bernoulli number.

Theorem 4a (Adams). Let X be a space with no torsion in $H_*(X; Z)$. Given $a \in H^{2k}(X; Z)$ choose $\xi \in K(X)$ such that $\mathrm{ch}\, \xi = a + a_{2k+2} + a_{2k+4} + \dots$. Then for each q the element $\mu(q) \circ a_{2k+2q}$ is in $H^{2k+2q}(X; Z)$ and $Sq^{-1}(a)_2 = \sum_{q=0}^{\infty} (\mu(q) \circ a_{2k+2q})_2$. A similar result holds for the Steenrod operation $^{-1}$.

Theorem 4b (Adams). Cohomology operations $\psi_1, \psi_2, \dots, \psi_k, \dots$ can be defined in K-theory in one and only one way such that if $\xi \in K(X)$ and $\mathrm{ch}\, \xi = \sum_{j=0}^{\infty} a_{ij}$ with $a_{ij} \in H^{2j}(X; Q)$ then $\mathrm{ch}(\psi_k \xi) = \sum_{j=0}^{\infty} k^j a_{2j}$.

Theorem 4c. Fix an integer n and let $k = 1, 2, 3, \dots$. The greatest common divisor d_n of $k^{2n}(k^{2n} - 1)$ is a divisor of the denominator of $\frac{Bn}{4n}$ which is of the form $2^{\rho_n + 3} \cdot$ odd number, where $n = 2^{\rho_n} \cdot$ odd number. In fact d_n or $2d_n$ equals the denominator of $\frac{Bn}{4n}$.

Remark. We shall need the Theorems 4a and 4c only as far as the prime 2 is concerned.

Theorem 4.1. If $l \neq 2, 4, 8$ there is no element of Hopf invariant one in $\pi_{2l-1}(S^l)$.

Proof. We may assume l even and set $l = 2n$. Let $f: S^{2l-1} \to S^l$ be a map of Hopf invariant one. Form the space

$X = S^l \cup_f e_{2l}$. $H^0(X; Z)$, $H^{2n}(X; Z)$, $H^{4n}(X; Z)$ are infinite cyclic and all other cohomology groups of X are zero. Let g_1 be a generator of $H^{2n}(X; Z)$. Then $g_2 = g_1 \cup g_1$ is a generator of $H^{4n}(X; Z)$. Choose $\xi \in K(X)$ such that $\operatorname{ch} \xi = g_1 + \lambda g_2$ with $\lambda \in Q$. Let 2^s be the largest power of 2 dividing the denominator of λ. Since $Sq^{-1}(g_1)_2 = (g_1)_2 + (g_2)_2 = (g_1)_2 + (\mu(n) \circ \lambda g_2)_2$ we have $n = s$.

By theorem 4b $\operatorname{ch}(\psi_k \xi) = k^n g_1 + k^{2n} \lambda g_2$. By corollary 3.6 $\operatorname{ch}(\psi_k \xi - k^n \xi) = \lambda(k^{2n} - k^n) g_2$ is an integral cohomology class. So $\lambda(k^{2n} - k^n)$ is an integer for all k. By theorem 4c the denominator of λ divides the denominator of $\frac{Bn}{4n}$. Let 2^n be the largest power of 2 dividing n. Thus $n \leq \rho_n + 3$, which implies $n \leq 4$.

The case $n = 3$ is also easy to exclude.

21

VECTOR FIELDS ON SPHERES

BY J. F. ADAMS[1]

Communicated by Deane Montgomery, October 9, 1961

Let us write $n=(2a+1)2^b$, where a and b are integers, and let us set $b=c+4d$, where c and d are integers and $0 \leqq c \leqq 3$; let us define $\rho(n)=2^c+8d$. Then it follows from the Hurwitz-Radon-Eckmann theorem in linear algebra that there exist $\rho(n)-1$ vector fields on S^{n-1} which are linearly independent at each point of S^{n-1} (cf. [**4**]).

THEOREM 1.1. *With the above notation, there do not exist $\rho(n)$ linearly independent vector fields on S^{n-1}.*

This theorem asserts that the known positive result, stated above, is best possible. Like the theorems given below, it is copied without change of numbering from a longer paper now in preparation.

Theorem 1.1 may be deduced from the following result (cf. [**1**]).

THEOREM 1.2. *The truncated projective space $RP^{m+\rho(m)}/RP^{m-1}$ is not coreducible; that is, there is no map $f: RP^{m+\rho(m)}/RP^{m-1} \to S^m$ such that the composite*

$$S^m = RP^m/RP^{m-1} \xrightarrow{i} RP^{m+\rho(m)}/RP^{m-1} \xrightarrow{f} S^m$$

has degree 1.

Theorem 1.2 is proved by employing the "extraordinary cohomology theory" $K(X)$ of Atiyah and Hirzebruch [**2**; **3**]. If our truncated projective space X were coreducible, then the group $K(X)$ would split as a direct sum, and this splitting would be compatible with any "cohomology operations" that one might define in the "cohomology theory" $K(X)$.

THEOREM 5.1. *It is possible to define operations*

$$\Psi^k_\Lambda: K_\Lambda(X) \to K_\Lambda(X)$$

for any integer k (positive, negative or zero) and for $\Lambda = R$ (real numbers) or $\Lambda = C$ (complex numbers). These operations have the following properties.

(i) *Ψ^k_Λ is natural for maps of X.*

(ii) *Ψ^k_Λ is a homomorphism of rings with unit.*

(iii) *The following diagram is commutative.*

[1] Supported in part by the National Science Foundation under grant G14779.

$$\begin{array}{ccc} K_R(X) & \xrightarrow{\Psi_R^k} & K_R(X) \\ c\downarrow & & \downarrow c \\ K_C(X) & \xrightarrow{\Psi_C^k} & K_C(X) \end{array}$$

(Here the homomorphism c is induced by "complexification" of real bundles.)

(iv) $\Psi_\Lambda^k \Psi_\Lambda^l = \Psi_\Lambda^{kl}$.

(v) Ψ_Λ^1 *and* Ψ_R^{-1} *are identity functions.* Ψ_Λ^0 *assigns to each bundle over* X *the trivial bundle with fibres of the same dimension.* Ψ_C^{-1} *assigns to each complex bundle over* X *the "complex conjugate" bundle.*

(vi) *If* $\xi \in K_C(X)$ *and* $\mathrm{ch}_q \xi$ *denotes the* $2q$*-dimensional component of the Chern character* ch ξ, *then*

$$\mathrm{ch}^q(\Psi_C^k \xi) = k^q \, \mathrm{ch}^q \xi.$$

This theorem is proved using virtual representations of groups. By (iv), (v) it is sufficient to define Ψ_Λ^k for $k>0$. One can define polynomials Q_n^k by setting

$$(x_1)^k + (x_2)^k + \cdots + (x_n)^k = Q_n^k(\sigma_1, \sigma_2, \cdots, \sigma_n)$$

where σ_i is the ith elementary symmetric function of $x_1, x_2, \cdots, x_n$. One can define a virtual representation of $GL(n, \Lambda)$ by setting

$$\psi_n^k = Q_n^k(E_\Lambda^1, E_\Lambda^2, \cdots, E_\Lambda^n)$$

where E_Λ^i denotes the ith exterior power representation. The operations Ψ_Λ^k are induced by the virtual representations ψ_n^k.

The values of our groups $K(X)$ and of our operations in them are given by the following result. In order to state it, we define $\phi(n, m)$ to be the number of integers s such that $m<s\leqq n$ and $s\equiv 0$, 1, 2 or 4 mod 8.

THEOREM 7.4. *Assume* $m \not\equiv -1$ mod 4. *Then* $\tilde{K}_R(RP^n/RP^m) = Z_{2^f}$, *where* $f=\phi(m, n)$. *If* $m=0$ *then the canonical real line-bundle* ξ *yields a generator* $\lambda=\xi-1$, *and the polynomials in* λ *are given by the formula*

$$\lambda Q(\lambda) = Q(-2)\lambda,$$

where Q *is any polynomial with integer coefficients. Otherwise the projection* $RP^n \rightarrow RP^n/RP^m$ *maps* $\tilde{K}_R(RP^n/RP^m)$ *isomorphically onto the subgroup of* $\tilde{K}_R(RP^n)$ *generated by* λ^{g+1}, *where* $g=\phi(m, 0)$. *We write* $\lambda^{(g+1)}$ *for the element in* $\tilde{K}_R(RP^n/RP^m)$ *which maps into* λ^{g+1}.

In the case $m \equiv -1 \bmod 4$ *we have*

$$\tilde{K}_R(RP^n/RP^{4t-1}) = \tilde{K}_R(RP^n/RP^{4t}) + Z;$$

here the first summand is embedded by an induced homomorphism and the second is generated by a suitable element $\bar{\lambda}^{(g)}$, *where* $g = \phi(4t, 0)$.

The operations are given by the following formulae.

(i) $$\Psi_R^k \lambda^{(g+1)} = \begin{cases} 0 & (k \text{ even}), \\ \lambda^{(g+1)} & (k \text{ odd}); \end{cases}$$

(ii) $$\Psi_R^k \bar{\lambda}^{(g)} = k^{2t}\bar{\lambda}^{(g)} + \begin{cases} (1/2)k^{2t}\lambda^{(g+1)} & (k \text{ even}), \\ (1/2)(k^{2t}-1)\lambda^{(g+1)} & (k \text{ odd}). \end{cases}$$

This theorem is proved by deducing results in the following order:

(i) Results on complex projective spaces for $\Lambda = C$.

(ii) Results on real projective spaces for $\Lambda = C$.

(iii) Results on real projective spaces for $\Lambda = R$.

References

1. M. F. Atiyah, *Thom complexes*, Proc. London Math. Soc. (3), vol. 11 (1961) pp. 291–310.

2. M. F. Atiyah and F. Hirzebruch, *Riemann-Roch theorems for differentiable manifolds*, Bull. Amer. Math. Soc. vol. 65 (1959) pp. 276–281.

3. ———, *Vector bundles and homogeneous spaces*, Proc. Sympos. Pure Math. Vol. 3, pp. 7–38 Amer. Math. Soc., Providence, R. I., 1961.

4. I. M. James, *Whitehead products and vector-fields on spheres*, Proc. Cambridge Philos. Soc. vol. 53 (1957) pp. 817–820.

22 ON THE GROUPS $J(X)$—IV

J. F. ADAMS

(*Received* 6 *July* 1965)

§1. INTRODUCTION

FROM ONE POINT of view, the present paper is mainly concerned with specialising the results on the groups $J(X)$, given in previous papers of this series [3, 4, 5], to the case $X = S^n$. It can, however, be read independently of the previous papers in this series; because from another point of view, it is concerned with the use of extraordinary cohomology theories to define invariants of homotopy classes of maps; and this machinery can be set up independently of the previous papers in this series. We refer to them only for certain key results.

From a third point of view, this paper represents a very belated attempt to honour the following two sentences in an earlier paper [2]. "However, it appears to the author that one can obtain much better results on the J-homomorphism by using the methods, rather than the results, of the present paper. On these grounds, it seems best to postpone discussion of the J-homomorphism to a subsequent paper." I offer topologists in general my sincere apologies for my long delay in writing up results which mostly date from 1961/62.

I will now summarise the results which relate to the homotopy groups of spheres. For this one needs some notation. The stable group $\lim_{n\to\infty} \pi_{n+r}(S^n)$ will be written π_r^S. The stable J-homomorphism is thus a homomorphism

$$J : \pi_r(SO) \to \pi_r^S.$$

THEOREM 1.1. *If $r \equiv 0 \bmod 8$ and $r > 0$ (so that $\pi_r(SO) = Z_2$), then J is a monomorphism and its image is a direct summand in π_r^S.*

Before considering the case $r \equiv 1 \bmod 8$, we need a preliminary result. Suppose that $r \equiv 1$ or $2 \bmod 8$. Then any map $f: S^{q+r} \to S^q$ induces a homomorphism

$$f^* : \tilde{K}_R^q(S^q) \to \tilde{K}_R^q(S^{q+r}),$$

where the functor $\tilde{K}_R^*$ is that due to Grothendieck–Atiyah–Hirzebruch [10, 11, 2]. We have

$$\tilde{K}_R^q(S^q) = Z, \qquad \tilde{K}_R^q(S^{q+r}) = Z_2.$$

THEOREM 1.2. *Suppose that $r \equiv 1$ or $2 \bmod 8$ and $r > 0$. Then π_r^S contains an element μ_r, of order 2, such that any map $f: S^{q+r} \to S^q$ representing μ_r induces a non-zero homomorphism of $\tilde{K}_R^q$.*

The elements μ_r may be described more precisely than is done in this theorem. We have $\mu_1 = \eta$ and $\mu_2 = \eta\eta$, where η is (as usual) the generator of π_1^S. The elements μ_r constitute a systematic family of elements, generalising η and $\eta\eta$; they have interesting properties, which I hope to discuss on another occasion. I am indebted to M. G. Barratt for ideas about systematic families of elements.

THEOREM 1.3. *Suppose that* $r \equiv 1 \bmod 8$ *and* $r > 1$ (*so that* $\pi_r(SO) = Z_2$). *Then J is a monomorphism and* π_r^S *contains a direct summand* $Z_2 + Z_2$, *one summand being generated by* μ_r *and the other being* Im J.

The case $r = 1$ is exceptional, in that the two summands coincide.

THEOREM 1.4. *Suppose that* $r \equiv 2 \bmod 8$ *and* $r > 0$. *Then* π_r^S *contains a direct summand* Z_2 *generated by* μ_r.

THEOREM 1.5. *Suppose* $r = 4s - 1 \equiv 3 \bmod 8$, *so that* $\pi_r(SO) = Z$. *Then the image of J is a cyclic group of order* $m(2s)$, *and is a direct summand in* π_r^S.

In this theorem, $m(t)$ is the numerical function discussed in [4, §2]. More explicitly, let B_s be the sth Bernoulli number; then $m(2s)$ is the denominator of $B_s/4s$, when this fraction is expressed in its lowest terms.

The direct sum splitting will be accomplished by defining (§7) a homomorphism

$$e'_R : \pi_r^S \to Z_{m(2s)}$$

such that

$$e'_R J : \pi_r(SO) \to Z_{m(2s)}$$

is an epimorphism.

THEOREM 1.6. *Suppose* $r = 4s - 1 \equiv 7 \bmod 8$, *so that* $\pi_r(SO) = Z$. *Then the image of J is a cyclic group of order either* $m(2s)$ *or* $2m(2s)$. *Moreover, there is a homomorphism*

$$e'_R : \pi_r^S \to Z_{m(2s)}$$

such that

$$e'_R J : \pi_r(SO) \to Z_{m(2s)}$$

is an epimorphism.

It follows that if the order of Im J is $m(2s)$, then Im J is a direct summand; this happens (for example) if $r = 7$ or 15. In any event, the subgroup of elements of odd order in Im J is a direct summand in π_r^S.

It will not be proved in this paper, but by more delicate arguments one can show that even for $r \equiv 7 \bmod 8$, the group π_r^S splits as (Ker e'_R) + $Z_{m(2s)}$; however, I do not know how the subgroup Im J lies with respect to this splitting.

The invariants (such as e'_R) which we shall introduce have convenient properties, and lend themselves to a variety of calculations; examples will be given in §§11, 12. They are not restricted to maps between spheres. The following result provides rather a striking example. We take p to be an odd prime, $g : S^{2q-1} \to S^{2q-1}$ to be a map of degree p^f, and Y to be the Moore space $S^{q-1} \cup_g e^{2q}$. Thus $\tilde{K}_C(Y) = Z_{p^f}$. $S^{2r}Y$ will mean the $2r$-fold suspension of Y; we take $r = (p-1)p^{f-1}$.

THEOREM 1.7. *For suitable q there is a map*

$$A : S^{2r}Y \to Y$$

which induces an isomorphism

$$A^* : \tilde{K}_C(Y) \to \tilde{K}_C(S^{2r}Y).$$

Therefore the composite

$$A.S^{2r}A.S^{4r}A.\dots S^{2r(s-1)}A : S^{2rs}Y \to Y$$

induces an isomorphism of $\tilde{K}_C$, and is essential for every s.

For $f = 1$ this result is related to Toda's sequence of elements $\alpha_s \in \pi^S_{2(p-1)s-1}$ [16,17], as will be explained in §12.

From the point of view of history or motivation, the sequence of ideas in this paper may be ordered as follows. Suppose given a map $f: X \to Y$. We may form the mapping cone $Y \cup_f CX$; by studying the group $K_C(Y \cup_f CX)$ and the homomorphism

$$ch: K_C(Y \cup_f CX) \to H^*(Y \cup_f CX; Q)$$

we may sometimes succeed in distinguishing $Y \cup_f CX$ from $Y \vee SX$; thus we may sometimes show that f is essential. This method was presumably known to Atiyah and Hirzebruch (*ca.* 1960/61); it is given in [6] (for the case in which X and Y are spheres) and was published by Dyer [13]. See also [19]. We touch on it in §7 of this paper.

One next realises that in the preceding construction, the possible Chern characters that can arise are severely limited by the fact that $K_C(Y \cup_f CX)$ admits operations Ψ^k. This observation leads to a proof of the non-existence of elements of Hopf invariant one (mod 2 and mod p); this proof was given in [6], and was first published by Dyer [13]. We touch on it in §8 of this paper. It should be said, however, that the most elegant proof by K-theory of the non-existence of elements of Hopf invariant one is somewhat different; see [8].

One next realises that the essential phenomenon we have to study is the short exact sequence

$$\tilde{K}_C(Y) \leftarrow \tilde{K}_C(Y \cup_f CX) \leftarrow \tilde{K}_C(SX)$$

of groups admitting operations Ψ^k. The class of this short exact sequence yields an element of a suitable group

$$\mathrm{Ext}^1(\tilde{K}_C(Y), \tilde{K}_C(SX)).$$

This element gives an invariant of f. If $K_C(Y \cup_f CX)$ is torsion-free this approach is equivalent to that using the Chern character; if $K_C(Y \cup_f CX)$ has torsion this approach is better than that using the Chern character. We therefore adopt this as our basic approach. It has been sketched in [7], and will be fully explained in §3.

In the above, we can of course use $\tilde{K}_R$ instead of $\tilde{K}_C$. The use of $\tilde{K}_R$ and the use of spaces with torsion gives the extra power needed to prove results such as Theorems 1.1, 1.3.

Once we realise that our invariants should take values in suitable Ext^1 groups, certain properties of the invariants become very plausible. Our invariants carry composition products (of homotopy classes) into composition products (in Ext) (§3); they carry Toda brackets (in homotopy) into Massey products (in Ext) (§§4,5). These products enable one to perform many calculations.

The arrangement of the paper is as follows. Since we make constant use of cofibre sequences

$$X \xrightarrow{f} Y \to Y \cup_f CX \to SX \dots,$$

we devote §2 to them. In §3 we define our invariants and give their basic properties. §§4, 5 are devoted to their properties on Toda brackets, as indicated above. So far the work has been done for a quite general cohomology theory; in §§6, 7 we specialise to the case of $\tilde{K}_C$ and $\tilde{K}_R$. §7 contains the main theorem about the cases in which X and Y are spheres and $\tilde{K}$ is torsion-free. §8 contains the relationship between the invariants of §7 and the classical Hopf invariant in the sense of Steenrod. §9 considers the case needed for Theorems 1.1, 1.3, in which X and Y are spheres but $\tilde{K}$ is not torsion-free. In §10 we discuss the value of our invariants on the image of J. In §11 we work out the general theory of §§4, 5 (about Toda brackets) for the special cases which most concern us. In §§12 we prove Theorem 1.7 and discuss related matters; since the same machinery serves to discuss certain 2-primary phenomena, we also prove Theorem 1.2 there. In §12 we also give a number of examples and applications; the reader's attention is particularly directed to these, since they provide essential motivation.

Since drafting the body of this paper, I have become aware of Toda's paper [19], which has a considerable overlap with the present paper. I am very grateful to Toda for a letter about his results.

Toda defines an invariant

$$CH^{n+k} : \pi_{2n+2k-1}(S^{2n}) \to Q/Z$$

which is presumably the same as the invariant e_C discussed in this paper. He also defines an invariant CH_*^{4m+2h}, which is presumably the same (up to a certain constant factor) as the invariant e'_R discussed in this paper.

To give Toda proper credit for his priority, I offer the following concordance of results. Corollary 7.7 of this paper is to be found in Toda's paper, and is the essential step in the proof of his Theorems 6.3, 6.5(i) and (ii) which give restrictions on the values that can be taken by his invariants (compare 7.14, 7.15 of this paper). Proposition 7.20 of this paper is Theorem 6.5 (iii) of [19]. Corollary 8.3 of this paper is Theorem 6.7 of [19]. The case $\Lambda = C$ of Theorem 11.1 of this paper is Theorem 6.4 of [19]. Theorem 12.11 of this paper is contained in 6.8 of [19].

§2. COFIBERINGS

As explained in the introduction, this paper will make much use of sequences of cofiberings. We shall therefore devote this section to summarising some material about cofibre sequences, following [15]. We need only deal with "good" spaces; for the applications, it would be sufficient to consider finite CW-complexes.

Let $f: X \to Y$ be a map. We can construct from it a cofibering

$$X \xrightarrow{f} Y \xrightarrow{i} Y \cup_f CX.$$

Here i is an injection map; and $Y \cup_f CX$ is the space obtained from Y by attaching CX, the cone on X, using f as attaching map.

Iterating this construction, we can construct

$$Y \xrightarrow{i} (Y \cup_f CX) \xrightarrow{j} (Y \cup_f CX) \cup_i CY$$

and (setting $Z = Y \cup_f CX$)

$$Z \xrightarrow{j} (Z \cup_i CY) \xrightarrow{k} (Z \cup_i CY) \cup_j CZ.$$

Now the space $(Y \cup_f CX) \cup_i CY$ is homotopy-equivalent to the suspension SX; and similarly, the space $(Z \cup_i CY) \cup_j CZ$ is homotopy-equivalent to SY. In order to avoid errors of sign in what follows, it is desirable to use the "same" homotopy equivalence in the two cases. If we do this, then the map

$$k: (Y \cup_f CX) \cup_i CY \to (Z \cup_i CY) \cup_j CZ$$

corresponds to

$$-Sf: SX \to SY.$$

(This is easy to check; or see [15, p. 309, Satz 4].) We shall therefore take the following as our basic cofibre sequence.

$$X \xrightarrow{f} Y \xrightarrow{i} Y \cup_f CX \xrightarrow{j} SX \xrightarrow{-Sf} SY \ldots$$

This construction has various obvious properties, which we record for use later.

PROPOSITION 2.1. *If $f \sim g$, then we can construct the following homotopy-commutative diagram, in which all the vertical arrows are homotopy equivalences.*

$$\begin{array}{ccccccccc}
X & \xrightarrow{f} & Y & \xrightarrow{i} & Y \cup_f CX & \xrightarrow{j} & SX & \xrightarrow{-Sf} & SY \\
\downarrow 1 & & \downarrow 1 & & \downarrow & & \downarrow 1 & & \downarrow 1 \\
X & \xrightarrow{g} & Y & \xrightarrow{i'} & Y \cup_g CX & \xrightarrow{j'} & SX & \xrightarrow{-Sg} & SY
\end{array}$$

PROPOSITION 2.2. *Given a commutative diagram*

$$\begin{array}{ccc}
X & \xrightarrow{f} & Y \\
\downarrow h & & \downarrow k \\
X' & \xrightarrow{f'} & Y'
\end{array}$$

we can construct the following commutative diagram.

$$\begin{array}{ccccccccc}
X & \xrightarrow{f} & Y & \xrightarrow{i} & Y \cup_f CX & \xrightarrow{j} & SX & \xrightarrow{-Sf} & SY \\
\downarrow h & & \downarrow k & & \downarrow & & \downarrow Sh & & \downarrow Sk \\
X' & \xrightarrow{f'} & Y' & \xrightarrow{i'} & Y' \cup_{f'} CX' & \xrightarrow{j'} & SX' & \xrightarrow{-Sf'} & SY'
\end{array}$$

These obvious and elementary propositions are special cases of the more general results proved in [15, pp. 311–316].

PROPOSITION 2.3. *Given*

$$X \xrightarrow{f} Y \xrightarrow{g} Z,$$

we can construct the following commutative diagram.

$$\begin{array}{ccccccccc}
X & \xrightarrow{f} & Y & \xrightarrow{i} & Y \cup_f CX & \xrightarrow{j} & SX & \xrightarrow{-Sf} & SY \\
\downarrow{\scriptstyle 1} & & \downarrow{\scriptstyle g} & & \downarrow & & \downarrow{\scriptstyle 1} & & \downarrow{\scriptstyle Sg} \\
X & \xrightarrow{gf} & Z & \xrightarrow{i'} & Z \cup_{gf} CX & \xrightarrow{j'} & SX & \xrightarrow{-S(gf)} & SZ \\
\downarrow{\scriptstyle f} & & \downarrow{\scriptstyle 1} & & \downarrow & & \downarrow{\scriptstyle Sf} & & \downarrow{\scriptstyle 1} \\
Y & \xrightarrow{g} & Z & \xrightarrow{i''} & Z \cup_g CY & \xrightarrow{j''} & SY & \xrightarrow{-Sg} & SZ
\end{array}$$

This follows from two applications of Proposition 2.2.

PROPOSITION 2.4. *For each r, we can construct the following homotopy-commutative diagram, in which all the vertical arrows are homotopy equivalences.*

$$\begin{array}{ccccc}
S^rY & \xrightarrow{S^ri} & S^r(Y \cup_f CX) & \xrightarrow{S^rj} & S^{r+1}X \\
\downarrow{\scriptstyle 1} & & \downarrow & & \downarrow{\scriptstyle (-1)^r} \\
S^rY & \xrightarrow{i'} & (S^rY) \cup_{S^r} C(S^rX) & \xrightarrow{j'} & S^{r+1}X
\end{array}$$

This proposition is easy to check, provided we use the "reduced" cone and suspension. The map $(-1)^r$ of $S^{r+1}X$ arises as a permutation of the suspension coordinates.

§3. DEFINITION AND ELEMENTARY PROPERTIES OF THE INVARIANTS d, e

In this section we shall define our basic invariants d and e. We shall also establish the elementary properties of these invariants.

We shall suppose given a half-exact functor in the sense of [12]. For example, the functor may be one component of a (reduced) extraordinary cohomology theory. More precisely, k is to be a contravariant functor defined on (say) the category of finite CW-complexes and homotopy classes of maps, and taking values in some abelian category [14], say A. If

$$X \xrightarrow{i} Y \xrightarrow{j} Z$$

is a cofibre sequence, then

$$k(X) \xleftarrow{i^*} k(Y) \xleftarrow{j^*} k(Z)$$

is to be an exact sequence in the abelian category A. It follows that we may identify $k(X \vee Y)$ with the direct sum $k(X) \oplus k(Y)$ in the category A; see [12, p. 1].

Now suppose given a map $f: X \to Y$ between (say) finite connected CW-complexes. We can consider the induced homomorphism

$$f^*: k(Y) \to k(X).$$

If we take $X = Y = S^n$ and take k to be $H^n(\ ;Z)$, then the invariant f^* gives us the degree of f. We therefore regard

$$f^*: k(Y) \to k(X)$$

as "the degree of f, measured by k-theory". We define

$$d(f) = f^* \in \operatorname{Hom}(k(Y), k(X)).$$

Here Hom(M, N) means the set of maps from M to N in the abelian category A.

The invariant $e(f)$ will be defined when $d(f) = 0$ and $d(Sf) = 0$. In this case we use the map $f: X \to Y$ to start the following cofibre sequence.

$$X \xrightarrow{f} Y \xrightarrow{i} Y \cup_f CX \xrightarrow{j} SX \xrightarrow{-Sf} SY$$

Since we assume that $f^* = 0$ and $(Sf)^* = 0$, the functor k yields the following short exact sequence in the abelian category A.

$$0 \leftarrow k(Y) \xleftarrow{i^*} k(Y \cup_f CX) \xleftarrow{j^*} k(SX) \leftarrow 0$$

In an abelian category we can define Ext^1 by classifying short exact sequences; therefore the short exact sequence above yields an element of

$$\text{Ext}^1(k(Y), k(SX)).$$

We call this element $e(f)$. The letter e stands for "extension", and goes well with d.

For example, let us consider the case in which $k = \tilde{H}^*(\quad; Z_2)$ and A is the category of graded modules over the mod 2 Steenrod algebra. Let us take $X = S^{m+n-1}$, $Y = S^m$. Given a map $f: S^{m+n-1} \to S^m$, we are led to consider the following short exact sequence.

$$0 \leftarrow \tilde{H}^*(S^m; Z_2) \leftarrow \tilde{H}^*(S^m \cup_f e^{m+n}; Z_2) \leftarrow \tilde{H}^*(S^{m+n}; Z_2) \leftarrow 0$$

As an extension of modules over the Steenrod algebra, this is completely determined by the Steenrod square

$$Sq^n: H^m(S^m \cup_f e^{m+n}; Z_2) \to H^{m+n}(S^m \cup_f e^{m+n}; Z_2).$$

We therefore recover Steenrod's approach to the mod 2 Hopf invariant.

The invariant $e(f)$ may thus be regarded as a "Steenrod–Hopf invariant" in which ordinary cohomology has been replaced by k-theory.

We have just defined

$$d(f) \in \text{Ext}^0(k(Y), k(X))$$

(if we interpret $\text{Ext}^0(M, N)$ as meaning $\text{Hom}(M, N)$), and

$$e(f) \in \text{Ext}^1(k(Y), k(SX)).$$

One would naturally hope to construct a third invariant, which should be defined when suitable d and e invariants vanish, and should take values in

$$\text{Ext}^2(k(Y), k(S^2X)).$$

Similarly for a fourth invariant, and so on. However, we will not pursue this line of thought any further here.

In later sections we will give examples and applications of the invariants d and e, and develop the resources to do practical calculations with them. For the moment we consider the elementary properties of these invariants.

PROPOSITION 3.1 (a). *If $f \sim g$, then $d(f) = d(g)$.*

(b) *If $f \sim g$ and $e(f)$ is defined, the $e(g)$ is defined and $e(f) = e(g)$.*

Proof. Part (a) is obvious. Part (b) is proved by applying the functor k to the diagram given in Proposition 2.1.

We now consider the situation in which we have two maps

$$X \xrightarrow{f} Y \xrightarrow{g} Z.$$

We aim to show that the invariants d and e send composition products (in homotopy) into composition products, i.e. Yoneda products, in Ext groups.

PROPOSITION 3.2 (a). *We have*

$$d(gf) = d(f)d(g).$$

(b) *If $e(f)$ is defined then so is $e(gf)$, and we have*

$$e(gf) = e(f)d(g).$$

(c) *If $e(g)$ is defined then so is $e(gf)$, and we have*

$$e(gf) = d(Sf)e(g).$$

Here statements (b) and (c) use the pairing of Ext^0 and Ext^1 to Ext^1.

Proof. All the statements about invariants d are obvious. For the rest, we apply the functor k to the diagram given in Proposition 2.3, and we obtain the following commutative diagram.

$$\begin{array}{ccccc}
k(Y) & \leftarrow & k(Y \cup_f CX) & \leftarrow & k(SX) \\
{\scriptstyle g^*}\uparrow & & \uparrow & & \uparrow{\scriptstyle 1} \\
k(Z) & \leftarrow & k(Z \cup_{gf} CX) & \leftarrow & k(SX) \\
{\scriptstyle 1}\uparrow & & \uparrow & & \uparrow{\scriptstyle (Sf)^*} \\
k(Z) & \leftarrow & k(Z \cup_g CY) & \leftarrow & k(SY)
\end{array}$$

If $e(f)$ is defined, it is represented by the top row; similarly for $e(gf)$ and the middle row; similarly for $e(g)$ and the bottom row. By definition of the products in Ext, this shows that

$$e(gf) = e(f) \cdot g^*$$

in case (b), and

$$e(gf) = (Sf)^* \cdot e(g)$$

in case (c). This completes the proof.

For our next proposition, we assume that X is a co-H-space, for example, a suspension. That is, we are provided with a map

$$\Delta : X \to X \vee X$$

of type (1, 1). This allows us to define the sum of two (base-point-preserving) maps

$$f, g : X \to Y;$$

by definition, $f + g$ is the composite

$$X \xrightarrow{\Delta} X \vee X \xrightarrow{f \vee g} Y \vee Y \xrightarrow{\mu} Y,$$

where μ is a map of type (1, 1) in the dual sense.

PROPOSITION 3.3 (a). *We have*

$$d(f + g) = d(f) + d(g).$$

(b) *If* $e(f)$ *and* $e(g)$ *are defined then so is* $e(f+g)$, *and*

$$e(f+g) = e(f) + e(g).$$

In part (b), the sum occurring on the right-hand side is, of course, the Baer sum in Ext^1.

Proof. All the statements about invariants d are obvious. For the rest, we may identify $k(Y \vee Y)$ with the direct sum $k(Y) \oplus k(Y)$, and $k(S(X \vee X))$ with $k(SX) \oplus k(SX)$. In this way we can identify the sequence

$$k(Y \vee Y) \leftarrow k((Y \vee Y) \cup_{f \vee g} C(X \vee X)) \leftarrow k(S(X \vee X))$$

with the direct sum of the sequences

$$k(Y) \leftarrow k(Y \cup_f CX) \leftarrow k(SX)$$

$$k(Y) \leftarrow k(Y \cup_g CX) \leftarrow k(SX).$$

That is: if $e(f)$ and $e(g)$ are defined, so is $e(f \vee g)$, and it can be identified with the "external" sum $e(f) \oplus e(g)$. According to Proposition 3.2, we have

$$\begin{aligned} e(f+g) &= e(\mu(f \vee g)\Delta) \\ &= (S\Delta)^* e(f \vee g)\mu^* \\ &= (S\Delta)^*(e(f) \oplus e(g))\mu^*. \end{aligned}$$

But with our identifications,

$$(S\Delta)^* : k(SX) \oplus k(SX) \to k(SX)$$

is a map of type (1, 1) in the category A, and

$$\mu^* : k(Y) \to k(Y) \oplus k(Y)$$

is a map of type (1, 1) in the dual sense. Thus the element

$$(S\Delta)^*(e(f) \oplus e(g))\mu^*$$

is the Baer sum of $e(f)$ and $e(g)$. This completes the proof.

We will now discuss the behaviour of our invariants under suspension. For this purpose we shall suppose that for some integer r, $k(S^r X)$ is known as a function of $k(X)$. For example, when we take $k(X) = \tilde{K}_C(X)$ [10, 11, 2], we shall take $r = 2$; when we take $k(X) = \tilde{K}_R(X)$ we shall take $r = 8$. If we took $k(X) = \tilde{H}^*(X; Z_2)$ we could take $r = 1$. More formally, we shall suppose given a functor T, from the abelian category A to itself, which preserves exact sequences; and we shall suppose given an isomorphism

$$k(S^r X) \cong Tk(X)$$

natural for maps of X. We shall allow ourselves to identify $k(S^r X)$ and $Tk(X)$ under this isomorphism.

Since the functor T preserves exact sequences, it defines a function

$$T : \mathrm{Ext}^1(M, N) \to \mathrm{Ext}^1(TM, TN).$$

This function is actually a homomorphism.

PROPOSITION 3.4 (a). *We have*

$$d(S^r f) = Td(f).$$

(b) *If* $e(f)$ *is defined, then so is* $e(S^r f)$, *and we have*

$$e(S^r f) = (-1)^r Te(f).$$

Proof. All the statements about the invariant d are obvious. For the rest, we apply the functor k to the diagram given in Proposition 2.4 and use the fact that $kS^r = Tk$.

We now define stable track groups by

$$\mathrm{Map}_S(X, Y) = \operatorname*{Dir\,Lim}_{n\to\infty} \mathrm{Map}(S^{nr}X, S^{nr}Y).$$

We also define stabilised Hom groups in the abelian category A by iterating T and taking direct limits; thus,

$$\mathrm{Hom}_S(M, N) = \operatorname*{Dir\,Lim}_{n\to\infty} \mathrm{Hom}(T^n M, T^n N).$$

Similarly, we define stabilised Ext^1 groups by iterating the homomorphism $(-1)^r T$ and taking direct limits; thus,

$$\mathrm{Ext}^1_S(M, N) = \operatorname*{Dir\,Lim}_{n\to\infty} \mathrm{Ext}^1(T^n M, T^n N).$$

PROPOSITION 3.5 (a). *The invariant* d *defines a homomorphism from* $\mathrm{Map}_S(X, Y)$ *to* $\mathrm{Hom}_S(k(Y), k(X))$.

(b) *The invariant* e *defines a homomorphism from the subgroup* $\mathrm{Ker}\, d \cap \mathrm{Ker}(dS)$ *of* $\mathrm{Map}_S(X, Y)$ *to* $\mathrm{Ext}^1_S(k(Y), k(SX))$.

This follows immediately from Propositions 3.1, 3.3, 3.4.

The pairing of Ext groups used in Proposition 3.2 are evidently compatible with the operations T on Ext^0 and $(-1)^r T$ on Ext^1; therefore these pairings pass to the limit. With this interpretation, Proposition 3.2 continues to give the value of the invariants d, e on a composite gf of stable homotopy classes.

§12. EXAMPLES

In this section we will give various examples and illustrations of our general methods, and prove certain results whose proof was deferred in earlier sections. To begin with, our work is directed towards proving Theorem 1.7.

We can actually make Theorem 1.7 a little more complete. As in §1, let p be an odd prime, let $g : S^{2q-1} \to S^{2q-1}$ be a map of degree p^f, and let Y be the Moore space $S^{2q-1} \cup_g e^{2q}$. Thus $\tilde{K}_C(Y) = Z_{p^f}$.

THEOREM 12.1. *There is a map*

$$A : S^{2r}Y \to Y$$

(for suitable q) such that the image of

$$A^* : \tilde{K}_C(Y) \to \tilde{K}_C(S^{2r}Y)$$

is Z_{p^t} *(where* $1 \leq t \leq f$*), if and only if r is divisible by* $(p-1)p^{t-1}$.

It is clear that this includes Theorem 1.7 (take $t = f$). We will show how to deduce Theorem 12.1 from Theorem 1.7.

First, suppose that there is a map $A : S^{2r}Y \to Y$ such that the image of A^* is Z_{p^t}. Then A^* commutes with the operations Ψ^k, which are given in Y and $S^{2r}Y$ by the formulae

$$\Psi^k x = k^q x, \qquad \Psi^k x = k^{q+r} x.$$

Therefore we have $k^{q+r} \equiv k^q \bmod p^t$; so r is divisible by $(p-1)p^{t-1}$.

Secondly, suppose that r is divisible by $(p-1)p^{t-1}$ and Theorem 1.7 is true. Set $Y' = S^{2q-1} \cup_h e^{2q}$, where h is a map of degree p^t. Then by Theorem 1.7 there is a map

$$A' : S^{2r}Y' \to Y'$$

inducing an isomorphism of $\tilde{K}_C$. We have only to take A to be the composite

$$S^{2r}Y \xrightarrow{S^{2r}i} S^{2r}Y' \xrightarrow{A'} Y' \xrightarrow{j} Y$$

where i, j are obvious maps such that $j^* : \tilde{K}_C(Y) \to \tilde{K}_C(Y')$ is an epimorphism and $i^* : \tilde{K}_C(Y') \to \tilde{K}_C(Y')$ is a monomorphism.

This completes the deduction of Theorem 12.1 from Theorem 1.7. We proceed with lemmas needed for the proof of Theorem 1.7. First we consider the cofibering

$$S^{2n-1} \xrightarrow{f} S^{2n-1} \xrightarrow{i} S^{2n-1} \cup_f e^{2n},$$

where f is a map of degree m. If $\Lambda = R$, we assume that n is even; thus we shall certainly have $d_R i = 0$, $d_R(Si) = 0$.

PROPOSITION 12.2. *$e_\Lambda i$ is the class of the extension*

$$0 \leftarrow Z_m \leftarrow Z \xleftarrow{-m} Z \leftarrow 0,$$

in which all the abelian groups have operations Ψ^k defined by

$$\Psi^k x = k^n x.$$

Proof. If we continue the cofibre sequence, it becomes

$$S^{2n-1} \cup_f e^{2n} \xrightarrow{j} S^{2n} \xrightarrow{-Sf} S^{2n};$$

we have only to apply $\tilde{K}_\Lambda$.

For the next proposition, we suppose given a diagram of the following form,

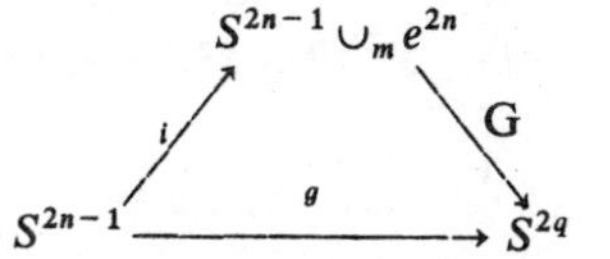

(Here we have written $S^{2n-1} \cup_m e^{2n}$ instead of $S^{2n-1} \cup_f e^{2n}$, where f is a map of degree m.) If $\Lambda = R$, we assume that n and q are even. Thus $\tilde{K}_\Lambda(S^{2q}) = Z$ and $\tilde{K}_\Lambda(S^{2n-1} \cup_m e^{2n}) = Z_m$; we can regard $d_\Lambda(G)$ as an integer mod m. We can also regard $e_\Lambda(g)$ as a rational mod 1; since $mg \sim 0$, $me_\Lambda(g)$ is an integer mod m.

PROPOSITION 12.3. *We have*

$$d_\Lambda(G) = -me_\Lambda(g) \qquad \text{mod } m$$

or equivalently

$$e_\Lambda(g) = -\frac{1}{m} d_\Lambda(G) \qquad \text{mod } 1.$$

Proof. This proposition is a special case of Proposition 3.2 (b), which states that

$$e(Gi) = e(i)\, d(G).$$

The element $e(i)$ has been given in Proposition 12.2; one has only to compute the product $e(i)\, d(G)$, which is an easy exercise in homological algebra.

LEMMA 12.4. *Let p be an odd prime, $m = p^f$, and $r = (p-1)p^f$. Then there is an element $\alpha \in \pi^S_{2r-1}$ satisfying the following conditions.*

(i) $m\alpha = 0$.

(ii) $e_C \alpha = -\dfrac{1}{m}$.

(iii) *The Toda bracket $\{m, \alpha, m\}$ is zero mod $m\pi^S_{2r}$.*

Proof. For $f = 1$ the result is easy; we have only to take α to be an element of Hopf invariant one mod p in π^S_{2p-3}. Then (i), (ii) are given by Corollary 8.4 and (iii) follows from the fact that the p-component of π^S_{2p-2} is zero.

For any f we can take α to be a suitable element in Im J, using Theorem 1.5 or 1.6 to obtain (i), (ii). Condition (iii) follows from the fact that $\{m, \alpha, m\}$ is an element of order 2 [18, p.26 (2.4) (i), p.33 (3.9) (i)].

Lemma 12.4 supplies the data for the following lemma, which we shall also use with $m = 2$.

LEMMA 12.5. *Suppose given* $\alpha \in \pi^S_{2r-1}$ *and* $m \in Z$ *such that*

(i) $m\alpha = 0$,

(ii) $e_C\alpha = -\dfrac{1}{m}$,

(iii) $\{m, \alpha, m\} = 0 \bmod m\pi^S_{2r}$.

Then for suitably large q there exist maps A which make the following diagram homotopy-commutative; and for any such A we have $d_C(A) = 1$.

$$\begin{array}{ccc} S^{2q+2r-1} \cup_m e^{2q+2r} & \xrightarrow{A} & S^{2q-1} \cup_m e^{2q} \\ \uparrow i & & \downarrow j \\ S^{2q+2r-1} & \xrightarrow{\alpha} & S^{2q} \end{array}$$

Proof. Conditions (i), (iii) enable one to construct the diagram. By Proposition 12.3 and condition (ii) we have $d_C(jA) = 1$. Hence $d_C(A) = 1$.

Theorem 1.7 follows immediately from Lemmas 12.4, 12.5. Since A induces an isomorphism of $\tilde{K}_C$, so does the composite

$$A . S^{2r}A . S^{4r}A . \dots S^{2r(s-1)}A : S^{2rs}Y \to Y;$$

Indeed we have

$$d_C(A . S^{2r}A . S^{4r}A . \dots S^{2r(s-1)}A) = 1.$$

Therefore this composite is essential for every s.

Under the assumptions of Lemma 12.5, we construct a map

$$\alpha_s : S^{2q+2rs-1} \to S^{2q}$$

by the following diagram.

$$\begin{array}{ccc} S^{2q+2rs-1} \cup_m e^{2q+2rs} & \xrightarrow{A \cdot S^{2r}A \cdots S^{2r(s-1)}A} & S^{2q-1} \cup_m e^{2q} \\ \uparrow i & & \downarrow j \\ S^{2q+2rs-1} & \xrightarrow{\alpha_s} & S^{2q} \end{array}$$

We have $\alpha_1 = \alpha$. The map α_s has order dividing m, since it can be extended over $S^{2q+2rs-1} \cup_m e^{2q+2rs}$. The maps α_s satisfy the equation

(12.6) $$\alpha_{s+t} \in \{\alpha_s, m, \alpha_t\}.$$

The case in which m is an odd prime p and $r = p - 1$ has been studied by Toda [16, 17].

PROPOSITION 12.7. *Under the assumptions of Lemma* 12.5, *the maps* α_s *are all essential; indeed we have*

$$e_C(\alpha_s) = -\frac{1}{m} \quad \mathit{mod}\ 1.$$

This improves and generalises a result of Toda [17]. Presumably the present proof is related to Toda's proof; however, it is hoped that the presentation given here may be found more conceptual.

Proofs. (i) Apply Proposition 12.3 to the diagram which defines α_s. (ii) Alternatively, apply Theorem 11.1 to equation (12.6) and use induction.

EXAMPLE 12.8. *We note that in* [16, 17] *Toda's elements* α_s *depend on the choice of* α_1, *which Toda does not fix; similarly, there is a choice for his element* α'_p. *However, we may take the choices so that*

$$e_C(\alpha_s) = -\frac{1}{p}, \qquad e_C(\alpha'_{rp}) = -\frac{1}{p^2}.$$

Then the coefficient δ *in Corollary* 11.6 *explains the coefficients which arise in Toda's formulae for*

$$\{\alpha_t, \alpha_s, p\} \quad and \quad \{\alpha'_{rp}, \alpha_s, p\}$$

[16, Theorem 4.17 (ii)].

We now pass on to study 2-primary phenomena. To begin with we prove the following result.

Theorem 12.13. *For each $s \geq 0$ there is an element μ_{8s+1} of order 2 in π_{8s+1}^S such that $e_C(\mu_{8s+1}) = \frac{1}{2}$ mod 1.*

Proof. Let α be the element of order 2 in π_7^S. Since $e_R' : \pi_7^S \to Z_{240}$ is an isomorphism, we have $e_C(\alpha) = \frac{1}{2}$ mod 1. Also, by a delicate result of Toda [18, p.31 Corollary 3.7] we have

$$\begin{aligned} \{2, \alpha, 2\} &= \alpha\eta \qquad \text{mod } 2 \\ &= 0 \qquad \text{mod } 2, \end{aligned}$$

since α is divisible by 2 and $2\eta = 0$. Thus we can apply Lemma 12.5 to construct a map A. Now we have the following diagram.

$$\begin{array}{ccccc} S^{2q+8s-1} \cup_2 e^{2q+8s} & \xrightarrow{A \cdot S^8 A \cdots + S^{8(s-1)}A} & S^{2q-1} \cup_2 e^{2q} & & \\ \uparrow i & \swarrow j & \uparrow i & \searrow \bar{\eta} & \\ S^{2q+8s-1} & \xrightarrow{\alpha_s} & S^{2q} \qquad S^{2q-1} & \xrightarrow{\eta} & S^{2q-2} \end{array}$$

We define μ_{8s+1} to be the composite

$$\bar{\eta} . A . S^8 A . \ldots S^{8(s-1)} A . i.$$

We have $\mu_1 = \eta$. The map μ_{8s+1} has order dividing 2, since it can be extended over $S^{2q+8s-1} \cup_2 e^{2q+8s}$. Since $e_C(\eta) = \frac{1}{2}$ mod 1, Proposition 12.3 shows that $d_C(\bar{\eta}) = 1$ mod 2. Hence

$$d_C(\bar{\eta} . A . S^8 A . \ldots S^{8(s-1)} A) = 1 \qquad \text{mod } 2.$$

A second application of Proposition 12.3 now yields

$$e_C(\mu_{8s+1}) = \tfrac{1}{2} \qquad \text{mod } 1.$$

Alternatively, we can obtain the same result by applying Theorem 11.1 to the equation

$$\mu_{8s+1} \in \{\eta, 2, \alpha_s\},$$

in which $e_C(\alpha_s) = \frac{1}{2}$ mod 1 by Proposition 12.7.

Proof of Theorem 7.18. Suppose $r \equiv 1$ mod 8. Then by Theorem 12.13 the homomorphism

$$e_C : \pi_r^S \to Z_2$$

is an epimorphism. But we also have

$$d_R : \pi_r^S \to Z_2$$

and Ker $d_R \subset$ Ker e_C by Lemma 7.21. Therefore $d_R = e_C$. This proves Theorem 7.18.

We have just shown that

$$d_R \mu_{8s+1} \neq 0.$$

(It is possible to show this directly from the construction of μ_{8s+1}, but this is unnecessary.)

Proposition 12.14. *If $r \equiv 1$ mod 8 and $s \equiv 1$ mod 8 then the composite $\mu_r \mu_s$ is non-zero; indeed*

$$d_R(\mu_r \mu_s) \neq 0.$$

This proposition generalises the behaviour of the composite $\eta\eta$. The proof is immediate.

Proof of Theorem 7.2. Let us define μ_{8s+2} to be one of the composites considered in Proposition 12.14, for example, $\eta\mu_{8s+1}$. Then we have shown that for $r \equiv 1, 2 \bmod 8$ and $r > 0$ we have $d_R\mu_r \neq 0$. Thus d_R is an epimorphism; and since μ_r is of order 2, π_r^S splits as a direct sum $Z_2 + \operatorname{Ker} d_R$, where the subgroup Z_2 is generated by μ_r.

EXAMPLE 12.15. *Suppose that* $\theta \in \pi_{8t-1}^S$ *is an element such that* $m(4t)e_R(\theta)$ *is odd. Then for* $r \equiv 1, 2$ *mod* 8 *the composite* $\theta\mu_r$ *is essential; indeed*

$$e_R(\theta\mu_r) \neq 0.$$

Proof. By Theorem 3.2 (c) we have

$$e_R(\theta\mu_r) = d_R(\mu_r)e_R(\theta).$$

Let us use the notation of §9; then $e_R(\theta)$ is a generator of the 2-component of $\operatorname{Ext}_S^1(M, N)$ and the homomorphism $d_R(\mu_r)$ may be identified with the quotient map $N \to N'$. So according to the discussion in §9, $d_R(\mu_r) \,.\, e_R(\theta)$ represents a generator of $\operatorname{Ext}_S^1(M, N')$.

This example provides a second proof for Theorem 9.5. In fact, let γ be a generator for $\pi_{8u-1}(SO)$ $(u > 0)$. Then the generators for $\pi_{8u}(SO)$, $\pi_{8u+1}(SO)$ can be written as composites $\gamma\eta$, $\gamma\eta\eta$; and we have

$$J(\gamma\eta) = J(\gamma)\eta$$
$$J(\gamma\eta\eta) = J(\gamma)\eta\eta.$$

Thus Theorem 9.5 follows from Example 12.15.

EXAMPLE 12.16. *If* $r \equiv 1$ *mod* 8 *then* $\{2, \mu_r, 2\}$ *is non-zero; indeed* $d_R\{2, \mu_r, 2\} \neq 0$.

This example generalises the behaviour of $\{2, \eta, 2\}$. The reader will find that it is an easy application of Theorem 5.3 (i). Alternatively, of course, one can quote [18, p.31 Corollary 3.7] to show that $\{2, \mu_r, 2\} = \mu_r\eta$ mod 2 and use Proposition 12.14.

PROPOSITION 12.17. If $r \equiv 2$ *mod* 8 *and* $s \equiv 1$ mod 8 *then the composition* $\mu_r\mu_s$ *is non-zero; indeed*

$$e'_R(\mu_r\mu_s) = \tfrac{1}{2} \qquad \text{mod } 1.$$

This proposition generalises the behaviour of the composite $\eta\eta\eta$.

Proof. Let

$$f: S^{2n-1} \to S^{2t}, \qquad g: S^{2t} \to S^{2q}$$

be maps representing μ_s, μ_r, where $2q \equiv 0$ mod 8, $2t \equiv 2$ mod 8, $2n - 1 \equiv 3$ mod 8. We have to consider the invariant $e_R(f)$. We have the following diagram.

$$\begin{array}{ccccc}
Z_2 = \tilde{K}_R(S^{2t}) & \leftarrow & \tilde{K}_R(S^{2t} \cup_f e^{2n}) & \leftarrow & \tilde{K}_R(S^{2n}) = Z \\
\uparrow{\scriptstyle epi} & & \uparrow{\scriptstyle r} & & \uparrow{\scriptstyle iso} \\
Z = \tilde{K}_C(S^{2t}) & \leftarrow & \tilde{K}_C(S^{2t} \cup_f e^{2n}) & \leftarrow & \tilde{K}_C(S^{2n}) = Z
\end{array}$$

Let ξ, η be generators in $\tilde{K}_C(S^{2t} \cup_f e^{2n})$. Then since $e_C(f) = \frac{1}{2}$ mod 1 we have (for a suitable choice of ξ)

$$\begin{aligned}\Psi^{-1}\xi &= (-1)^t\xi + \tfrac{1}{2}((-1)^n - (-1)^t)\eta \\ &= -\xi + \eta.\end{aligned}$$

Now in $\tilde{K}_R(S^{2t} \cup_f e^{2n})$ we have $r\Psi^{-1} = r$; thus we have $2r\xi = r\eta$. Thus $e_R(f)$ is the non-trivial extension

$$0 \leftarrow Z_2 \leftarrow Z \overset{2}{\leftarrow} Z \leftarrow 0$$

in which all the groups are given operations Ψ^k by the formula $\Psi^k x = k^n x$.

We must now compute the product $e_R(f)\, d_R(g)$, where

$$d_R(g) \colon \tilde{K}_R(S^{2q}) \rightarrow \tilde{K}_R(S^{2t})$$

is the epimorphism $Z \rightarrow Z_2$. We easily find that $e_R(f)\, d_R(g)$ is the extension corresponding to the rational $\frac{1}{2}$ mod 1.

PROPOSITION 12.18. *If* $r \equiv 1$ *mod* 8 *and* $s \equiv 1$ *mod* 8 *then any representative of the Toda bracket* $\{\mu_r, 2, \mu_s\}$ *is an element of order* 4; *indeed* $e'_R\{\mu_r, 2, \mu_s\} = \frac{1}{4}$ *mod* $\frac{1}{2}$.

This proposition generalises the behaviour of $\{\eta, 2, \eta\}$.

Proof. We have just shown that the indeterminacy of $\{\mu_r, 2, \mu_s\}$ consists at least of the integers $\frac{1}{2}$ mod 1. By Theorem 11.1 we have

$$\begin{aligned} e_C\{\mu_r, 2, \mu_s\} &= -\tfrac{1}{2}.2.\tfrac{1}{2} && \text{mod } 1 \\ &= \tfrac{1}{2} && \text{mod } 1. \end{aligned}$$

By Proposition 7.14 this is equivalent to

$$e'_R\{\mu_r, 2, \mu_s\} = \tfrac{1}{4} \qquad \text{mod } \tfrac{1}{2}.$$

On the other hand, we have

$$2\{\mu_r, 2, \mu_s\} = \{2, \mu_r, 2\}\mu_s \qquad \text{mod } 0.$$

This actually gives $\eta\mu_r\, \mu_s$; but at all events it is an element of order 2 at most, so $\{\mu_r, 2, \mu_s\}$ has order dividing 4. This completes the proof.

REFERENCES

1. J. F. Adams: On Chern characters and the structure of the unitary group, *Proc. Camb. Phil. Soc.* **57** (1961), 189–199.
2. J. F. Adams: Vector field on spheres, *Ann. Math.* **75** (1962), 603–632.
3. J. F. Adams: On the groups $J(X)$—I, *Topology* **2** (1963), 181–195.
4. J. F. Adams: On the groups $J(X)$—II, *Topology* **3** (1965), 137–171.
5. J. F. Adams: On the groups $J(X)$—III, *Topology* **3** (1965), 193–222.
6. J. F. Adams: Lectures on $K^*(X)$, (mimeographed notes, Manchester, 1962).
7. J. F. Adams: Cohomology operations (mimeographed notes, Seattle, 1963).
8. J. F. Adams and M. F. Atiyah: *K*-theory and the Hopf invariant, *Quart. J. Math.*, to appear.
9. J. F. Adams and G. Walker: Complex Stiefel Manifolds, *Proc. Camb. Phil. Soc.* **61** (1965), 81–103.
10. M. F. Atiyah and F. Hirzebruch: Riemann–Roch theorems for differentiable manifolds, *Bull. Amer. Math. Soc.* **65** (1959), 276–281.
11. M. F. Atiyah and F. Hirzebruch: Vector bundles and homogeneous spaces, *Proc. of Symposia in Pure Mathematics* 3, Differential Geometry, Amer. Math. Soc., 1961, pp. 7–38.
12. A. Dold: Half exact functors and cohomology, (mimeographed notes, Seattle, 1963).
13. E. Dyer: Chern characters of certain complexes, *Math. Z.* **80** (1963), 363–373.
14. S. MacLane: *Homology*, Springer, Berlin, 1963.
15. D. Puppe: Homotopiemengen und ihre induzierten Abbildungen. I *Math. Z.* **69** (1958), 299–344.
16. H. Toda: *p*-primary components of homotopy groups. IV. Compositions and toric constructions, *Mem. Coll. Sci. Kyoto*, Ser. **A 32** (1959), 297–332.
17. H. Toda: On unstable homotopy of spheres and classical groups, *Proc. Nat. Acad. Sci., Wash.* **46** (1960), 1102–1105.
18. H. Toda: Composition methods in homotopy groups of spheres, *Ann. Math. Stud. No.* 49, Princeton 1962.
19. H. Toda: A survey of homotopy theory, *Sûgaku* **15** (1963/4), 141–155.

23

The next piece is a summary on complex cobordism, written by me especially for the present work.

23
A SUMMARY ON COMPLEX COBORDISM

J. F. Adams

The subject of complex cobordism has two aspects: a geometrical side, on which it links up with the theory of complex manifolds, and a homotopy-theoretic side, on which it links up with generalised homology and cohomology theories and the study of spectra. I begin by sketching this. (Afterwards I will present the calculation of $\pi_*(MU)$, and finish by sketching some topics from the further development of the subject.)

Let M_1^m and M_2^m be two smooth manifolds, of dimension m, compact, without boundary and both of the same sort: that is, both non-oriented, or both oriented, or both with whatever extra structure is to be considered. Then we say that M_1^m and M_2^m are cobordant if there is a smooth manifold W^{m+1} of dimension $m + 1$, compact, with boundary, and of the same sort, such that the boundary of W^{m+1} is the disjoint union of M_1^m and M_2^m. Here the notion of 'boundary' is taken in the sense appropriate to manifolds of the sort considered, so that we include a condition on any extra structure we may have. For example, if we are working with oriented manifolds, then we ask that the boundary of the orientation class on W^{m+1} should be plus the orientation class on M_2^m, minus the orientation class on M_1^m. Similarly for other forms of extra structure. Cobordism is an equivalence relation, and divides m-manifolds of the sort considered into equivalence classes. The set of equivalence classes becomes an abelian group if we use the disjoint union of manifolds as the group operation.

In particular, let M^m be a smooth manifold, compact and without boundary. We will say that it is weakly almost-complex, or stably almost-complex, if it can be embedded in a sphere S^{m+2n} of sufficiently high dimension with a normal bundle which is a U(n)-bundle. (It is also possible to say this in terms of the tangent bundle of M^m, but the definition given is the most basic.) In terms of such manifolds we define the complex cobordism group Ω^U_m. These groups make up a graded ring if we use the Cartesian product of manifolds as the product operation. (This also happens with the other sorts of 'extra structure' usually considered.)

We may say that it is difficult to classify manifolds into isomorphism classes because there are so many different isomorphism classes; cobordism is a cruder equivalence relation than isomorphism, so that the cobordism classes are larger than the isomorphism classes, but there are fewer of them; for many purposes it is sufficient to know the cruder classification into cobordism classes, and therefore the calculation of cobordism groups becomes important.

The fundamental step in the calculation of cobordism groups is the introduction of Thom complexes, as in the fundamental paper of Thom ('Quelques propriétés globales de variétés différentiables', Comment. Math. Helvetici 28 (1954), 17-86). Let ξ be a U(n)-bundle over a CW-complex X. Let E be the associated bundle whose fibre is the unit disc D^{2n} in complex n-space C^n; let E_0 be its boundary, that is, the associated bundle whose fibre is the unit sphere S^{2n-1} in C^n. Then the Thom complex $M(\xi)$ is the quotient space E/E_0. Alternatively, let V be the associated bundle whose fibre is the sphere $C^n \cup (\infty)$ and let V_∞ be the section at ∞; then $M(\xi)$ is the quotient space V/V_∞. In particular, let ξ_n be the universal U(n)-bundle over BU(n);

then MU_n is defined to be $M(\xi_n)$.

The fundamental fact, then, is that we have an isomorphism

$$\Omega^U_m \cong \lim_{n\to\infty} \pi_{m+2n}(MU_n) .$$

This isomorphism allows one to calculate the groups Ω^U_m by applying homotopy-theory, following Milnor and Novikov. (See Milnor, 'On the cobordism ring Ω^* and a complex analogue', Amer. Jour. Math. 82 (1960), 505-521; this paper by Milnor is highly recommended. See also Thom, 'Travaux de Milnor sur le cobordisme', Seminaire Bourbaki no. 180, 1958/59, and Novikov, 'Some problems in the topology of manifolds connected with the theory of Thom spaces', Doklady Akad. Nauk SSSR 132 (1960), 1031-1034.) However, the notation $\lim_{n\to\infty}$ implies that we are given both the groups $\pi_{m+2n}(MU_n)$ and certain homomorphisms between them; we have not yet defined these homomorphisms, so we must return to the details.

Over the space $BU(n) \times BU(m)$ there is the external Whitney sum $\xi_n \times \xi_m$. This bundle admits a classifying map to ξ_{n+m}, and of course this induces a map from $M(\xi_n \times \xi_m)$ to $M(\xi_{n+m})$. But we can check that the Thom complex $M(\xi_n \times \xi_m)$ is homeomorphic to the 'smash' product $M(\xi_n) \wedge M(\xi_m)$, where $X \wedge Y = X \times Y / X \vee Y$; so we have a product map

$$MU_n \wedge MU_m \to MU_{n+m} .$$

Also we have a map

$$S^2 \xrightarrow{i} MU_1$$

given by the injection of one fibre; we we can construct

$$S^2 \wedge MU_m \xrightarrow{i \wedge 1} MU_1 \wedge MU_m \longrightarrow MU_{m+1} .$$

These maps make the sequence of spaces MU_n into a spectrum MU. The spectrum MU is a ring-spectrum, because of the product maps constructed above.

Let us now return to the geometrical situation. Take a manifold M^m embedded in S^{m+2n} with normal bundle ν which is a U(n)-bundle. Then we have a classifying map from ν to ξ_n, and this induces a map of Thom complexes from $M(\nu)$ to $M(\xi_n)$. But $M(\nu)$ is obtained from the sphere S^{m+2n} by taking the complement of a tubular neighbourhood of M^m and identifying that complement to a point. So we get a map $S^{m+2n} \to MU_n$. This is the classical Pontryagin-Thom construction, and it is this construction which induces the isomorphism

$$\Omega^U_m \xrightarrow{\cong} \lim_{n\to\infty} \pi_{m+2n}(MU_n) = \pi_m(MU) .$$

This isomorphism preserves the products.

So we see that the homotopy groups of the spectrum MU admit a geometrical interpretation. As a matter of fact the homology theory determined by the spectrum MU (see paper no. 13) also admits a geometrical interpretation. It is called complex bordism. Suppose given a space X; to construct $MU_m(X)$ we take maps $f: M^m \to X$, where M^m runs over the stably almost-complex manifolds, and classify these maps into suitable equivalence classes. The process is faintly reminiscent of the way in which one constructs singular homology by considering maps $f: \sigma^m \to X$. See Atiyah, 'Bordism and cobordism', Proc. Camb. Phil. Soc. 57 (1961),

200-208; also Conner and Floyd (1).

Similarly, the Thom spectrum MU gives rise to a cohomology functor $MU^*(X)$ or $\Omega^*_U(X)$, and this is called complex cobordism.

Of course, in order to make use of a generalised homology or cohomology functor one has to know the coefficient groups, and so we are forced back to computing the homotopy ring $\pi_*(MU)$ of the spectrum MU. The result is as follows.

Theorem 1 (Milnor, Novikov). $\pi_*(MU)$ _is a polynomial ring (over_ Z) _on generators of dimension_ 2, 4, 6, 8, ...

Let Q be the rationals; then it is certainly clear that $\pi_*(MU) \otimes Q$ is a polynomial ring over Q on generators of dimensions 2, 4, 6, 8, ...; for we have

$$\pi_*(MU) \otimes Q \cong H_*(MU) \otimes Q ,$$

as may be seen from Serre's C-theory (§8); and we may suppose that $H_*(MU)$ is known. To show that $\pi_*(MU)$ has no torsion, one may use the following spectral sequence, which is due to the present writer (see §10).

Lemma 2. _There is a spectral sequence_

$$\mathrm{Ext}_A^{s,t}(H^*(MU; Z_p), Z_p) \underset{s}{\Longrightarrow} \pi_{t-s}(MU) .$$

Here p is supposed to be a prime; A is the Steenrod algebra of operations on mod p cohomology; and $H^*(MU; Z_p)$ and Z_p are considered as modules over A. Of course the validity of the lemma is not restricted to the spectrum MU; but in this case the spectral sequence is a spectral sequence of rings; all the products arise from the fact that MU is a ring-spectrum.

The ring $\pi_*(MU)$ has a unit element 1 (represented by the point considered as a 0-manifold, or by the injection of a fibre considered as a map $S^{2n} \to MU_n$). The element $p1 \in \pi_0(MU)$ has filtration 1 in the spectral sequence; it defines an element $\{p1\}$ in $E_r^{1,1}$ for all r.

The calculation of the E_2-term of the spectral sequence is purely computational, and we present the answer.

Lemma 3. $\mathrm{Ext}_A^{s,t}(H^*(MU; Z_p), Z_p)$ *is a polynomial algebra over* Z_p *on generators of the following bidegrees.*

(i) $s = 0,\ t = 2n$ *whenever* $n > 0$ *and* $n + 1$ *is not a power of* p.

(ii) $s = 1,\ t = 2p^f - 1$ *for each* $f \geq 0$, *the generator for* $f = 0$ *being* $\{p1\}$.

Since $E_2^{s,t}$ is zero when $t - s$ is odd, all the differentials in the spectral sequence are zero, and the spectral sequence is trivial.

Corollary 4. *The torsion subgroup of* $\pi_*(MU)$ *is zero.*

Sketch proof. Multiplication by $\{p1\}$ is mono on E_2^{**} (since $\{p\,1\}$ is one of the polynomial generators), therefore on E_∞^{**}; therefore multiplication by p is mono on $\pi_*(MU)$. This holds for each p.

Milnor's published proof of Corollary 4 is slightly different. He considers 'homotopy with coefficients', defined in terms of maps from Y to MU, where Y is the complex $S^1 \cup_p e^2$ in which the attaching map is of degree p. There is a spectral sequence similar to that of Lemma 2, which is again trivial and shows that $[Y, MU]_r = 0$ if r is odd. If there were any p-torsion in $\pi_*(MU)$, then an obvious exact sequence (the 'universal coefficient theorem for homotopy with coefficients') would show that there were non-zero elements in $[Y, MU]_*$ in two consecutive dimensions - a contradiction.

Corollary 5. $\pi_*(MU) \otimes Z_p$ is a polynomial algebra over Z_p on generators of dimension 2, 4, 6, 8, ...

It is easy to believe that this follows from lemma 3, since the effect of tensoring with Z_p is simply to identify the element p1 with zero. The precise proof, however, requires the following technical lemma on the convergence of the spectral sequence (or at least the special case $f = 1$).

Lemma 6. Given q and f, there exists $s = s(q, f)$ such that every element x of filtration s in $\pi_q(MU)$ has the form $x = p^f y$ for some y in $\pi_q(MU)$.

The validity of this lemma is not restricted to the spectrum MU; it is a general lemma. It may be proved by the argument given in Comment. Math. Helv. 32 (1958), 191-192.

(I remark at this point that I owe John Milnor an apology. He once asked me for a lemma in the same direction as lemma 6, but I failed to get my thoughts clear enough to provide it at that time. I assume that Milnor had in mind a line of argument similar to that presented here.)

We now introduce a lemma from pure algebra.

Lemma 7. Let R_* be a graded anticommutative ring with unit, such that R_q is a finitely-generated abelian group for each q, and such that for each prime p, $R_* \otimes Z_p$ is a polynomial algebra over Z_p on generators of dimension 2, 4, 6, 8, ... Then R_* is a polynomial algebra over Z on generators of dimension 2, 4, 6, 8, ...

In this lemma, the particular dimensions of the generators are irrelevant; all that matters is that the dimensions of the generators should be the same for all p. The proof is easy.

We know from Serre's C-theory (§8) that $\pi_q(MU)$ is finitely-generated for each q. So lemma 7 applies to $R_* = \pi_*(MU)$; this completes the proof of Theorem 1.

At the risk of increasing the level of sophistication, and certainly of adding material I do not think compulsory for all students, I add a few words about the further development of the subject.

In general, if we are asked to compute a ring, it is not good enough just to know that it is a polynomial algebra; we would like to know definite elements which provide generators. It is easy to name generators for $\pi_*(MU) \otimes Q$ geometrically; the complex projective spaces CP^n will do. Unfortunately, they do not provide generators for $\pi_*(MU)$; for example, in $\pi_6(MU)$ there is an element which in terms of projective spaces can only be written as $\frac{1}{2}(CP^3 - (CP^1)^3)$. No canonical choice of polynomial generators for $\pi_*(MU)$ is yet known to me.

However, Milnor showed that it is possible to obtain generators by taking Z-linear combinations of certain manifolds $H_{i,j}$. To construct $H_{i,j}$, one takes in $CP^i \times CP^j$ the subset of pairs $((w_0, w_1, \ldots, w_i), (z_0, z_1, \ldots, z_j))$ such that

$$w_0 z_0 + w_1 z_1 + \ldots + w_k z_k = 0 ,$$

where $k = \min(i, j)$. The manifold $H_{i,j}$ is a 'hypersurface of type (1, 1)'.

The real reason why generators can be obtained in this way is linked with the theory of 'formal groups'. The study of formal groups may be approached through an analogy. Let G be a commutative real Lie group of dimension 1; and let us choose a chart round the identity e in which e corresponds to the real number 0. Then the product in G may be given by a power-series

$$\mu(x, y) = x + y + \sum_{i,j \geq 1} a_{ij} x^i y^j ;$$

this power-series will be convergent for x and y sufficiently close to 0, and will have properties such as

$$\mu(x, y) = \mu(y, x)$$

$$\mu(x, \mu(y, z)) = \mu(\mu(x, y), z) .$$

A 'formal product' is a formal power-series

$$\mu(x, y) = x + y + \sum_{i,j \geq 1} a_{ij} x^i y^j$$

having the same formal properties; but now the coefficients a_{ij}

are supposed to lie in some abstract ring R.

The connection between cobordism and the theory of 'formal groups' was observed by S. P. Novikov (The methods of algebraic topology from the viewpoint of cobordism theories, Izvestiya Akad. Nauk SSSR 31 (1967), 855-951) and exploited by D. Quillen (to be published). Briefly, the Thom complex MU_1 is equivalent to complex projective space CP^∞. This equivalence defines an element ω in the cobordism group $MU^2(CP^\infty)$. The cobordism ring $MU^*(CP^\infty)$ is a ring of formal power-series, in one variable ω, over $\pi_*(MU)$ as the ring of coefficients. Similarly, $MU^*(CP^\infty \times CP^\infty)$ is a ring of formal power-series over $\pi_*(MU)$ on two generators, ω_1 and ω_2, induced from ω by the projection of $CP^\infty \times CP^\infty$ onto its two factors. Now CP^∞ is an H-space; we have a product map

$$g : CP^\infty \times CP^\infty \to CP^\infty .$$

Therefore we can form the element $g^*\omega$ in $MU^2(CP^\infty \times CP^\infty)$, and it can be written as a formal power-series

$$g^*\omega = \omega_1 + \omega_2 + \sum_{i,j \geq 1} a_{ij}\, \omega_1^i\, \omega_2^j .$$

Here each coefficient a_{ij} lies in $\pi_{2(i+j-1)}(MU)$. This formal power-series is a formal product.

It can be shown that the ring $\pi_*(MU)$ is generated by the coefficients a_{ij} which arise in this power-series. Moreover, the proof can be made 'clean', at least in the sense that it relies mainly on theory and does not involve computations except such as arise inevitably from the general theory. (To sketch this proof would involve more about formal groups than seems appropriate here.)

We have said that the elements a_{ij} generate $\pi_*(MU)$. But the element a_{ij} coincides with H_{ij}, up to sign, and modulo decomposable elements; therefore the elements H_{ij} generate $\pi_*(MU)$. (Actually the last assertion about a_{ij} is true only if we exclude low values of i and low values of j; but these can be taken into account separately.)

We turn next to summarise the connection between complex bordism or cobordism and complex K-theory. Here the first theorem is that of Conner and Floyd (2). As in 'Lectures on generalised cohomology', Springer-Verlag, Lecture Notes in Mathematics no. 99, p. 38, we state it for the covariant functors to avoid finiteness assumptions on X. The theorem says that the complex bordism of X determines its complex K-homology.

Theorem 8. We have

$$BU_*(X) \cong MU_*(X) \otimes_{\pi_*(MU)} \pi_*(BU) .$$

For further details and proof, see the references cited above.

The second theorem in this direction is that of Hattori and Stong (Hattori, 'Integral characteristic numbers for weakly almost complex manifolds', Topology 5 (1966), 259-280; Stong, 'Relations among characteristic numbers I', Topology 4 (1965), 267-281). Recall that it follows from Corollary 4 that the Hurewicz homomorphism

$$h:\pi_*(MU) \to H_*(MU)$$

is mono. The result was originally conceived as a characterisation of the image of h, and as a tool for practical calculation; and indeed it is effective as such. However, the most elegant formulation has a rather theoretical appearance; it goes as follows.

Theorem 9. The Hurewicz homomorphism in complex K-homology,

$$k : \pi_*(MU) \to BU_*(MU)$$

is the injection of a direct summand.

For further details and proof, see the reference to Hattori cited above.

If one wishes to describe the image of k, that can also be done. It is shown in 'Lectures on generalised cohomology', Springer-Verlag, Lecture Notes in Mathematics no. 99, pp. 56-76, that $BU_*(MU)$ is a comodule with respect to the coalgebra $BU_*(BU)$, so that we have a structure map

$$\psi : BU_*(MU) \to BU_*(BU) \otimes_{\pi_*(BU)} BU_*(MU) .$$

An element x in $BU_*(MU)$ is said to be 'primitive' if $\psi x = 1 \otimes x$. The set of primitive elements is written $P(BU_*(MU))$.

Proposition 10. The Hurewicz homomorphism in complex K-homology gives an isomorphism

$$k : \pi_*(MU) \xrightarrow{\cong} P(BU_*(MU)) .$$

It is easy to deduce this from Theorem 9.

Finally, the algebra of cohomology operations on the cohomology functor $MU^*(\ )$ has been completely determined by Novikov (Izvestiya Akad. Nauk SSSR 31 (1967), 855-951). The method may be explained as follows. We have to compute $MU^*(MU)$. Now in ordinary cohomology we have a 'Thom isomorphism'

$$H^q(BU(n)) \rightarrow \tilde{H}^{q+2n}(MU_n) .$$

Similarly, here we have an isomorphism

$$MU^*(BU) \xrightarrow{\cong} MU^*(MU) .$$

But to compute $MU^*(BU)$ amounts to studying characteristic classes, definex on unitary bundles, with values in $MU^*(\ )$; and these have been studied by Conner and Floyd (2). The result follows.

24

The final piece is an excellent survey article by Novikov. As with many other survey articles, the reader's first object in reading it should be to gain a general understanding of what is going on rather than a grasp of the technical details behind each sentence.

24

NEW IDEAS IN ALGEBRAIC TOPOLOGY (K-THEORY AND ITS APPLICATIONS)

S.P. NOVIKOV

Contents

Introduction

In recent years there has been a widespread development in topology of the so-called *generalized homology theories*. Of these perhaps the most striking are *K*-theory and the *bordism* and *cobordism* theories. The term *homology theory* is used here, because these objects, often very different in their geometric meaning, share many of the properties of ordinary homology and cohomology, the analogy being extremely useful in solving concrete problems. The *K-functor*, which arose in algebraic geometry in the well-known work of Grothendieck, has been successfully applied by Atiyah and Hirzebruch to differential topology and has led quickly to the solution of a number of delicate problems.

Among the results obtained strictly with the help of K-theory the work of Atiyah and Singer on the problem of the index of elliptic operators and of Adams on vector fields on spheres and the Whitehead J-homomorphism are outstanding. More or less influenced by the K-functor other functors have appeared, with importance for topology – the *J-functor*, *bordism theories* and *Milnor's microbundle k-functor*. These have thrown new light on old results and have led to some new ones. Note, for example, the results of Milnor, Mazur, Hirsch, Novikov and others on the problem of the relation between smooth and combinatorial manifolds, based on Milnor's k_{PL}-functor, and the theorems of Browder and Novikov on the tangent bundles of manifolds of the same homotopy type, successfully treated with the help of Atiyah's J-functor. Particularly interesting applications of bordism theory have been obtained by a number of authors (Conner and Floyd, Brown and Peterson, Lashof and Rothenberg).

In this survey I shall try roughly to describe this work, though the account will be very far from being complete. In order to describe recent results I shall of course have to devote a large part of the survey to presenting material that is more or less classical (and is in any case no longer new). This material is collected in the first two chapters[1].

Chapter I

CLASSICAL CONCEPTS AND RESULTS

To begin with we recall the very well-known concepts of fibre bundle, vector bundle, etc. We shall not give strict definitions, but confine ourselves to the intuitive ideas.

§1. The concept of a fibre bundle

Let X be a space, the *base*, and to each point $x \in X$ let there be associated a space F_x, the *fibre*, such that in a good topology the set $\bigcup_x F_x = E$, the *space*, is projected continuously onto X, each point of the fibre F_x being mapped to the corresponding point x. To each path $g : I \to X$, where I is the interval from 0 to 1, there corresponds a map $\lambda(g) : F_{x_0} \to F_{x_1}$, where $x_0 = g(0)$ and $x_1 = g(1)$. The map $\lambda(g)$ has to depend continuously on the path g and satisfy the following conditions:

[1] The distribution of the literature over the various sections is indicated on p. 60/61.

a) $\lambda(g^{-1}) = \lambda(g)^{-1}$, where g^{-1} is the inverse path,[1]

b) $\lambda(fg) = \lambda(f)\lambda(g)$, where fg denotes the composition of paths.

If we suppose that all the fibres F_x are homeomorphic and that, for each closed path g, $\lambda(g)$ is a homeomorphism belonging to some subgroup G of the group of homeomorphisms of the "standard fibre" F, then the group G is called the *structure group of the bundle*. These concepts *standard fibre* and *structure group* can be defined rigorously. Thus, the concept of fibre bundle includes the "space" E, the "base" X, the projection $p: E \to X$, the "fibre" F, homeomorphic to $p^{-1}(x)$, $x \in X$, and the group G.

We illustrate this by examples.

EXAMPLE 1. The line R is projected onto the circle S^1, consisting of all complex numbers $|z| = 1$, where $p: \varphi \to e^{i\varphi}$.

Here $E = R$, $X = S^1$, F is the integers, G the group of "translations" of the integers $g_n: m \to m + n$.

EXAMPLE 2. The sphere is projected onto the real projective space RP^n so that a single point $x \in RP^n$ is a pair of points of the sphere - vectors α and $-\alpha$ if the sphere is given by the equation $\Sigma x_i^2 = 1$. Here $E = S^n$, $X = RP^n$, the fibre F is a pair of points and G is a group of order 2.

EXAMPLE 3. The Möbius band Q is a skew product with base the circle $X = S^1$, fibre the interval from -1 to 1 and $E = Q$. The group G is of order 2, because the fibre - the interval - when taken round a contour - the base - is mapped to itself by reflection with respect to zero.

EXAMPLE 4. If H is a Lie group and G a closed subgroup, then we obtain a fibre bundle by setting $X = H/G$, $E = H$ and $G = G$, with p the natural projection $H \to H/G$ onto the space of cosets. For example, if H is the group of rotations of three-dimensional space SO_3 and G the group of rotations of a plane, then H/G is the sphere S^2 and we have a fibration $p: SO_3 \to S^2$ with fibre $S^1 = SO_2$. There are a great number of fibre bundles of this type.

EXAMPLE 5. A Riemannian metric induces on a closed manifold the concept of the parallel "transport" of a vector along a path. This shows that the tangent vectors form a fibre bundle (the "tangent bundle"), the base being the manifold itself and the fibre all the tangent vectors at a point (a Euclidean space). The group G is SO_n if there are no paths changing orientation ("the manifold is oriented") and O_n otherwise.

Such a fibre bundle, with fibre R^n and group $G = O_n$ or SO_n, we shall call a "real vector bundle".

EXAMPLE 6. If a manifold M^n is smoothly embedded in a manifold W^{n+k}, then a neighbourhood E of it in W^{n+k} may be fibred by normal balls. The neighbourhood is then also a fibre bundle with fibre D^k (a ball) or R^k, group SO_k or O_k and base M^n. This also is a vector bundle ("the normal bundle").

The concept of a "complex vector bundle" with fibre C^n and group U_n or SU_n is introduced in a similar way.

It is useful to relate to any fibre bundle its "associated" principal bundle, a principal bundle being one in which the fibre coincides with the

[1] If the maps λ are not homeomorphisms, but only homotopy equivalences, then a) has to be weakened by postulating only that $\lambda(e) = 1$, where e is the constant path.

group G and all the transformations are right translations of the group.

A principal bundle can be described as follows: the group G acts without fixed points on the space E, the base X is the set of orbits E/G and the projection $E \to E/G$ is the natural one.

The fact is that an arbitrary fibre bundle with arbitrary fibre may be uniquely defined by the choice of the maps $\lambda(f) \in G$ for the closed paths f from a single point. One can therefore by this same choice construct a principal fibre bundle with the same base X, but with fibre G, a transformation of the fibre being replaced by the translation of G induced by the same element $\lambda(f)$. One can also change one fibre into another if the same group acts on it.

§2. A general description of fibre bundles

1. Bundle maps. If two bundles $\eta_1 = (E_1, X_1, F, G)$ and $\eta_2 = (E_2, X_2, F, G)$ are given with common group and fibre, then a map $E_1 \to E_2$ is said to be a *bundle map* if it sends fibres into fibres homeomorphically and if it commutes with the action of G on F. The bundle map induces a map $X_1 \to X_2$.

2. Equivalence of bundles. Two bundles with the same base X, fibre F and group G are said to be *equal* if there is a bundle map of the one to the other such that the induced map of the base is the identity.

3. Induced bundles. If one has a bundle $\eta = (E, X, F, G)$, a space Y and a map $f : Y \to X$, then over Y there is a unique bundle "induced by f" with the same fibre and group and with base Y, mapped to the first bundle η and including on the base the map $f : Y \to X$. This bundle is denoted by $f^*\eta$. That is, to a bundle over X there uniquely corresponds a bundle over Y (bundles are mapped contravariantly just as functions are).

4. Examples. *EXAMPLE* 1. If a bundle over X is equivalent to a "trivial" bundle, that is, is such that $G = e$ and $E = X \times F$, then the same is true of any bundle induced by it – this implies that an induced bundle $f^*\eta$ cannot be more complicated than the the original one.

EXAMPLE 2. Let M^k be a manifold embedded in R^{n+k}. Consider the manifold $G_{k,n}$ of k-planes in R^{n+k} passing through the origin. Over $G_{k,n}$ (as base) there is a fibre bundle η with fibre R^k: to be precise, the fibre over $x \in G_{k,n}$ is the plane x itself, of dimension k. The group G here is O_k One can translate the tangent plane at a point $m \in M^k$ to the origin. We get a map $f : M^k \to G_{k,n}$; the tangent bundle to M^k, clearly, is $f^*\eta$, where η is the bundle over $G_{k,n}$.

One may assume here that n is very large. There is an important classifying

THEOREM. *Every fibre bundle with finite-dimensional base X and group O_k is induced by a map $X \to G_{k,n}$, unique up to homotopy, provided n is sufficiently large.*

The set of fibre bundles over X with group O_k is therefore identical with the set of homotopy classes of maps of X to $G_{k,n}$ (denoted by $\pi(X, G_{k,n})$).

Because of this theorem the bundle η constructed in Example 2 is said to be "universal". Such universal bundles can also be constructed in a similar way for $G = SO_k$ (one has to take oriented k-planes in R^{n+k}) and for $G = U_k$ (complex k-planes in C^{n+k}). It is easy to construct similar universal bundles for an arbitrary Lie group G.

The standard notation for the base of a universal bundle for a group G is BG. For example $G_{k,n} = BO_k$ for n infinitely large.

In the sequel we shall only be concerned with vector bundles, real or complex.

§3. Operations on fibre bundles

1. Sum (Whitney). Let η_1 and η_2 be two vector bundles with bases X_1 and X_2. Then over $X_1 \times X_2$ there is a bundle $\eta_1 \times \eta_2$, whose fibre is the direct product of the fibres of η_1 and η_2. If $X_1 = X_2 = X$, then there is a "diagonal map" $\Delta : X \to X \times X$, where $\Delta(x) = (x, x)$. Set

$$\eta_1 \oplus \eta_2 = \Delta^* (\eta_1 \times \eta_2).$$

We get a bundle $\eta_1 \oplus \eta_2$ over X, the "sum" of the bundles η_1 and η_2 with common base X.

2. Product (tensor). Let η_1 and η_2 be two bundles with common base X. Then over $X \times X$ there is a bundle $\eta_1 \hat{\otimes} \eta_2$ whose fibre is the tensor product of the fibres (over R, if they are real, or over C if they are complex bundles). Set, as before,

$$\eta_1 \otimes \eta_2 = \Delta^* (\eta_1 \hat{\otimes} \eta_2);$$

$\eta_1 \otimes \eta_2$ is a bundle with the same base space X.

It is easy to prove bilinearity:

$$\eta_1 \otimes (\eta_2 \oplus \eta_3) = (\eta_1 \otimes \eta_2) \oplus (\eta_1 \otimes \eta_3).$$

The unit for tensor multiplication is the trivial (line) bundle with one-dimensional fibre R or C.

3. Representation of the structure group. Let G be the structure group of the bundle η with fibre F and base X and for simplicity let $h : G \to O_N$ or $h : G \to U_N$ be a faithful orthogonal or unitary representation. By the method indicated above one can "change" the fibre F to R_N or C_N by means of the representation h. We get a bundle denoted by h. The following cases are the most interesting:

1) h is the natural inclusion $O_N \subset U_N$ ("complexification", denoted by c).

2) h is the natural inclusion $U_N \subset O_{2N}$ ("realization", denoted by r).

3) If any $h : G \subset O_N$ or $G \subset U_N$ is given, we can take its "exterior power" $\Lambda^i h$, $0 \leqslant i \leqslant N$, where $\Lambda^1 h = h$ and $\Lambda^N h = \Lambda^0 h = e$. For example, differential forms are "sections" of the exterior powers of the (co-) tangent bundle.

Note that $cr\eta = \eta \oplus \bar{\eta}$ and $rc\eta = \eta \oplus \eta$, where $\bar{\eta}$ denotes the complex conjugate bundle to η.

4) If two representations h_1 and h_2 are given, then we can form their sum and product, which we shall denote simply by the symbols for addition (with positive integral coefficients) and multiplication. All these operations enable us to construct new bundles with the same base.

Chapter II

CHARACTERISTIC CLASSES AND COBORDISMS

§4. The cohomological invariants of a fibre bundle. The characteristic classes of Stiefel-Whitney, Pontryagin and Chern

To each real fibre bundle there are associated a collection of invariants, the Stiefel and Pontryagin classes. Let η be a bundle with fibre R^n and base X. Classes $W_i \in H^i(X, Z_2)$ are defined, with $W_0 = 1 \in H^0(X, Z_2)$ and $W_i = 0$, $i > n$. They are called the Stiefel-Whitney classes. There are also Pontryagin classes $p_i \in H^{4i}(X, Z)$ and the Euler-Poincaré class $\chi \in H^n(X, Z)$ if $n = 2k$. Moreover, $p_k = \chi^2$ for $n = 2k$ and $p_i = 0$ for $i > k$. If the classes p_i and χ are taken mod 2, then they are expressible in terms of the Stiefel classes; to be precise, $p_i = W_{2i}^2$ mod 2 and $\chi = W_n$ mod 2. We form the "Stiefel polynomial" $W = 1 + W_1 + W_2 + \ldots$ and the "Pontryagin polynomial" $P = 1 + p_1 + p_2 + \ldots$

In a similar way one can associate to a complex bundle ζ, with fibre C^n, Chern classes $c_i \in H^{2i}(X, Z)$ and a Chern polynomial $C = 1 + c_1 + c_2 + \ldots$ Note that c_i mod 2 is equal to $W_{2i}(r\zeta)$, where r is the operation of making the bundle ζ real, and $c_{2i}(c\eta) = (-1)^i p_i(\eta)$, where c is the operation of complexifying the bundle η.

Properties of these classes are:

a) Bundle maps map classes to classes.

b) The Whitney formulae $W(\eta_1 \oplus \eta_2) = W(\eta_1)\, W(\eta_2)$,

$$C(\zeta_1 \oplus \zeta_2) = C(\zeta_1)\, C(\zeta_2),$$

up to elements of order 2: $P(\eta_1 \oplus \eta_2) = P(\eta_1)\, P(\eta_2)$.

c) Let us factorize the Chern polynomial $C(\zeta)$ formally: $C = \prod_{i=1}^{n} (1 + b_i)$, $\dim b_1 = 2$, and form the series $\operatorname{ch} \zeta = \sum_{i=1}^{n} e^{b_i} \in H^*(X, Q)$, where Q is the rational numbers. Clearly $\operatorname{ch} \zeta$ is expressible in terms only of the elementary symmetric functions of the b_i which are the $c_k(\zeta)$. Therefore $\operatorname{ch} \zeta = \sum_{k \geqslant 0} \operatorname{ch}^k \zeta$ is meaningful in the cohomology ring. Similarly, for a real bundle η, let

$$\operatorname{ch} \eta = \operatorname{ch} c\eta = \sum_{k=0}^{\infty} \operatorname{ch}^{2k} c\eta, \qquad \operatorname{ch}^{2k+1} c\eta = 0.$$

$\operatorname{ch} \zeta$ is called the "Chern character". It has the following properties:

(24)

$$\operatorname{ch} \zeta_1 \oplus \zeta_2 = \operatorname{ch} \zeta_1 + \operatorname{ch} \zeta_2,$$
$$\operatorname{ch} \zeta_1 \otimes \zeta_2 = \operatorname{ch} \zeta_1 \operatorname{ch} \zeta_2.$$

d) For the Euler class χ we have $\chi(\eta_1 \oplus \eta_2) = \chi(\eta_1)\,\chi(\eta_2)$. If η is the tangent bundle of a smooth closed manifold M^n, then the scalar product $(\chi(\eta), [M^n])$ is equal to the Euler characteristic of M^n. If M^k is a complex manifold, then $\chi(r\eta) = c_{\frac{k}{2}}(\eta)$.

e) If η is a complex U_n - bundle with base X and $r\eta$ its real form, then $P(r\eta) = \prod_{i=1}^{n} (1 + b_i^2)$, where $c(\eta) = \prod_{i=1}^{n} (1 + b_i)$, $\dim b_i = 2$,

For example $p_1(r\eta) = c_1^2 - 2c_2$.

It has already been remarked that $W(r\eta) = C(\eta) \bmod 2$.

EXAMPLE 1. The natural one-dimensional normal bundle η_1 of RP^n in RP^{n+1} (the "Möbius band") has Stiefel polynomial $W(\eta_1) = 1 + x$, where x is the basis element of the group $H^1(RP^n, Z_2)$. Similarly the one-dimensional complex normal bundle ζ_1 of CP^n in CP^{n+1} (the "complex Möbius band") has Chern polynomial $C(\zeta_1) = 1 + x$, where x is the basis element of the group $H^2(CP^n, Z) = Z$.

EXAMPLE 2. If $\tau(M)$ is the tangent bundle of the manifold (complex if it is complex), then for $M = RP^n$ or CP^n we have the formulae

$$\tau(RP^n) \oplus 1_R = \eta_1 \oplus \ldots \oplus \eta_1 \qquad (n+1 \text{ terms}),$$
$$\tau(CP^n) \oplus 1_C = \zeta_1 \oplus \ldots \oplus \zeta_1 \qquad (n+1 \text{ terms}),$$

where 1_R and 1_C are the one-dimensional trivial bundles. Therefore $W(RP^n) = (1 + x)^{n+1}$, $x \in H^1(RP^n, Z_2)$ and $C(CP^n) = (1 + x)^{n+1}$, $x \in H^2(CP^n, Z)$. By property e) it is easy to deduce that $P(CP^n) = (1 + x^2)^{n+1}$; here $W(M)$, $C(M)$, $P(M)$ denote the polynomials of the manifold, that is, the polynomials W, C, P of the tangent bundle. In particular $C(CP^1) = C(S^2) = 1 + 2x$, $C(CP^2) = 1 + 3x + 3x^2$ and $P(CP^2) = 1 + 3x^2$, $\operatorname{ch}\tau(CP^n) = (n + 1)e^x - 1$.

§5. The characteristic numbers of Pontryagin, Chern and Stiefel. Cobordisms

If M^n is a closed manifold, one can consider polynomials of degree n in the Stiefel and Pontryagin classes and take their scalar products with the fundamental cycle of the manifold. We get numbers (or numbers mod 2 for Stiefel classes). If M is a complex or quasicomplex manifold, one can do the same with the Chern classes of its tangent bundle. We get Stiefel, Pontryagin and Chern numbers. The following important theorem holds:

THEOREM OF PONTRYAGIN-THOM. 1) *A manifold is the boundary of a compact manifold with boundary if and only if its Stiefel numbers are zero.*

2) *An oriented manifold is the boundary of an oriented compact manifold with boundary if and only if its Pontryagin and Stiefel numbers are zero.*[1]

[1] Statement 2) was finally proved only at the end of the 50's by Milnor, Rokhlin, Averbukh and Wall.

The corresponding theorem (Milnor, Novikov) for quasicomplex manifolds has a more complicated formulation, since one first has to define what it means to "be a boundary", but after this it runs analogously, involving the Chern numbers.

One defines the "cobordism rings" $\Omega^* = \Sigma\Omega^i$ roughly as follows: the "sum" of two manifolds of some class or other (orientable, quasicomplex, etc.) is defined by forming their disjoint union and the "product" by forming their direct product; one also defines what it means to "be a boundary" for each such class. The correctness of the definition has to be verified. In this way there arise the "cobordism rings" Ω^* with the operations of addition and multiplication. Such a ring arises naturally for each of the classical series of Lie groups $\{O_n\}$, $\{SO_n\}$, $\{U_n\}$, $\{SU_n\}$, $\{Sp_n\}$, $\{Spin_n\}$ and the unit group. The following types of cobordism occur:

$N = \Omega^*_O$ – non-oriented manifolds,

$\Omega = \Omega^*_{SO}$ – oriented manifolds,

Ω^*_U – quasicomplex manifolds (a complex structure on the stable tangent bundle),

Ω^*_{SU} – special quasicomplex manifolds (the first Chern class c_1 is equal to zero),

Ω^*_{Spin} – spinor manifolds,

Ω^*_{Sp} – symplectic manifolds,

Ω^*_e – the Pontryagin framed manifolds, Ω^i_e being isomorphic to the stable homotopy group of spheres of index i, $\Omega^i_e \approx \pi_{N+i}(S^N)$.

The following results on cobordism rings are known.

1°. Any type of cobordism is completely defined by characteristic numbers after factoring by torsion. After tensoring with the field of rational numbers all the cobordism rings become polynomial rings (theorems arising from results of Cartan-Serre in homotopy theory and Thom's work on cobordism).

2°. Ω^*_O is a polynomial algebra over the field Z_2 (Thom). $\Omega^*_{SO} \otimes Z_2$ is a subalgebra of Ω^*_O; Ω^*_{SO} does not contain elements of order 4 or elements of odd order; the quotient of Ω^*_{SO} by 2-torsion is a polynomial ring over the integers (Rokhlin, Wall, Averbukh, Milnor).

3°. Ω^*_U is a polynomial ring. $\Omega^*_{Sp} \otimes K$ and $\Omega^*_{Spin} \otimes K$ are polynomial algebras, where the characteristic of the field K is not equal to 2. (Milnor, Novikov).

4°. $\Omega^*_{SU} \otimes K$ is a polynomial algebra if the characteristic of the field K is not equal to 2. The ring Ω^*_{SU} has no elements of odd order. The whole subgroup $\Omega^{odd}_{SU} = \Sigma\Omega^{2k+1}_{SU}$ belongs to the ideal generated by $\Omega^1_{SU} = Z_2$. that is, $\Omega^{2k+1}_{SU} = \Omega^1_{SU}\Omega^{2k}_{SU}$ (Novikov). It seems that these are all the results on cobordisms that have been known for some time (several years, at least).

To these results it is useful to add the following: for a complex (real) manifold to represent a polynomial generator of the ring Ω^*_U or Ω^*_{SO} it is necessary and sufficient that (in the complex case) the components $\mathrm{ch}^n\eta$ of the tangent bundle should be such that

$$|(n!\ \mathrm{ch}^n\eta,\ [M^n])| = \begin{cases} 1, & n+1 \neq p^i, \\ p, & n+1 = p^i, \end{cases}$$

where p is an arbitrary prime number. For example, for $M^n = CP^n$ we have $(n!\ \mathrm{ch}^n \eta,\ [CP^n]) = n + 1$, from which it follows that CP^n is a polynomial generator only for dimensions of the form $p - 1$, where p is prime. In particular, CP^1, CP^2, CP^4, and CP^6 are generators, while CP^3, CP^5, CP^7 and CP^8 are not. Milnor has exhibited a system of "genuine" geometrical generators. For Ω^*_{SO} the situation is analogous, but one is then only concerned with CP^{2k}.

§6. The Hirzebruch genera. Theorems of Riemann-Roch type

The simplest and oldest invariants of a complex Kähler (in particular, algebraic) manifold are the "dimensions of the holomorphic forms of rank Q", denoted by $h^{q,0}$. An important invariant is the "arithmetic genus" $\chi = \Sigma(-1)^q h^{q,0}$. Apart from ordinary forms one can also speak of forms with values in an algebraic vector bundle ζ over M^n; we then obtain

$$\chi(M^n, \zeta) = \Sigma(-1)^q h^{q,0}(\zeta).$$

Another invariant (in the real case) is the "signature" of a manifold of dimension $4k$, this being simply the signature of the quadratic form $(x^2, [M^{4k}])$, where $x \in H^{2k}(M^{4k}, R)$. This signature is denoted by $\tau(M^{4k})$.

We set:

1) $L = \sum\limits_{i\geqslant 0} L_i = \prod\limits_j \frac{\mathrm{tg}\, h b_j}{b_j}$, $\dim b_i = 2$ and $P = \prod\limits_j (1 + b_j^2)$. Here $L_i = L_i(p_1, \ldots, p_i)$ and $\dim L_i = 4i$.

2) $T = \sum\limits_{i\geqslant 0} T_i = \prod\limits_j \frac{b_j}{1 - e^{-b_j}}$, $T_i = T_i(c_1, \ldots, c_i)$, $\dim T_i = 2i$, $C = \prod\limits_j (1 + b_j)$.

3) $A = \sum\limits_{i\geqslant 0} A_i = \prod\limits_j \frac{\frac{b_j}{2}}{\sin h \frac{b_j}{2}}$, $A_i = A_i(p_1, \ldots, p_i)$, $\dim A_i = 4i$.

Note that $T = e^{\frac{1}{2}\sum b_j} A$ and also that

$$L_1 = \frac{1}{3} p_1, \quad L_2 = \frac{1}{45}(7p_2 - p_1^2),$$

$$T_1 = \frac{1}{2} C_1, \quad T_2 = \frac{1}{12}(C_2 + C_1^2),$$

$$A_1 = \frac{1}{24} p_1, \quad A_2 = \frac{p_1^2 - 4L_1}{32 \cdot 28}.$$

Hirzebruch proved the following facts which are the basis of cobordism theory:

$$(L_k(p_1, \ldots, p_k),\ [M^{4k}]) = \tau(M^{4k}) \quad \text{(real case)} -$$

for $k = 1$ this becomes the theorem of Rokhlin and Thom;
$(\mathrm{ch}\,\zeta\, T(M^n),\ [M^n]) = \chi(M^n, \zeta)$, where ζ is an algebraic bundle and M^n is an algebraic manifold, - for $n = 1$ this becomes the classical theorem of Riemann-Roch and for $n = 2$ the formulae of Noether and Kodaira.

The following generalizations of these Hirzebruch formulae turn out to be extremely important:

1°. If $W_1 = W_2 = 0$, for a real manifold M^{4k}, then $(A_k,\ [M^{4k}])$ is an integer. For odd k this number is divisible by 2. For $k = 1$ this reduces to the theorem of Rokhlin.

2°. For a quasicomplex manifold M^n the Todd genus $(T_n,\ [M^n])$ is an integer (Borel-Hirzebruch, Milnor).

Note that in the algebraic case the first integrality theorem was known, when $c_1 = 0$, since $T = e^{c_1/2}A$.

§7. Bott periodicity

As has already been said, the set of G-bundles with base X is identica with the set of homotopy classes $\pi(X, BG)$,. where BG is the base of the universal bundle. Let $G = O_n$, U_n or Sp_n. How many bundles are there with base S^k? The set of these bundles is $\pi(S^k, BG) = \pi_k(BG)$. By a theorem of Cartan-Serre we know that $\pi_k(BG) \otimes Q$ is determined by cohomological invariants (Pontryagin and Chern classes). However, we do not know the torsion. For example $\pi_2(BSO_n) = Z_2$ for $n > 2$, $\pi_4(BU_2) = Z_2$.

How are we to compute fully the homotopy groups of the spaces BG? The theorems of Bott completely compute $\pi_i(BO_n)$ for $i < n$, $\pi_i(BU_n)$ for $i < 2n$, and $\pi_i(BSp_n)$ for $i < 4n$.

THEOREM. *There exists a canonical isomorphism between the sets*
1) $\pi(X, BU_n)$ *and* $\pi(E^2X, BU_n)$ *for* $\dim X < 2n - 2$,
2) $\pi(X, BO_n)$ *and* $\pi(E^8X, BO_n)$ *for* $\dim X \leqslant n - 9$,
3) $\pi(X, BSp_n)$ *and* $\pi(E^8X, BSp_n)$ *for* $\dim X \leqslant 4n - 8$.

For $x = S^k$ *these are isomorphisms of cohomology groups.* EX *here denote the suspension of the space* X, *and* $E^kX = EE^{k-1}X$.

For $X = S^k$ the Chern character of the bundle generating the group $\pi_{2k}(BU_n)$ is equal to $\mathrm{ch}^k\,\eta = x$, where x is the basis element of the group $H^{2k}(S^{2k}, Z) \subset H^{2k}(S^{2k}, Q)$. for $X = S^{4k}$ the Chern character of the bundle generating the group $\pi_{4k}(BO_n)$ is equal to $\mathrm{ch}^{2k}\,\eta = a_k x$, where x is as in the complex case and $a_k = \begin{cases} 1, & \text{if } k \text{ is even.} \\ 2, & \text{if } k \text{ is odd.} \end{cases}$

Here, of course, the dimension of the sphere is supposed to satisfy the conditions of Bott's theorem. Using the relation between the Chern classes and the Chern character we get the fact that the Chern class of a complex bundle with base S^{2k} is divisible by $(k - 1)!$, while the Pontryagin class of a real bundle over S^{4k} is divisible by $a_k(2k - 1)!$.

One can also add to Bott's theorem information on the first dimensions not satisfying the "stability" conditions:
1) $\pi_{2k+1}(BU_k) = Z_{k!}$; 2) the kernel of the map $\pi_n(BO_n) \to \pi_n(BO_{n+1})$ is

always Z_2 for odd $n+1 \neq 2, 4, 8$; from this there follows at once the nonparallelizability of the spheres S^k for $k \neq 1, 3, 7$ and the nonexistence of finite-dimensional division algebras over R, except for the dimensions 2, 4, 8. (This has also been proved by Adams with other arguments; as a corollary of Bott's Theorem it was proved by Milnor and Kervaire.)

§8. Thom complexes

Let X be the base of a fibre bundle η with fibre R^n. The following space will be called the Thom complex $T_\eta : T_\eta = E_\eta/A_\eta$, where E_η is the bundle space of η and A_η is the subspace consisting of all vectors of length $\geqslant 1$ in each fibre.

The Thom isomorphism is defined to be a certain map

$$\varphi: H^i(X) \longrightarrow H^{n+i}(T_\eta), \quad i \geqslant 0$$

(cohomology over Z_2 for O_n-bundles and over Z for SO_n-bundles).

Note that there is a natural inclusion

$$j: X \subset T_\eta.$$

EXAMPLE. $X = BG$ and η the universal G-bundle, where G is a subgroup of O_n. Then η has R^n as fibre. For example $G = O_n$, SO_n, $U_n \subset O_{2n}$, $SU_n \subset O_{2n}$, $Sp_n \subset O_{4n}$, $e \subset O_n$. There is an inclusion $Spin_n \subset O_N$. The space T_η is denoted in this case by MG. For $G = e \subset O_n$ we have $Me = S^n$.

THOM'S THEOREM (for $G = e$, Pontryagin). *The groups $\pi_{n+k}(MG)$ are isomorphic to the cobordism groups Ω_G^k for large n and $G = O_n$, SO_n, $U_{n/2}$, $SU_{n/2}$, $Sp_{n/4}$, e.* [1]

On the basis of this fact cobordisms can be computed. Note that the ring structure of the cobordisms has a natural homotopy interpretation in terms of the Thom complex MG (Milnor, Novikov).

§9. Notes on the invariance of the classes

The Stiefel classes can be defined in terms of the Thom complex. Let

$$W_i(\eta) = \varphi^{-1} Sq^i \varphi(1),$$

where $\varphi : H^k(X) \to H^{k+n}(T_\eta)$. This formula is easily verifiable for the universal bundle. If it is applied to the tangent (or normal) bundle η of a smooth manifold M^n, it makes it possible to compute the Stiefel classes of the manifold in terms of the cohomological invariants of the manifold – the Steenrod squares and the cohomology ring (the formulae of Thom and Wu). In a similar way it is sometimes possible to compute the Pontryagin classes up to some modulus or other (for example, p_i mod 3). However, unlike the Stiefel classes, the Pontryagin classes (rational,

---

[1] In Thom's work this is formulated only for $G = 0$, SO. For $G = e$ this theorem is contained in older work of Pontryagin. In the remaining cases the proof is similar.

integral) are not invariants of homotopy type (Dold). Moreover, except for Hirzebruch's formula, $(L_k, [M^{4k}]) = \tau(M^{4k})$, there are no "rational" homotopy-invariant relations for simply-connected manifolds (we shall say more about this later). On the basis of cobordisms and Hirzebruch's formula about L_k, Thom, Rokhlin and Shvarts have proved the combinatorial invariance of the rational classes p_k; the class $L_k(M^{4k+1})$ is even a homotopy invariant (Appendix). A fundamental problem is the topological meaning of the rational Pontryagin classes (we shall say more about this later).

Chapter III

GENERALIZED COHOMOLOGIES. THE K-FUNCTOR AND THE THEORY OF BORDISMS. MICROBUNDLES.

§10. Generalized cohomologies. Examples.

We first recall what is meant in general by cohomology theories. They are defined by the following properties:

1°. "Naturality" (functoriality) and homotopy invariance - to a homotopy class of maps of complexes $X \to Y$ there corresponds a homomorphism $H^*(Y) \to H^*(X)$, composition of maps corresponds to composition of homomorphisms and the identity map to the identity homomorphism.

2°. "Factorization" $H^*(X, Y) = \widetilde{H}^*(X/Y)$, $Y \subset X$.

3°. "Exact sequence of a pair".

4°. "Normalization" - one has to prescribe the cohomology of a point. (Absolute cohomology groups $H^i(X)$ should be considered as the corresponding $H^i(X \cup P, P)$.)[1] It is important, in particular, that for a point $X = P$ we have $H^i(P) = 0$, $i \neq 0$; $H^0(P) = G$ is called the "coefficient group".

Note the important property of cohomology that it is a "representable functor", that is

$$H^i(X, Y; G) = \pi(X/Y, K(G, i)),$$

where $K(G, i)$ is an Eilenberg-MacLane complex for which $\pi_i(K(G, i)) = G$ and $\pi_j(K(G, i)) = 0$, $i \neq j$.

In topology it is well-known how these general statements are used in solving concrete problems. Homotopy theory, however, makes it possible to construct an unlimited number of algebraic functors of a similar kind, possessing all the above properties with the exception of the normalization property (among them also their representability as homotopy classes of maps).

If the normalization condition is added, then by a theorem of Eilenberg and Steenrod we get nothing other than ordinary cohomology.

A "generalized cohomology" is a representable functor for which the cohomology of a point is non-trivial in arbitrary dimensions. One can also define "generalized homologies".

[1] Here and in what follows we shall be concerned with spaces with base point; in $X \cup P$ the base point is P.

A "cohomology operation" is a mapping of such a functor into itself or into another similar functor (not always natural with respect to additive homomorphisms).

Even for ordinary cohomology there is a very large number of "cohomology operations", for example the Pontryagin powers and the Steenrod powers. In the study of these operations and their applications their representability is very important. It seems that the first generalized cohomologies were introduced in particular cases by Atiyah and Hirzebruch in the solution of concrete problems, while the general interrelations were pointed out by Whitehead.

Consider a typical functor of this type. Let $F = \{X_n, f_n\}$ be a spectrum of spaces and maps $f_n : X_n \to \Omega X_{n+1}$ or $EX_n \to X_{n+1}$, where Ω is the Serre operation of taking loops on X_{n+1}.

Let K be an arbitrary complex. Maps

$$f_{n*}: \pi(K, X_n) \longrightarrow \pi(K, \Omega X_{n+1}) \approx \pi(EK, X_{n+1}).$$

are defined. Let $H^i_F(K, L) = \operatorname{dir}\lim_{n\to\infty} \pi(E^{n-i}K/L, X_n)$ (the limit being taken over f_{n*}). The groups $H^i_F(K, L)$ are defined for $-\infty < i < \infty$. We also set $H^i_F(K) = H^i_F(K \cup P, P)$, where P is a point.

What is the cohomology of a point in this case?

By definition

$$H^i_F(P) = H^i_F(P \cup P_1, P_1) = \operatorname{dir}\lim_{n\to\infty} \pi_{n+i}(X_n).$$

In all the "good" cases we have "stability", that is for large n $\pi_{n+i}(X_n)$ only depends on i.

Thus, in every such case the computation of the cohomology of a point is the solution of a problem (as a rule a difficult one) in homotopy theory.

We set $H^*_F(K, L) = \Sigma H^i_F(K, L)$.

In many cases in such "generalized cohomologies" one has also the structure of a graded ring. This has not yet been axiomatized.

In most cases we have to deal with generalized cohomology rings.

There is an easily proved fact which sometimes makes it possible to compute " generalized cohomologies" if they are known for a point.

There exists a spectral sequence $\{E_r, d_r\}$, where $E_r = \sum_{p,q} E^{p,q}_r$, $d_r : E^{p,q}_r \to E^{p+r,q-r+1}_r$ and $E^{p,q}_2 = H^p(K, L; H^q_F(P))$ (in the usual sense) and $\sum_{p+q=n} E^{p,q}_\infty$ is related to $H^n_F(K, L)$.

EXAMPLE 1. $X_n = K(G, n)$, $\Omega X_n = X_{n-1}$; we get the ordinary cohomology theory. Here cohomology is trivial for negative dimensions.

EXAMPLE 2. a) $X_{2n} = BU$, $X_{2n+1} = U$ $(U = \bigcup_n U_n)$. Bott periodicity gives $X_n = \Omega X_{n+1}$. Let $H^i_F(K, L) = K^i_C(K, L)$, $K^*_C = \Sigma K^i_C$.

b) $X_{8n} = BO$, $X_{8n-1} = O$ $(O = \bigcup_n O_n)$.

Then $X_{8n+4} = BSp$ and $X_{8n+3} = Sp$ ($Sp = \bigcup_n Sp_n$) (Bott periodicity). Let $H_F^i(K, L) = K_R^i(K, L)$, $K_R^* = \Sigma K_R^i$. The last two cohomology theories are calle the *K-theories* (*real and complex*). We shall give here an alternative definition (Grothendieck).

Denote by $L(Y)$ the set of all vector bundles with base Y and by $L(Y)$ the free abelian group (with composition sign +) generated by $L(Y)$. Let

$$\eta_1 + \eta_2 = \eta_1 \oplus \eta_2, \quad \eta_1, \eta_2 \in L(Y).$$

The quotient group of $L(Y)$ by this equivalence relation is denoted by $K(Y)$ or sometimes in the real case by $K_R(Y)$ and in the complex case by $K_C(Y)$. K_R and K_C are rings with respect to $\otimes$. There is a dimension homomorphism

$$(\dim) : K_\Lambda(Y) \to Z, \quad \Lambda = R, C.$$

Let $K_\Lambda^0 = \mathrm{Ker}(\dim)$. Let $K_\Lambda^{-i}(K, L) = K_\Lambda^0(E^i K/L)$, $\overline{K}_\Lambda^* = \Sigma K_\Lambda^i$, $i < 0$. By Bott periodicity $K_C^0(E^2Y) = K_C^0(Y)$ and $K_R^0(E^8Y) = K_R^0(Y)$, and therefore the definition extends to positive numbers $i > 0$, by periodicity.

Now let $K_\Lambda^* = \Sigma K_\Lambda^i$, $-\infty < i < \infty$. It can be shown that this new definition of K_R^*, K_C^* coincides with the original one.

For the form of the functor at a point it is sufficient by Bott periodicity to consider the structure of the ring only for the sum $\sum_{i \leqslant 0} K^i = \overline{K}^*$. We give the following generators and relations.

a) $\overline{K}_C^*$ is generated by $u \in K_C^{-2}$ and $1 \in K_C^0$. The ring is a polynomial ring over u.

b) $\overline{K}_R^*$ is generated by $1 \in K_R^0$, $u \in K_R^{-1}$, $v \in K_R^{-4}$, $w \in K_R^{-8}$; the relation are $2u = 0$, $u^3 = 0$, $uv = 0$, $v^2 = 4w$.

"Chern characters" defined by

$$ch: K_C(K, L) \to H^*(K, L; Q),$$
$$ch: K_R(K, L) \to H^*(K, L; Q),$$

are ring homomorphisms. A new formulation of Bott periodicity is:

$$K_C(X \times S^2) = K_C(X) \otimes K_C(S^2),$$
$$K_R(X \times S^8) = K_R(X) \otimes K_R(S^8).$$

An important fact is that there is a Thom isomorphism for the complex K-functor. Let η be a complex vector bundle with base X and T_η its Thom space. Then there is a Thom isomorphism:

$$\varphi_K: K_C(X) \to K_C^0(T_\eta).$$

By contrast with ordinary cohomology, however, there are in this case many "Thom isomorphisms". This has important consequences later on.

EXAMPLE 3.

$$\{X_n = MO_n\}, \qquad \{X_{4n} = MSp_n\},$$
$$\{X_n = MSO_n\}, \qquad \{X_n = S^n = Me_n\},$$
$$\{X_{2n} = MU_n\}, \qquad e \subset O_n.$$
$$\{X_{2n} = MSU_n\}.$$

We call the corresponding cohomologies *cobordisms*. They admit also a good geometric interpretation, and there are very useful homology theories dual to them - *bordisms*.

For let us set $H_i^F(K, L) = \pi_{n+i}(X_n \divideontimes K/L)$, where $\divideontimes$ is the operation of multiplication followed by identifying to a point the coordinate axes. In particular, when $X_n = S^n$, we get the "stable" homotopy groups of the space K/L. For a manifold bordisms and cobordisms are related by a duality law of Poincaré type. The role of the "coefficient group", that is the ring of bordisms of a point, is here played by the ring Ω_G, of especially good structure when $G = O_n$ or U_n.

EXAMPLE 4. Let $H_n = (\Omega^n S^n)^0$. Then inclusions $H_n \subset H_{n+1}$ and $O_n \subset H_n$ are defined, since H_n consists of maps $S^n \to S^n$ of degree $+1$ preserving base point. H_n is an H-space.

Let[1] $BH_n = (\widehat{\Omega^{n-1} S^n})$. Then (natural) maps $BO_n \to BH_n$ are defined; let $H = \bigcup_n H_n$; also let $BH = \lim_n BH_n$. Then a natural map $BO \to BH$ is defined. Let $BH = X_0$ and $X_{-i} = \Omega^i BH$. The groups $\pi_{i+1}(BH)$ are isomorphic to the stable homotopy groups of spheres of index i.

Let $\Pi^{-i}(K, L) = \pi(E^i K/L, BH)$, $i \geqslant 0$. It is easy to verify that $\Pi^{-i}(K, L)$ is a group for all i. A natural map

$$J\colon K_R^{-i}(K, L) \to \Pi^{-i}(K, L).$$

is defined. The image JK_R^{-i} is called the *J-functor* $J(K, L)$. For $K = S^0$, $L = P$ we have

$$K_R^{-i}(K, L) = \pi_i(BO) \approx \pi_{i-1}(O),$$
$$\Pi^{-i}(K, L) = \pi_i(BH) = \pi_{N+i-1}(S^N),$$

N large.

We have the classical J-homomorphism of Whitehead:

$$J\colon \pi_{i-1}(O_N) \to \pi_{N+i-1}(S^N).$$

The J-functor is very important in many problems of differential topology.

In this example all the properties of generalized cohomologies are present except that the spectrum X_n is only defined for $n \leqslant 0$ and therefore, generally speaking, only the cohomology of negative degree is defined. This is more frequently, though less efficaciously, the case in homotopy theory. We give another example of the same type.

EXAMPLE 5. Consider "Milnor's microbundles" in the combinatorial

---

[1] The sign $\wedge$ on top here denotes the universal covering space.

sense. A complex K lies piecewise-linearly in a complex $L \supset K$. To each (small) neighbourhood $U \subset K$ there exists a neighbourhood $V \subset L$ such that $V \cap L$ such that $V \cap K = U$ and V is combinatorially equivalent to $U \times R^n$ so that a "projection" $p : V \to U$ is given. On the intersection of two such neighbourhoods U_1 and U_2 the projections $p_{U_1} : V_1 \to U_1$ and $p_{U_2} : V_2 \to U_2$ agree on $V_1 \cap V_2$ (but the "dimension" of the fibres is not supposed to be unique). Two microbundles $\eta_1 : K \subset L_1$ and $\eta_2 : K \subset L_2$ are to be regarded equal if there are neighbourhoods $Q_i \subset L_i$ of the complex K in L_i combinatorially equivalent and preserving the "fibre structure" round K.

For microbundles there is a "Whitney sum", a "structure group" PL_n and a universal bundle with base BPL_n. There are also inclusions $O_n \subset PL_n$ and $PL_n \subset PL_{n+1}$. Let $PL = \operatorname{dir}\lim_{n\to\infty} PL_n$ and define BPL analogously. A map $BO \to BPL$ is defined. Using the Whitney sum $\oplus$, one can introduce a "Grothendieck group" of microbundles $k_{PL}(K, L)$ and $k_{PL} = Z + k^0_{PL}(K, L)$. Analogously one introduces $k^{-i}_{PL}(K, L) = k^0_{PL}(E^iK/L)$. There is the Milnor homomophism

$$j\colon\ K_R(K, L) \longrightarrow k_{PL}(K, L);$$

there are also the concepts of "tangent" and "stable normal" microbundle of a combinatorial manifold.

The following facts are known:

a) A combinatorial manifold is smoothable if and only if its "tangent bundle" $\eta(M^n) \in k_{PL}(M^n)$ belongs to the image of j (Milnor).

b) The relative groups $\pi_i(PL, O)$ are isomorphic to Milnor's groups of differentiable structures on spheres Γ^i (Mazur, Hirsch). In particular, the image of $\pi_7(0) = Z$ is divisible by 7 in the group $\pi_7(PL) = k^0_{PL}(S^8)$ (Kervaire). From this it can easily be deduced that for a suitably chosen manifold the homomorphism $j : K_R(M^n) \to k_{PL}(M^n)$ has kernel isomorphic to Z_7. From this in its turn it follows that there exists smooth combinatorially identical manifolds with differing tangent bundles – and even differing 7-torsion in the Pontryagin class p_2 (Milnor). In this way the torsion of the Pontryagin classes is seen not to be a combinatorial invariant.

c) Homotopy type and the tangent bundle (or its invariants, the rational Pontryagin classes) determine a combinatorial manifold modulo "a finite number of possibilities" if $\pi_1 = 0$ and the dimension is >5 (Novikov).

One may consider the corresponding J-functor. A homomorphism of functors

$$\begin{array}{ccc} k^{-i}_{PL}(K, L) & \xrightarrow{J_{PL}} & \Pi^{-i}(K, L). \\ & \nwarrow j \quad J \nearrow & \\ & K^{-i}_R(K, L) & \end{array}$$

is defined. This triangle is commutative, that is $J = J_{PL} \circ j$. If $L = P$,

(24)

$K = S^0$, then $k_{PL}^{-i}(K, L) = \pi_i(BPL)$. We have

$$\begin{array}{ccc} \pi_{i-1}(PL) & \xrightarrow{J_{PL}} & \pi_{N+i-1}(S^N), \\ & {}_{j}\nwarrow \quad \nearrow_{J} & \\ & \pi_{i-1}(0) & \end{array}$$

where N is large. Note that the map j is a monomorphism for $K = S^0$, $L = P$ (Adams). One can consider a natural "fibration"

$$p\colon BH \to X_{SO}$$

with fibre BSO. There corresponds the exact sequence

$$K_R^{-1}(M^n) \xrightarrow{J} \Pi^{-1}(M^n) \xrightarrow{\varkappa} \Gamma_{SO}(M^n) \xrightarrow{\delta} K_R^0(M^n) \xrightarrow{J} \Pi^0(M^n).$$

Here M^n is a closed simply-connected smooth manifold, and $\Gamma_{SO}(M^n)$ denotes $\pi(M^n, \Omega X_{SO})$. We denote by $\pi^+(M^n)$ the group of homotopy classes of maps of M^n onto itself of degree $+1$ preserving the stable "tangent" element. $-n \in K_R^0(M^n)$, where $n \in K_R^0$ is the stable "normal" element. Then the group $\pi^+(M^n)$ acts somehow on the groups which enter here. Novikov's result on the diffeomorphism problem can be formulated as follows: for $n \neq 4k + 2$ the set of manifolds, having the same homotopy type and tangent bundle as a given manifold M^n, is the set of orbits of the group $\pi^+(M^n)$ on $\operatorname{Im} \varkappa = \Pi^{-1}(M^n)/JK_R^{-1}(M^n)$, if these manifolds are identified modulo the group of differential structures on the sphere $\theta^n(\partial\pi)$. For $n = 4k + 2$ the additive Arf-invariant $\varphi\colon \operatorname{Im} \varkappa \to Z_2$ is defined on the group $\operatorname{Im} \varkappa$ and in place of $\operatorname{Im} \varkappa$ one has to take $\operatorname{Ker} \varphi$. The role of the group $\operatorname{Ker} J = \operatorname{Im} \delta$ will be indicated in Chap. IV, §12. There is an analogous result in the piecewise linear case with k_{PL}^{-1} replacing K_R^{-1} and BPL replacing BSO. It is interesting to note that although this result is equivalent in form to the old one yet in other (non-stable) problems, concerning the type of n-dimensional knots to be precise, the corresponding statement only approximates the correct one, as J. Levine has shown for embeddings[1] of the sphere S^n in S^{n+k}. This result of Levine's is not yet published; the possibility of interpreting the answer as an approximation has also been pointed out to me by A.S. Shvarts. It is interesting to note that in the example of the homotopy type of S^n, where the diffeomorphism classification was earlier obtained by Milnor and Kervaire, the group structure arises on account of the fact that $\pi^+(M^n) = 1$, if we look at things from the point of view of the general theorem. The group $\Gamma_{SO}(M^n)$ brings together a number of problems on diffeomorphisms - the subgroup $\operatorname{Im} \varkappa \subset \Gamma_{SO}$, with the problem of normal (tangent) bundles (cf. Chap. IV, §12), being related to the set $\{n + \operatorname{Ker} J\} \subset K_R^0(M^n)$, $\operatorname{Ker} J = \operatorname{Im} \delta = \Gamma_{SO}/\operatorname{Im} \varkappa$.

---

1 Here in place of H_n one has to take G_n, consisting of the maps $S^{n-1} \to S^{n-1}$ of degree $+1$ without fixed point.

Chapter IV

SOME APPLICATIONS OF THE K- AND J-FUNCTORS AND BORDISM THEORIES

§11. Strict application of K-theory

Atiyah and Hirzebruch have proved a number of theorems generalizing to the case of differential manifolds Grothendieck's form of the Riemann-Roch theorem.

Consider two manifolds $M_1^{n_1}$ and $M_2^{n_2}$, $n_1 - n_2 = 8k$, and a map $f: M_1 \to M_2$. We shall suppose that the manifolds are oriented. Fix (if possible) elements $c_1 \in H^2(M_1, Z)$ and $c_1' \in H^2(M_2, \underline{Z})$ such that $c_1 \bmod 2 = w_2$ and c' $c_1' \bmod 2 = w_2$. Let $f^*c_1' = c_1$. Denote by $\overline{f}_*$ the map $Df_*D: H^*(M_1) \to H^*(M_2)$, D being the Poincaré duality operator. Let $\zeta \in K_R(M_1^{n_1})$. Then we have the following fact ("the Riemann-Roch theorem"):

There exists an additive map $f_! : K_R(M_1) \to K_R(M_2)$ such that

$$\overline{f}_* (\operatorname{ch} \zeta A (M_1)) = \operatorname{ch} f_! \zeta A (M_2).$$

There is an analogous theorem for maps of quasicomplex manifolds, but here one can dispense with the condition $n_1 - n_2 = 8k$, while the T-genus replaces the A-genus. One should note that an important part in the theorems is played by, first, Bott periodicity in terms of K-theory and, secondly, the Thom isomorphism of K-theory. The situation is that in K-theory there is a Thom isomorphism in the following cases:

a) for the complex K-functor

$$\varphi_K: K_C(X) \longrightarrow K_C^0(T_\eta),$$

where η is a U_N-bundle over X;

b) for the real K-functor

$$\varphi_K: K_R(X) \longrightarrow K_R(T_\eta),$$

where η is a $Spin$-bundle over X ($w_1 = w_2 = 0$).

However, this time there are many Thom isomorphisms. One should notice that the Thom isomorphism may be chosen such that the following conditions hold:

$$\operatorname{ch}(\varphi_K \alpha) = T(\eta)\, \varphi(\operatorname{ch} \alpha) \qquad \text{(case a)),}$$

$$\operatorname{ch}(\varphi_K \alpha) = A(\eta)\, \varphi(\operatorname{ch} \alpha) \qquad \text{(case b)),}$$

where φ_K is the chosen Thom isomorphism in K-theory and φ is the standard Thom isomorphism in ordinary cohomology. One should notice that it is sufficient to construct these Thom isomorphisms for universal bundles, depending only on the invariants of the Lie groups and their representations.

In this context the T- and A-genera therefore arise out of the commutativity law for the Chern character with the (chosen) Thom isomorphism. Other "universal" Thom isomorphisms may lead to other "multiplicative genera" and "Riemann-Roch theorems". The integrality of the A-genus is obtained here (as in the algebraic Grothendieck theorem) if M_2 is taken to be a point. Of course, these theorems are not given here in their general form, nor do we give a number of interesting corollaries.

Another important result has been obtained in a neighbouring field by Atiyah and Hirzebruch. This concerns the "index" of an elliptic operator on a manifold without boundary. To an elliptic operator (defined and taking its values on the sections of bundles F_1 and F_2 over X^{2l}) there corresponds a "symbol" $\sigma: \pi^*F_1 \overset{\approx}{\rightarrow} \pi^*F_2$, where $\pi : S_X \to X$ is the natural fibration of the manifold of unit tangent (co-)vectors over X and X is a Riemannian manifold. The index of the operator depends in fact only on the homotopy class of the symbol and is trivial if the isomorphism σ induces an isomorphism $(\pi\sigma) : F_1 \to F_2$ of fibre bundles over X (Vol'pert, Dynin). In an elegant way one constructs an "invariant" $\alpha(\sigma) \in K_R^0(T_\eta)$, where η is the tangent bundle on X and the index is a homomorphism

$$I\colon\ K_R^0(T_\eta) \rightarrow Z.$$

Moreover, by a theorem of Cartan-Serre the index depends only on $\operatorname{ch} \alpha(\sigma) \in H^*(T_\eta, Q)$. One constructs a "special Thom isomorphism"

$$\varphi_K^{-1}\colon\ K_R^0(T_\eta) \otimes Q \rightarrow K_R(X) \otimes Q,$$

and the index depends only on $\operatorname{ch}(\varphi_K^{-1}\alpha(\sigma))$.

The situation is therefore reduced to pairs consisting of a manifold X and a vector bundle ζ over X with a special operator ("the Hirzebruch index"), whose Chern character is easily computed – such operators give a "complete" set for the index problem.

Moreover, an important theorem, the "intrinsic homological invariance" of the index, has been proved in a convenient cobordism form, and the solution of the problem is easily accomplished on the basis of Thom's theory. The final formula is

$$I(\sigma) = ((\varphi^{-1} \operatorname{ch} \alpha(\sigma))\, T(c\eta),\ [X]).$$

For example, the A-genus is the index of the "Dirac operator" of a spinor structure.

Particularly interesting results have also been obtained by Atiyah in the theory of smooth embeddings. A striking example is the application of these general theorems to the embeddings of the complex projective spaces CP^n, which cannot be embedded in spaces of dimension less than (for example) $2(n - \beta(n))$, where $\beta(n)$ is the number of 1's in the binary representation of n. Interesting links have also been found between the K_C-functor, the cohomology of finite groups and the "representation rings". Indeed, in the most recent papers cohomology operations in the K-functor are used, though not quite explicitly.

§12. Simultaneous applications of the K- and J-functors. Cohomology operation in K-theory

We state to begin with some "general" theorems on the connection of the K- and J-functors with topological problems. It is easy to see that the normal bundle η of a smooth closed manifold X has a Thom complex T_η with spherical fundamental cycle (the Thom complex is "reduced"), while the Thom complex of a trivial bundle of dimension N over X is the bouquet of a sphere S^N and the suspension $E^N X$, that is, it is "coreduced". Moreover for any two elements $\alpha_1, \alpha_2 \in K(X)$ possessing either of these properties simultaneously $J(\alpha_1) = J(\alpha_2)$, where $J: K_R(X) \to J(X)$, that is, these properties are J-invariant. Atiyah noticed the essential fact that the Thom complexes of an element $\alpha + N$, $\alpha \in K_R^0$ and $(n - \alpha) + N_1$, where n is the representative in $K_R^0(X)$ of the normal bundle, are "S-dual" to each other; he proved that the "reducibility" of the Thom complex of the element α is equivalent to the statement that this element is equal to the normal bundle in the J-functor, that is, the set of "reducible" elements of $K_R^0(X)$ is exactly $J^{-1}J(n)$, where $n \in K_R^0(X)$ is the normal bundle. Similarly for the "tangent" bundle $(-n) \in K_R^0(X)$. For simply-connected odd-dimensional manifolds of dimension $\geqslant 5$ Novikov and Browder proved the converse "realization theorem", namely that every element $(-n_i) \in J^{-1}J(-n)$ is the "tangent bundle" of some manifold X_i^k of homotopy type X^k; for even k the formulation is more complicated; it is final only for $k \equiv 0 \bmod 4$, when the "tangent bundles" consist of all the elements $-n_i \in J^{-1}J(-n)$, satisfying the "Hirzebruch condition" $(L_q, [X^{4q}]) = \tau(X^{4q})$.

It is interesting to note that for a non-simply-connected manifold of dimension $4k + 1$ this theorem is not true by the formula for the rational class $L_k(M^{4k+1})$ given in the Appendix.

From the finiteness of the J-functor and from these theorems it is clear that for simply-connected manifolds one can vary the tangent bundle (and the Pontryagin classes) very freely within a given boundary type. There are analogous theorems for combinatorial manifolds also – one has to consider k_{PL}^0 in place of K_R^0 and J_{PL} in place of J. A combinatorial manifold has the homotopy type of a smooth manifold if (under analogous homotopy restrictions) its "normal microbundle" $n \in k_{PL}^0(X)$ is such that there is in the set $J_{PL}^{-1} J_{PL}(n)$ an element of the image $jK_R^0(X) \subset k_{PL}^0(X)$, if the dimension is odd (for even dimensions it is, as before, more complicated). For example for $M^6 = S^2 \times S^4$ the "tangent" elements are all $\tau_i \in K_R^0(M^6)$ such that $w_2 = 0$ and $p_1 = 48\lambda u$, where λ is an arbitrary integer and u is the basis element of $H^4(M^6) = Z$. There is a family of manifolds M^6 homotopically equivalent to $S^2 \times S^4$ and with class $p_1 = 48\lambda u$. It is interestin that it follows from the latest paper of Novikov that the class $p_1(M^6)$ is topologically invariant; we get a proof of the difference of homeomorphism and homotopy type of closed simply-connected manifolds. For $n > 3$ no example of this, even non-simply-connected, was known.

Finally, the following lemma due to Adams and Atiyah turns out to be very important in further applications: if $K_R^0(RP^k) = J^0(RP^k)$ for all dimensions, then on all spheres S^n there are exactly $\rho(n) - 1$ linearly independent vector fields and no more, where $\rho(n)$ is defined as follows: if $n + 1 = 2^b(2a + 1)$, $b = c + 4d$, $0 \leqslant c \leqslant 3$, then $\rho(n) = 2^c + 8d$. Note that this number of fields was already known classically and had been proved in a number of cases. In particular, Toda had also noted this lemma in another context, and had computed the number of fields where this was possible, enabling him to compute the homotopy groups of spheres and the classical J-homomorphism on this foundation (for example for $k \leqslant 19$ and $n \leqslant 2^{11}$). Adams has solved the j-functor problem completely in this case for all k. We give the basic outline of his method.

It had already been observed (Grothendieck, Atiyah) that there were "operations" in K-theory, related to the exterior powers. They are denoted by λ_i and possess the property

$$\lambda_i(x+y) = \sum_{j+l=i} \lambda_j(x)\lambda_l(y).$$

However, these operations are non-additive and it is therefore difficult to apply them. Adams was led to introduce operations $\psi^i_\Lambda\colon K_\Lambda(X) \longrightarrow K_\Lambda(X)$, (expressed in terms of the λ_l) which were ring homomorphisms and such that $\psi^i_\Lambda \psi^j_\Lambda = \psi^{ij}_\Lambda$, where $\psi^0_\Lambda = (\dim)$, ψ_C^{-1} — is complex conjugation and $\psi_R^{-1} = \psi_R^1 = 1$. Moreover, $\psi^i_\Lambda(x) = x^i$, if x is a one-dimensional Λ-bundle. Note that $\mathrm{ch}^n \psi_C^k \eta = k^n \mathrm{ch}^n \eta$. These operations are remarkable from the point of view of the usefulness of their application. It is rather easy to compute the K_C-functor with its operations for CP^n and CP^n/CP^{n-k-1}, rather more difficult to compute the K_R-functor with its operations for RP^n and RP^{n-k-1}. Adams did this and the desired result about $J(RP^k)$ followed almost immediately from the answer. From this it also follows that the classical J-homomorphism

$$J \otimes Z_2\colon \pi_{i-1}(O_N) \otimes Z_2 \longrightarrow \pi_{N+i-1}(S^N) \otimes Z_2$$

is always a monomorphism, and this implies the topological invariance of the tangent bundle of a sphere S^i for $i \equiv 1, 2 \bmod 8$ (for $i \not\equiv 1, 2 \bmod 8$ the invariance was known). In a number of cases (for example for $X = S^{4n}$) Adams has succeeded by other methods in giving an upper bound on the order of the J-functor: to be precise, if $x \in K_C(X)$ then $J(k^N(\varphi^K_C - 1)x) = 0$ for large N and for all k. This gives a complete or almost complete answer for S^{4n}. To obtain similar estimates in the general case is an interesting problem.

The introduction of the operations ψ^k made it possible to introduce new "characteristic classes" into K-theory: let $x \in K^0_R(X)$ and let the Thom complex of the bundle $\eta = x + N$ possess a Thom isomorphism $\varphi_K : K_R(X) \to K^0_R(T_\eta)$. Let $\rho_l(x) = \varphi_K^{-1}\psi^l_R\varphi_K(1)$ by analogy with the Stiefel classes. Since φ_K is not uniquely determined in K-theory these classes ρ_l are not uniquely defined but the degree of indeterminacy is easily computed. These classes are very useful for estimating the order of $J^0(X)$ from below (Adams, Bott).

§13. Bordism theory

In Chapters I and II we have already spoken of cobordism groups and rings and have pointed out the general cohomological and homological theories connected with them (the first cohomological approach here, as in

K-theory, was made by Atiyah). This theory was developed by Conner and Floyd in connection with problems on the fixed points of transformations; recently the theory has been used to obtain a number of other results, among which one should note the work of Brown-Peterson on the relations between the Stiefel-Whitney classes of closed manifolds and also the results on the groups Ω^i_{SU}, obtained simultaneously by a number of authors (Conner-Floyd, Brown-Peterson, Lashof-Rothenberg). In particular in the middle dimension Brown-Peterson and Lashof-Rothenberg have with the help of the group Ω^{8k+2}_{SU} solved the well-known problem about the Arf-invariant of Kervarie-Milnor in the theory of differential structures on spheres.

a) Brown and Peterson study the following ideal:

$$\lambda_i(x+y) = \sum_{j+l=i} \lambda_j(x)\,\lambda_l(y).$$

where $\tau_{M^n} : M^n \to BO$ is the classifying map for the (stable) tangent bundle of M^n. By using O-bordism theory it has been proved that this ideal consists only of those elements "trivially" belonging to it purely algebraically, according to the formulae of Thom and Wu. Analogous results, more complicated to formulate, have been obtained for the ideal $I_n(SO, 2)$. In the proof a "right" action of the Steenrod algebra in the category of complexes and bundles is introduced and studied; there is an elegant treatment of the Thom-Wu formulae and an isomorphism of cohomology theories

$$J \otimes Z_2\colon\ \pi_{i-1}(O_N) \otimes Z_2 \longrightarrow \pi_{N+i-1}(S^N) \otimes Z_2$$

is constructed, $N_*(X)$ being the bordism group; it is sufficient to verify this isomorphism for points only. It is then applied to $X = K(Z_2, m)$ and this gives the result.

b) In the work of Conner and Floyd U-bordism theory is applied to the study of the number of fixed points of involutions of a quasicomplex manifold onto itself. More precisely, they study a quasicomplex involution $T : M^{2n} \to M^{2n}$ such that its set of fixed points decomposes into a union $V = M^0 \cup M^2 \cup \dots \cup M^{2n-2}$ of quasicomplex submanifolds in whose normal spaces "reflection" (in the form of non-degeneracy) occurs. Cutting out an invariant small neighbourhood B of V from M^{2n} and setting

$$N = (M \setminus B)/T,$$

we see that

$$\partial N = \partial B/T$$

and that N is quasicomplex.

Consider now the "bordism" group $U_*(RP^\infty)$ to which we relate the group $\sum_{i,k} H_k(RP^\infty, \Omega^i_U)$. The ring Ω^*_U is therefore a polynomial ring over Z, having one generator in each dimension $2m$, this generator being CP^1 when $m = 1$. Thus,

$$I_n(O, 2) = \bigcap_{M^n} \operatorname{Ker} \tau^*_{M^n},$$

while the "geometrical" generator is $(RP^{2k-1} \times M^{2l}, f)$, where M^{2l} is the generator of Ω_U^{2L} and f is the projection $RP^{2k-1} \times M^{2l} \to RP^{\infty}$. The original involution determines a "relation"

$$\partial(N, F),\ F\colon N \to RP^{\infty},$$

where N was defined above and $F: N \to RP^{\infty}$ is obtained by factorizing with respect to T the map $M \setminus B \to S^{\infty}$, which commutes with the involutions. Thus, there is a correspondence between involutions on arbitrary manifolds M^{2n} and relations between the basis elements in $U_*(RP^{\infty})$, arising by pull-back from the group $\Sigma H_K(RP^{\infty}, \Omega_U^*)$ related to $U_*(RP^{\infty})$.

There is an elegant proof of the following relation:

$$2(RP^{2k-1} \otimes 1) = RP^{2k-3} \otimes CP^1,$$

where RP^{2i-1} are the basis cycles in RP^{∞} and 1 and CP^1 are the generators of Ω_U^0 and Ω_U^2 (in fact, all other relations follow from this one).

Since CP^1 is a polynomial generator in Ω_U^*, that is, the powers $(CP^1)^m$ are irreducible in Ω_U^*, it follows that if the fixed points of an involution are isolated, for example, for a complex manifold M^{2n} (the dimension is real, the involution quasicomplex) then their number is divisible by 2^n, since $\partial(N, F)$ in this case is

$$\sum RP^{2n-1} \otimes 1 = \lambda RP^{2n-1} \otimes 1,$$

and λ must be divisible by 2^n since

$$\lambda RP^{2n-1} \otimes 1 = 0.$$

This is the main result, but if one wishes to give a more general formulation, concerned not only with the zero-dimensional case, then in the case when the normal bundles of the manifolds $M^{2i} \subset M^{2n}$ are trivial it can be expressed as follows:

Let $[M^{2i}]$ denote the class of M^{2i} in Ω_U^{2i} and let x be the element $\frac{CP^1}{2} \in \Omega_U^2 \otimes Q$. Then the element

$$\frac{1}{2}(x^{n-1}[M^0] + x^{n-2}[M^2] + \ldots + [M^{2n-2}])$$

is "integral" in $\Omega_U^* \otimes Q$, that is, belongs to Ω_U^*.

c) Another beautiful application of the general theory of U- and SU-bordisms is the final computation of the 2-torsion in the ring Ω_{SU}^* (Conner-Floyd, Lashof-Rothenberg, Brown-Peterson). Here connections have successfully been found between the U- and SU-bordisms of different objects. From their results it follows, for example, that all the 2-torsion in Ω_{SU}^k is $\Omega_{SU}^1 \Omega_{SU}^{k-1}$; for even k this was not known and the old methods led to serious difficulties. As has already been pointed out, these methods of studying the ring Ω_{SU}^* led to the solution of the Arf-invariant problem in the middle dimension.

The most important unsolved problems of this theory are the study of the multiplicative structure of the ring Ω_{SU}^* and also of the ring Ω_{Spin}^*

on which little is known is "modulo 2". It would be useful to compute "bordisms" and "cobordisms" for a much wider class of spaces, as this would widen greatly the possibilities of application. Note, for example, that the U_*-theory is contained as a direct component of the K_C-theory.

The large number of interconnections introduced into topology by the new outlook on homology theory is apparently such that it would be impossible to describe beforehand the circle of problems that will be solved in this region even in the relatively near future.

APPENDIX

The Hirzebruch formula and coverings

Novikov has found an analogue to the Hirzebruch formula, relating the Pontryagin classes to the fundamental group. Let M^{4k+n} be a smooth (or PL) manifold and let $x \in H_{4k}(M^{4k+n}, Z)$ be an irreducible element, such that $Dx = y_1, y_2, \ldots, y_n$ mod Tor, $y_i \in H^1$, D being Poincaré duality. Consider a covering $p : \hat{M} \to M^{4k+n}$, for which there are paths γ, covered by closed by paths, such that $(\gamma, y_i) = 0$ $(i = 1, \ldots, n)$. Let $\hat{x} \in H_{4k}(M, Z)$ be an element such that $p_* \hat{x} = x$, $\hat{x}$ being invariant with respect to the monodromy group of the covering. It is unique up to an additive algebraic restriction on this element. Let $\tau(\hat{x})$ be the signature of the quadratic form $(y^2, \hat{x})$, $y \in H^{2k}(\hat{M}, R)$, the non-degenerate part of which is finite-dimensional.

For $n = 1$, and also for $n = 2$ provided that $H_{2k+1}(\hat{M}, R)$ is finite-dimensional, it has been proved that $(L_k(M^{4k+n}), x) = \tau(\hat{x})$, which already in these cases leads to a number of corollaries and also has application to the problem of the topological invariance of the Pontryagin classes even for simply-connected objects.

Note in proof. The author has recently completed the proof of the topological invariance of the rational Pontryagin classes (see [46]).

Translator's note. In [46] Novikov states that the topological invariance is an easy consequence of the following fundamental lemma of which he sketches the proof:

LEMMA: *Suppose that the Cartesian product $M^{4k} \times R^m$ has an arbitrary smooth structure, turning the product into an open smooth manifold W, M^{4k} being a compact closed simply-connected manifold. Then* $(L_k(W), [M^{4k}] \otimes 1) = \tau(M^{4k})$, *where the $L_k(W)$ are the Hirzebruch polynomials for the manifold W and $\tau(M^{4k})$ is the signature of the manifold M^{4k}.*

Some pointers to the literature

The literature is distributed over the various sections in the following manner:

§§1-3 – [1],
§4 – [2], [15],
§5 – [3]-[13],

§6 – [14], [15],
§7 – [16] - [21]
§8 – [3], [6], [11], [12], [28].
§9 – [2], [22]-[24], [33],
§10 – [25]-[33], [35], [40],
§11 – [29], [34]-[38],
§12 – [28], [33], [35], [39]-[41],
§13 – [42]-[45]
Appendix – [39], [46],

References

[1] N. Steenrod, The topology of fibre bundles, University Press, Princeton 1951.
Translation: *Topologiya kosykh proizvedenii*, Gostekhizdat, Moscow 1953.

[2] J. Milnor, Lectures on characteristic classes, *Lektsii o kharakteristicheskikh klassakh*. Matematika 5 : 2 (1959).

[3] R. Thom, Quelques propriétés globales des variétés différentiables, Math. Helv. 28 (1954), 17-86.
Translation in the collection (Fibre spaces) *Raccloennye prostranstva*, IL, Moscow 1958.

[4] V.A. Rokhlin, New results in the theory of four-dimensional manifolds, Dokl. Akad. Nauk SSSR *84* (1952), 221-224.

[5] V.A. Rokhlin, The theory of intrinsic homologies, Uspekhi Mat. Nauk *14* (1959), no. 4, 3-20.

[6] J. Milnor, On the cobordism ring Ω^* and a complex analogue. I, Amer. J. Math. *82* (1960), 505-521.

[7] J. Milnor, A survey of cobordism theory. Enseign. Math. *8*, (1962), 16-23.

[8] C.T.C. Wall, Determination of the cobordism ring, Ann. of Math. *72* (1960), 292-311.

[9] C.T.C. Wall, Note on the cobordism ring, Bull. Amer. Math. Soc. *65*, (1959), 329-331.

[10] B.G. Averbukh, The algebraic structure of the intrinsic homology groups, Dokl. Akad. Nauk SSSR *125* (1959), 11-14.

[11] S.P. Novikov, Some problems in the topology of manifolds, connected with the theory of Thom spaces, Dokl. Akad. Nauk *132*, (1960), 1031-1034.
= Soviet Math. Dokl. 1, 717-720.

[12] S.P. Novikov, Homotopy properties of Thom complexes, Mat. Sb. *57* (*99*), (1962), 407-442.

[13] S.P. Novikov, Differential topology in the collection "Algebra. Topology" 1962, VINITI (1963).

[14] F. Hirzebruch, Neue topologische Methoden in der algebraischen Geometrie, Springer, Berlin 1956; second edition, 1962.

[15] A. Borel and F. Hirzebruch, Characteristic classes and homogeneous spaces. Amer. J. Math. *80*, 458-538 (1958); *81* (1959), 315-382; *82* (1960), *491*-501.

[16] R. Bott, The stable homotopy of the classical groups, Ann. of Math. *70* (1959), 313-337.

[17] J. Milnor, Some consequences of a theorem of Bott, Ann. of Math. 68 (1958), 444-449.
= Matematika *3*, 3 (1959).

[18] M. Kervaire, and J. Milnor, Bernoulli numbers, homotopy groups and a theorem of Rohlin, Proc. Int. Cong. Math. 1958. Camb. Univ. Press 1960.

[19] J. Adams. On Chern characters and the structure of the unitary group, Proc. Camb. Phil. Soc. *57* (1961), 189-199.

[20] J. Adams, On the non-existence of elements of Hopf invariant one, Bull. Amer Math. Soc. *64* (1958), 279-282.

[21] J. Milnor, *Lektsii po teorii Morsa* (Lectures on Morse theory), "Nauka", Moscow 1965.

[22] R. Thom, Les classes charactéristique de Pontryagin des variétés triangulées Symp. Int. Top. Alg., Mexico 1958.

[23] V.A. Rokhlin, On the Pontryagin characteristic classes, Dokl. Akad. Nauk SSSR *113* (1957), 276-279.

[24] V.A. Rokhlin and A.S. Shvarts, The combinatorial invariance of the Pontryagin classes, Dokl. Akad. Nauk *114* (1957), 490-493.

[25] S. Eilenberg and N. Steenrod, Foundations of algebraic topology, Univ. Press, Princeton 1952.
Translation: *Osnovaniya algebraicheskoi topologii*, Fizmatgiz, Moscow 1956.

[26] G. Whitehead, Generalized homology theories, Trans. Amer. Math. Soc. *102* (1962), 227-283.

[27] M.F. Atiyah, Bordism and cobordism, Proc. Cambridge Phil. Soc. *57* (1961), 200-208.

[28] M.F. Atiyah. Thom complexes, Proc. London Math. Soc. *11* (1960), 291-310.

[29] M.F. Atiyah and F. Hirzebruch, Riemann-Roch theorems for differentiable manifolds, Bull. Amer. Math. Soc. *65* (1959), 276-281.

[30] J. Milnor, Microbundles, I, Topology *3*, Suppl. 1 (1964), 53-80.

[31] M. Hirsch, Obstruction theories for smoothing manifolds and maps, Bull. Amer. Math. Soc. *69* (1963), 352-356.

[32] S.P. Novikov, The Topology Summer Institute (Seattle 1963), Uspekhi Mat. Nauk *20*, 1 (1965), 147-170.
= Russian Math. Surveys 20, 1 (1965), 145-167.

[33] S.P. Novikov, Homotopy equivalent smooth manifolds, I. Appendices 1, 2, Izv. Akad. Nauk SSSR. Ser. Mat. *28*, 2 (1964).

[34] M.F. Atiyah and I. Singer, The index of elliptic operators, Bull. Amer. Math. Soc. *69* (1963), 422-433.

[35] R. Bott, Lectures on *K*-theory (mimeographed), 1963.

[36] R. Bott, M.F. Atiyah, I. Singer, Seminar (mimeographed) 1964.

[37] M.F. Atiyah and F. Hirzebruch, Vector bundles and homogeneous spaces, Proc. Symp. Pure Math., Amer. Math. Soc. Vol. III (1961), 7-38.

[38] M.F. Atiyah and F. Hirzebruch, Quelques théorèmes de non-plongement pour les variétés differentiables, Bull. Soc. Math. France 87 (1959), 383-396.
= Matematika *5*, 3 (1961).

[39] S.P. Novikov, On the homotopy and topological invariance of certain rational Pontryagin classes, Dokl. Akad. Nauk SSSR *162*, 6 (1965).

[40] J. Adams, Vector fields on spheres, Ann. Math. *75* (1962), 603-632.

[41] J. Adams, On the groups *J(X)*. I, Topology *2* (1963), 181-195.

[42] E. Brown and F. Peterson, Relations among characteristic classes. I, Topology *3* (1964), 39-52.

[43] P. Conner and E. Floyd, Differentiable periodic maps, Springer, Berlin 1964.

[44] P. Conner and E. Floyd, Periodic maps which preserve a complex structure, Bull. Amer. Math. Soc. *70* (1964), 574-579.

[45] P. Conner and E. Floyd, The *SU*-bordism theory, Bull Amer. Math. Soc. *70* (1964), 670-675.

[46] S.P. Novikov, The topological invariance of the rational Pontryagin classes, Dokl. Akad. Nauk SSSR *163* (1965), 298-300.

Translated by I.R. Porteous